Inventive Engineering

Knowledge and Skills for Creative Engineers

Inventive Engineering

Knowledge and Skills for Creative Engineers

Tomasz Arciszewski

George Mason University, USA

CRC Press
Taylor & Francis Group
Boca Raton   London   New York

CRC Press is an imprint of the
Taylor & Francis Group, an **Informa** business

A SPON PRESS BOOK

CRC Press
Taylor & Francis Group
6000 Broken Sound Parkway NW, Suite 300
Boca Raton, FL 33487-2742

© 2016 by Taylor & Francis Group, LLC
CRC Press is an imprint of Taylor & Francis Group, an Informa business

No claim to original U.S. Government works

Printed on acid-free paper
Version Date: 20160323

International Standard Book Number-13: 978-1-4987-1124-1 (Paperback)

Library of Congress Cataloging-in-Publication Data

Names: Arciszewski, Tomasz, author.
Title: Inventive engineering : knowledge and skills for creative engineers / Tomasz Arciszewski.
Description: Boca Raton : Taylor & Francis, a CRC title, part of the Taylor & Francis imprint, a member of the Taylor & Francis Group, the academic division of T&F Informa, plc, [2016] | Includes bibliographical references and index.
Identifiers: LCCN 2016000351 | ISBN 9781498711241 (pbk. : alk. paper)
Subjects: LCSH: Engineering--Vocational guidance. | Creative ability in technology. | Soft skills. | Inventions.
Classification: LCC TA157 .A785 2016 | DDC 620--dc23
LC record available at http://lccn.loc.gov/2016000351

Visit the Taylor & Francis Web site at
http://www.taylorandfrancis.com

and the CRC Press Web site at
http://www.crcpress.com

Contents

Foreword

Ours is a rapidly evolving profession—one that demands a high level of specialized knowledge and a voracious intellectual appetite capable of keeping pace with technological advances. Perhaps the most critical attributes to seek and nurture are our ability to lead effectively, act ethically, communicate clearly, and manage efficiently. Engineers who possess these skills will not only enjoy greater personal success but will also make more meaningful contributions to their businesses and to society as a whole.

Beyond *what* we know, Arciszewski invites readers to consider *why* our expertise matters, and *how* we approach these changing demands. Knowledge is the bedrock, but fertile imaginations are the topsoil where innovation takes root. Creativity and the ability to apply knowledge in novel ways become differentiating factors in the talent marketplace.

Arciszewski continues to advance this vital conversation about the future of engineering with clear writing, compelling ideas, and practical advice. At the same time, he continues to fearlessly color outside the lines, inviting us to see the big picture from a new perspective. He invites engineers to be as creative as we are analytical. He embraces a refreshingly holistic approach, asking us to focus first on rewarding and creative work that fuels personal satisfaction and genuine happiness. This, in turn, leads to career advancement. Too many of us have that equation reversed, and too many young people start their engineering careers focused on the wrong outcomes. While Arciszewski's previous book was written for instructors and academic administrators, this edition speaks directly to engineering students. The two create a unique synergistic unity.

Arciszewski coined the name "Inventive Engineering" in the 1990s while teaching courses by the same name at George Mason University. The name reflects a new way of framing deeply rooted concepts for a modern era. In his typical style, Arciszewski offers a comprehensive explanation of how this term found its way into the modern lexicon of professional engineering, examining its roots in cognitive, philosophical, and biological underpinnings. He traces the lineage of inventive engineering from ancient Asia to the European Renaissance and from the writings of Greek philosophers to the works of futurist Richard Florida.

In a demonstration of boundary-spanning thought leadership, Arciszewski invites readers to think more broadly about our profession. He challenges us to demand more of ourselves and those we prepare for careers in engineering, and he expands our lexicon with language and ideas that need to become more commonplace in engineering.

Everyone who reads this will interpret the words and ideas with different points of view. All will be invited to lift their noses out of the blueprints, seek opportunities for experimentation, and think about how we can seek better ways to serve humanity.

Jeffrey S. Russell
University of Wisconsin-Madison

Preface

I have been studying, practicing, and teaching Inventive Engineering for the last 45 years. As a result of that, I have discovered a secret, which is the key to becoming an inventive engineer. This secret has compelled me to write this book. Learning about the methods is necessary but grossly insufficient. A student must undergo a transition from an analyst (specializing in solving analytical problems) to a person also capable of developing new engineering ideas, or inventions. This transformation requires understanding the process and acquiring a balanced body of knowledge, including not only knowledge about methods but also my unique integrated knowledge with roots in several domains. This second kind of knowledge is absolutely critical to motivate the student and to prepare him or her for the most important transition in life.

The book answers three fundamental questions: Why learn inventive engineering? (Chapter 1), What is the fundamental knowledge behind inventive engineering and what is our new science? (Chapters 2 through 5), and finally, How to practice inventive engineering? (Chapters 6 through 10). The book has its roots in engineering, systems science, cognitive psychology, heuristics, history, and political science. The presented knowledge is integrated as a result of my years of interdisciplinary studies and cooperation with many scholars in various areas.

Acknowledgments

This book magically emerged over a period of about a year (summer 2014–spring 2015). Its history is unusual and reflects a fundamental life transformation I underwent during this time. I retired from George Mason University, renovated and sold my house in Fairfax, built a new house in the Shenandoah Valley, taught a course in Poland, moved twice, and worked on the book at the same time, often in the middle of construction when contractors were playing very loud country music. Writing this book was probably the only sanity in my life when everything else was rapidly changing and creating incredible challenges: emotional, logistic, and financial. For all those reasons, I would like to acknowledge the constant support of Tony Moore and Amber Donley, both of Taylor and Francis, whose contributions are numerous and important, and Michelle van Kampen of Deanta Global Publishing Services. Even more, I would like to acknowledge the support of my wife Eva and my daughters Joanna and Milena. Their support was absolutely essential for survival over the last year and brought love and stability to my life at a time when it was most desired.

The cooperation with Professor Frank Fish of West Chester University in Philadelphia, the Director of the Liquid Life Laboratory, is also gratefully acknowledged. He shared with the Author his unique understanding of biology and of ocean life and provided various private pictures, which were used in the book.

Finally, I would like to acknowledge the editorial assistance of Nikky Lynn Pugh. She was the editor of my previous book (*Successful Education. How to Educate Creative Engineers*) and helped me with adopting and updating the background knowledge presented in Chapters 2 and 3 ("Lessons from the Past" and "Lessons from Modern Science").

Tomasz Arciszewski
Middlebrook, Virginia

Author

Tomasz Arciszewski is a professor emeritus at George Mason University (GMU) in Virginia, the United States. He is a global scholar with experience that spans five continents and 24 countries. His life mission is to inspire and educate engineers in becoming more creative, thus enabling them to serve humanity in an expanded capacity, not only as analysts but also as inventors. This mission has led him to study not only engineering but also cognitive psychology, history, political science, and even art. The Author is like a Renaissance scholar and believes that the synergy of the sciences leads to the emergence of modern transdisciplinary knowledge, which is key to inventions.

The Author's formal academic background is in the area of structural engineering and mechanics. His early interests included steel skeleton structures in tall buildings and steel space structures. Over the course of his academic career he became increasingly interested in inventive engineering, automated knowledge acquisition (machine learning), evolutionary computation and designing, and education. His interdisciplinary signature courses on Inventive Engineering at GMU attracted graduate and undergraduate students for more than 15 years. Also, his expertise in this emerging area took him round the globe, lecturing and inspiring new initiatives at universities in Asia, Europe, and the United States.

Dr. Arciszewski has published more than 175 research and technical articles in various journals, books, and conference proceedings publications. In 2009, he published the book *Successful Education. How to Educate Creative Engineers.* This book has been translated into Mandarin and Polish. The Author is also an inventor, with patents in Canada, the United States, and Poland in the areas of tall buildings and space structures.

In the professional arena, Dr. Arciszewski was the founding chair of the Executive Committee of the American Society of Civil Engineers (ASCE) Global Center of Excellence in Computing. Among other functions within the ASCE, he served as an associate editor and a technical editor of the *Journal of Computing in Civil Engineering* and as a corresponding member of the International Activities Committee and the Body of Knowledge II Academic Fulfillment Committee. He also served on 70 technical committees at various international conferences. At present, the Author is advisor to the president of the Polish Chamber of Commerce and is associated with the GMU's Center for the Advancement in Well-Being and with the Intelligent Agents Center. In 2004, he received the ASCE Computing Award, and in 2006 he received the Intelligent Computing in Engineering Award from the European Group for Intelligent Computing in Engineering.

Dr. Arciszewski earned all his degrees from the Warsaw University of Technology. Before joining George Mason University in 1993, he was a member of the faculty at Wayne State University for 10 years. Prior to 1984, he held teaching positions in the Department of Civil Engineering at the University of Nigeria and in the Department of Metal Structures at the Warsaw University of Technology. In Nigeria, his students gave him the honorary title of chief, and the Author highly values this title, which is rarely given to foreigners.

The Author lives with his family in a mountaintop retreat in the Shenandoah Valley in Virginia, where he continues his studies, writings, and lecturing all over the world. He also advises senior executives on how to improve their competitive advantage by becoming more creative.

Chapter 1

Why learn inventive engineering

Ten reasons

A letter from the author to the reader of this book and potential inventive engineer:

Dear Future Inventive Engineer,

Your future begins today. The academic decisions you make today will shape your future professionally and personally for many years to come. They will shape your opportunities, your failures, and your successes. I know this statement can be easily forgotten when you are in the middle of never-ending projects, exams, and internships—all the "stuff" it takes to be an engineering student—not to mention your everyday and personal life. I understand and respect your feelings because I've been there. However ...

Did you know that this period of study, intense as it may be, is really only a brief moment in your life? It is a critical period for your future, though, and perhaps one of the most important junctures you will ever come to. From this crossroads, you get to choose one direction or another. One road is marked with stepping stones that will lead to ultimate well-being, success, and satisfaction. The other will lead you to mediocrity and limited opportunities. What awaits for you down that "road of mediocrity"? Do not let its smoothly paved surface fool you. Take this road in today's competitive market and the result will be a loss of professional opportunities—opportunities lost possibly to other engineers who decided to go in the direction of success. The road of mediocrity is dangerous, to be sure, and it is ultimately full of disappointment and unwanted consequences. As you may have guessed, I want to lead you down the stepping stones of ultimate success and well-being. The journey along this road will be challenging but also incredibly satisfying, and the pay off will be immeasurable. There is one caveat to the path to true success, however. The gate to this path is locked. You need a key. I offer you this key. I call it "inventive engineering."

Let me share a secret with you. It is a mystery discovered by me as a result of my 45 years as an academic and inventive engineer. It is also a priceless piece of experience accumulated through years of living all over the world, including Africa, Europe, and North America. And, by the way, this secret is only for you, so please do not share it with others ...

I discovered a long time ago that thinking only about current goals and challenges is like looking at a very small piece of the big mosaic that is your life. It is "local thinking" instead of thinking "globally" (Figure 1.1). Global thinkers look at the "big picture" of how their current activities are connected to their dreams and to their vision of the future. "Local thinkers," on the other hand, are never fully successful; their focused and limited understanding of the world simply prevents them from developing their full potential. Even more importantly, because they do not see or even think about the future, they are not motivated by their dreams. This sad state of affairs drastically reduces their ability to work hard, to focus, and to be creative. Let's face it—to say the world for fledgling engineers is competitive is an understatement. Being a local thinker in this world leads directly to reduced opportunities and loss.

After teaching for nearly half a century, I have finally realized that engineering education is not just about knowledge and professional abilities and skills. So what *is* the ultimate goal of your studies if not these things? It is this: to transform yourself from a mere student following his or her instructor or master into a truly successful engineer, a professional who is not only able to think practically and analytically but also creatively or inventively, and to translate that creativity and inventiveness into innovative ideas and projects that stand out from all the others. This transformation is not only intellectual; it is also psychological. You need to open your mind, develop a set of new abilities, and become a different person in many ways. This book will help you understand this transformation. Most importantly, this book will prepare you for it and guide you along the way.

My dream is that you will become a successful engineer. Even more, I dream that you will have a chance to win at that global competition that

Figure 1.1 Local and global thinking. (Based on Arciszewski, T., *Successful Education. How to Educate Creative Engineers*, Successful Education LLC, Fairfax, VA, 2009.)

you are about to enter into. And this will happen for you, I am sure, because of the unique combination of traditional engineering knowledge and inventive engineering knowledge that you will soon possess—and because you are motivated to realize your dreams.

I know how we engineers think. We are all about precise calculations and structured information gathering. So let me provide you with ten powerful and rational reasons why you should learn inventive engineering and ultimately become a truly successful inventive engineer.

1.1 TEN REASONS WHY YOU SHOULD BECOME AN INVENTIVE ENGINEER

1.1.1 Reason No. 1. The new era is coming

The history of Western civilization can be seen as an evolutionary process. In this process, our societies gradually undergo a series of transformations as they move through various eras. Each era can be characterized by the most important human activities that occur at a particular time and that provide the material support for society while also shaping that society's culture and priorities.

During the *nomadic era*, tribes traveled constantly, following migrating wild-animal herds, and hunting was the most important human activity. The next period, the *agricultural era*, ended only recently, in the nineteenth century. In this period, the dominant human activity was the cultivation of crops. Control over land was the key to power. In the next period, the *industrial era,* manufacturing became the main focus, and the ownership of industry was the source of power. To a certain degree, we are still living in the industrial era. In the 1960s, however, the *knowledge era* emerged. The critical human activity since then has been knowledge acquisition, processing, storage, and utilization. "Knowledge is power" is the best way to describe this time period. Recently, at the beginning of this century, the picture has changed again. Knowledge has become a commodity and is easily available over the Internet. The center of gravity has been moving from knowledge to innovation and creativity, which has become the most important human activity, deciding the future not only of individual people but also of their nations and ultimately of the whole of civilization. American political scientist George Pink (2006) calls this period the *era of innovation and creativity*. Today the most important human activity is creativity, particularly engineering creativity, which advances technology and is the driving force behind the evolution of society (Figure 1.2).

Today, you and I stand at the very beginning of the era of innovation and creativity. And you should know that you will be always working within this context. To find out why, let us go back to the 1960s and to the very

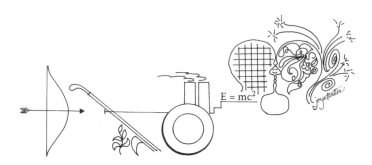

Figure 1.2 Evolution of Western civilization. (Based on Pink, D.H., *A Whole New Mind. Why Right-Brainers Will Rule the Future*, Riverhead Books, New York, 2006. With permission. Drawn by Joy E. Tartter.)

beginning of the knowledge era. At that time, all engineering students were confronted with a simple life-changing question:

> "Is it possible to succeed as an engineer if I do not know how to use computers?"

I guess by now, we all know what the correct answer is. Today, however, I have an equally important question for you:

> "Is it possible today to survive as an engineer if I am not creative?"

And I hope that by now you know the answer to this question too. You got it. The answer is a very big *no*—and that is Reason No. 1 why you should learn inventive engineering.

But there are other reasons. Read on to find out what they are.

1.1.2 Reason No. 2. Emerging global challenges

Our world is changing rapidly. Today, we are confronted by a number of complex, interwoven, and fundamental challenges. These challenges, if not properly addressed, may not only change our lives; they may threaten the very existence of our countries and even of our entire civilization. The global pollution boom, for example, has already created nearly unbearable living conditions in many metropolitan areas in Asia. People living there suffer from a myriad of pollution-related health problems, and pollution and its fluctuating levels affect everyday life for millions of people there. In 2013, I spent a mere ten days teaching in Beijing, China. That was more than a year ago, and I am still struggling with chronic irritation in my eyes. My itching eyes are a constant and very personal reminder to me of the harmful impact of global pollution.

So, up and coming engineer, how would *you* solve this global pollution problem? Would you use numerical optimization methods? Would you find a solution in a textbook? Unfortunately, using these methods is not going to get you the right answer. To solve this kind of problem, you will need to practice so-called out-of-the-box thinking, or creative thinking, which will produce entirely new solutions and most likely new inventions to boot.

But I have some bad news for you: You will probably not be taught how to develop inventions, and thus, you will not even be able even to begin to address this challenge. Now here is the good news: If you become an active student of inventive engineering, it is most likely that you *will* be able to develop inventions, and these inventions may become your stepping stones to addressing the global pollution challenge as well as other fundamental challenges of our times. And that is Reason No. 2 for learning inventive engineering.

1.1.3 Reason No. 3. Emerging engineering challenges

Technological progress is usually driven by systematic research and development. However, there is another powerful driver: government regulations. In many countries, both federal and local governments (e.g., the state of California in the United States) impose various regulations and requirements (regarding, for example, car safety or energy efficiency) that are often impossible to comply with. Usually, there are imposed deadlines for satisfying these new requirements and serious penalties for missing deadlines. Yet the necessary engineering solutions that would lead to compliance are still to be found.

Let me tell you a story about a former Ph.D. student of mine, now a very good friend. He is a talented mechanical engineer and an inventor with several patents to his name. He works in the area of crash and safety engineering for a major car manufacturing company in the United States. When he was a junior engineer, he was given the task of redesigning the so-called B-pillar in a new car. The car was ready for production, but it could not satisfy the new safety requirements regarding side collisions. The B-pillar was initially designed to satisfy rollover requirements and was made very rigid in order to carry the weight of the car. Unfortunately, such a rigid pillar was also dangerous in the case of a side collision, when a human head may hit it.

The challenge for my mechanical engineer friend was to design a B-pillar in such a way that it would be "rigid" for a rollover and "soft" for a side collision. His superiors decided that the answer to this challenge was to redesign the cross section of the B-pillar. Each day of delay was bringing huge financial losses, not to mention lost prestige and the giving away of the market share. Before my former student was assigned this challenge, a team of the best and most experienced crash engineers had been given the redesign task. They failed miserably, using a purely analytical approach with the most sophisticated analytical models and state-of-the-art computers.

When my former student learned about his challenge, he was devastated. He was sure that he would also fail and face terrible professional consequences. Fortunately, after a day or two of despair, he recalled my lectures on inventive engineering and began thinking creatively. First, he quickly realized that the answer to the dilemma lay not in the reshaping of the cross section but was actually outside the B-pillar altogether; an entirely new system had to be developed. His out-of-the-box thinking (i.e., outside of the imposed problem formulation) led him to the clear realization that his challenge was to protect the driver or a passenger during a side collision, not to simply redesign the cross section. Following this line of reasoning, he soon realized that an energy-absorbing element should be placed somewhere between the moving human head and the B-pillar during a side collision. He investigated a number of such elements and discovered that the perfect solution was in the form of a small metal shell element shaped like a tub. This element would be placed on the cabin side of the B-pillar and under a plastic cover. In the case of a side collision, when hit by a human head, the element would deform and absorb the energy that would otherwise be absorbed by the human skull and brain. The invented element was immediately produced and installed in all new cars of the given model, not only making their production possible but, more importantly, potentially saving hundreds of human lives. That is what I call inventive engineering in action!

This real-life example demonstrates that there are many engineering challenges that simply cannot be addressed using traditional deductive and analytical approaches alone. There are hundreds of such challenges that develop every year, and inventive engineers are absolutely needed to tackle them in order for progress to occur. I should add that my former student got his first patent from his invention of the tub-like energy-absorbing element, and this was only the beginning of his career as an inventor.

If *you* want to become an inventor who advances progress in engineering, possibly saving human lives in the process, you have your Reason No. 3 for learning inventive engineering.

1.1.4 Reason No. 4. The competition and you

During your long professional career, you will have to compete with people who are more talented than you are, who are better educated than you are, and who (no matter how much energy you have) work much harder than you do. You want to win against these people, right? So what to do? Obviously, you need a competitive advantage that will give you at least a fair chance to rise about the others. I propose that your chance to rise to the top will be in the form of your *inventive abilities*. And that is Reason No. 4 for learning inventive engineering.

1.1.5 Reason No. 5. The exporting of engineering jobs

You want to become an engineer because you love creating new devices and you wish to serve society. You study hard for this reason and also because you believe that your engineering career will bring to you and your family stability and a constant stream of income. By becoming a professional engineer, this steady income should be a given, right? Although that may have been true at one point in history, times are changing. There is an emerging challenge that may fundamentally change the situation of professionals in the United States. What is this challenge? Let me explain the issue in simple terms.

Your education is focused on learning how to do regular engineering work, that is, how to analyze and design typical bridges and not how to invent *new* types of bridges. Following this analogy, it makes sense that in the future you will participate in global competition with all engineers who know how to analyze and design typical bridges. But, again, I have some bad news for you: There are many such engineers in this country as well as all over the world. Some engineers outside of the United States are less prepared for this global competition than you are, but many are better prepared, particularly in India. Students in India study in English and use the same textbooks and software you use, but they take many more courses than you are required to take. What is more, their engineering programs are often much more demanding than yours in terms of the amount of work, the difficulty of problems, the toughness of instructors, and so on. As a result of all this, many of your competitors will simply know more than you and will be better prepared for their work than you will be. Even worse, their entry salaries will be several times lower than yours. And this is a *very* dangerous situation to be in as an American engineer.

In the modern world, manufacturing jobs as well as analytical and routine design jobs usually flow to where the best value is offered, that is, the best combination of available engineering knowledge at the lowest cost of labor. If you demand an American salary but you are less educated than your competitor abroad (who, by the way, is paid two or four times less than you), it is only a matter of time before your job will fall to the best value and will for you, I am sad to say, simply disappear.

If you are truly concerned about your future, you should know that the only way for you to win at today's global competition game (and to keep your job) is to create your own unique competitive advantage. Assuming that you cannot extend your studies by two or three years and that you do not want to reduce your salary by 75%, you have no choice. In fact, your only chance at survival in your given profession is to become creative and to become an inventive engineer capable of inventing new engineering systems. And that is Reason No. 5 for learning inventive engineering.

1.1.6 Reason No. 6. Systems and you

In the late 1940s, the American scientist Norbert Wiener (1950) proposed the science of Cybernetics. It deals with the development of abstract models of purposeful living organisms, artificial objects, and processes in nature and engineering. Such models he called "systems."

You, my dear friend, can also be considered a "system" living in a specific environment. In order to survive and grow, you must adapt to this environment. You communicate with this environment using a large number of signals that you regularly receive through feedbacks (i.e., interrelationships) that exist between you and your surroundings. Your environment may be interpreted as your job, your profession, and also your society, or even as the overall global community. Feedbacks help you adjust your everyday decisions to the changing environment. Unfortunately, however, such everyday operational decisions are not sufficient for you to survive and grow in the long run. For this purpose, the system (i.e., you) needs to undergo both quantitative and qualitative changes. This first type of change (quantitative) may be interpreted as *incremental improvements*. For example, it can be defined as acquiring additional knowledge or learning new but complementary professional skills. Such changes may require a lot of effort, but they will merely be a continuation of your engineering studies. Obviously, you are well prepared to initiate them as most professionals do throughout the course of their careers. We have an entirely different situation in the case of qualitative changes, however. These changes are also interpreted as fundamental, revolutionary, or structural improvements, but unfortunately you are not ready to manage the process of acquiring them. Qualitative changes require a fundamental transformation of your knowledge and your priorities. That is, they usually require acquiring an entirely new and different body of knowledge and skills, not to mention changing your attitudes and undergoing a complete psychological transformation.

Please do not be frustrated in discovering the gross inadequacies of your traditional engineering education because there *is hope*. I have a solution for you: Learn inventive engineering and you will know how to create entirely new concepts of complex systems, including the system called "you." The process of learning inventive engineering will help you to reinvent yourself and grow as your career does. And that is Reason No. 6 for learning inventive engineering.

1.1.7 Reason No. 7. Critical contradiction

TRIZ is the Russian acronym for the theory of inventive problem solving. The Russian inventor and patent expert Genrich Altshuller originally proposed this theory. He began working on it in the 1940s, and today there are many schools established by his followers, who continue to expand on the concept of TRIZ and adapt it for various markets and applications.

One of the key notions behind TRIZ is the concept of a *contradiction*. "Every great invention (i.e., unknown yet feasible and potentially patentable design concept) is the result of resolving a contradiction," claims Altshuller. He distinguished two types of contradictions: technical and physical. A technical contradiction is an interrelated pair of technical, or abstract, contradictory characteristics of an engineering system (e.g., "weight" and "speed"). A physical contradiction, on the other hand, takes place when a given physical characteristic needs to increase and at the same time decrease to satisfy conflicting requirements. For example, the power of an engine must increase to improve acceleration and at the same time decrease to reduce fuel consumption.

Let us use the concept of a contradiction in discovering our next reason to learn inventive engineering. Imagine that you are in a hypothetical situation. As a student of engineering, ideally you would want to study as little as possible. Yet at the same time you also desire to make a lot of money after you graduate. We have here a clear contradiction. Usually a person's income in the engineering profession is proportional to the amount of knowledge acquired, not inversely proportional, as your dream would require. To succeed, you need to resolve this absolutely critical contradiction, which just may determine the course of your life.

On the surface, you cannot resolve your contradiction. However, if you decide to learn inventive engineering, you will acquire an unusual body of knowledge and skills that may help you actually realize your dream. Yes, if you follow your hypothetical trajectory of studying very little, you still will know very little as far as traditional engineering knowledge and skills are concerned. However, you will have at least one competitive advantage with respect to other similarly intellectually challenged engineers. Therefore, inventive engineering may become your key to the good life and that is more than a good Reason No. 7 to learn inventive engineering.

1.1.8 Reason No. 8. A creative and happy life

People doing the same routine work again and again, no matter how sophisticated the work is, inevitably become bored, uninterested in their work, and generally frustrated by the monotony of their lives as time goes on. This state of being often results in a variety of mental as well as medical problems and makes life just plain miserable. On the other hand, being creative and doing nonroutine work is plenty of fun and brings a lot of satisfaction. It makes common sense that people doing creative work are, in general, more happy and enthusiastic about their lives than people doing exclusively routine work day after day.

What's the conclusion? If you want to be happy in your life, create an opportunity to do creative work. Learn inventive engineering. And that is Reason No. 8.

1.1.9 Reason No. 9. Creativity and wealth

Richard Florida, an American political scientist, has discovered that the rate of growth of wealth as measured by the gross national product per capita is not uniformly distributed throughout the country. This rate changes from region to region and is proportional to the number of patents per capita in an individual region. This number is also a good measure of the level of engineering creativity in a given area. Therefore, it is safe to say that the growth of creativity is directly related to the growth of wealth. In other words, if you want to contribute to the wealth of your region, the best way is to become an inventor and to produce patented inventions. First off, however, you have to learn how to actually produce inventions. And that is Reason No. 9 for learning inventive engineering.

1.1.10 Reason No. 10. Software and industrial robots

We are currently witnessing "the Robot Revolution," which is the process of gradually replacing workers by industrial robots of increasing sophistication. Such robots are programmable, highly flexible, and capable of performing long and complex sequences of various operations. You may not be aware, however, that several years ago, another revolution began. It was called the Software Robots Revolution, and this revolution will have a tremendous impact on your future and may even eliminate your job altogether.

Software robots are computer programs that automatically perform complex tasks. In the last century, these tasks were the exclusive domain of humans. For software robots, we are talking about the taking over of such activities as payroll preparation and engineering design. In the case of engineering design, even today we have computer programs that do much of the routine engineering work once done by people. Examples of these tasks include the analysis, design, and optimization of structural systems. There is a light at the end of the software tunnel, however. These programs do not perform so-called creative engineering work; that is, they do not invent new engineering systems. That task still requires a human, although engineering creativity is the subject of intensive design research and this situation may soon change.

The emergence of the software robot revolution means that if you are an engineer and you do not know how to invent and how to conduct creative work, you are in a difficult position. It is only matter of time before a software robot will be capable of doing all the routine work you now do and you will be asked to find a different job, most likely one that is not as interesting and definitely not as well paid as engineering.

So how does one avoid this danger? Simply learn how to invent—and that is our Reason No. 10 for learning inventive engineering.

SUMMING UP

Even though I have presented you with ten solid reasons, I believe that even one reason should be more than enough for you to decide to learn inventive engineering. I wish you good luck in your studies, and I am sure that, should you decide to follow the path of inventive engineering, which leads to true success and satisfaction, you will not regret your decision. On the contrary, you will consider it a life-changing event for the better. In the process of learning inventive engineering, you will become a different person, and this transformation will allow you to have a more successful and fulfilling life. And that, obviously, is anyone's ultimate goal.

Yours truly,
The Author

Chapter 2

Lessons from the past

2.1 WHY LEARN FROM THE RENAISSANCE?

We are living in a period of rapid technological progress that requires the constant evolution of how we think about engineering. To understand how we can ensure the evolution of our field, it is helpful to look back at the historical factors that led to the da Vincian era, that is, the Renaissance. Inventive engineers must develop a strong understanding of why engineers in general were so successful, respected, and important in the past. That brings us to the *Renaissance man*, an individual (who can be of either gender) who personified various Renaissance-era ideas that led to the emergence of several generations of exceptionally successful creators. These individuals were also engineers, and their fifteenth- and sixteenth-century creations still impress and inspire us today.

2.2 RENAISSANCE EMERGENCE

The evolution of Western civilization in the last millennium can be described as a three-stage process. In this process, the dominant sociocultural paradigm gradually evolved from *the Dark Ages* (or the Middle Ages) through the Renaissance to the present modern age.

The Dark Ages ended in the fourteenth or the fifteenth century, depending on the country. This era was a time when the Roman Catholic Church was a dominant influence in the lives of European societies. During this time, life was highly restricted in terms of expressions of human creativity, and there was a forced focus mostly on the religious side of existence. As a result, very little changed in Europe over a period of several centuries. Even science in the Middle Ages was mostly driven by the official theology of the Roman Catholic Church. It was a period of scholastic studies that mostly concentrated on dogmatic and deductive considerations without any empirical verification of results which would involve induction or out-of-the-box thinking. During this period, it appears that the main purpose of science was to validate the existence of God. To make matters worse, the problems that were considered by scientists were often fantastical or of little practical importance. In the Dark Ages, a subject of "important scientific inquiry" may, for example, have been "How many devils can fit on a pinhead?" The science of the Middle Ages produced very few significant results, and this period is usually considered one of scientific stagnation and even decline.

A good reflection of the Middle Ages and their philosophy as mirrored in art can be found in the famous painting of Lord Hilarion, which presents the saint as an ascetic living in a cave and entirely focused on his spiritual life (Figure 2.1). Lord Hilarion is an ideal man of the Middle Ages. Academic learning in the Dark Ages was also a reflection of the times. Usually, learning was simply the memorization of various texts, mostly religious, which were read by a "professor" and repeated by students several times (Figure 2.2). Since this teaching reflected the official church doctrine, there was very little room for any questioning of provided knowledge, not

Figure 2.1 Lord Hilarion. (With permission. Drawn by Michael Sikes.)

Figure 2.2 Scholastic learning. (With permission. Drawn by Michael Sikes.)

to mention for discussion or creativity. Such teaching only reinforced the intellectual stagnation of the Middle Ages.

The word *Renaissance* is a combination of the French verb *renaître* (to revive) and of the noun *naissance* (birth). It means "revival" (Gelb 1998, 1999, 2004). Usually, the Renaissance is understood as the revival of the classical ideas of human power and of unlimited human potential, which were entirely rejected in the Dark Ages. The Renaissance can be described as a transition from the ascetics of the Middle Ages to the consumerism and enjoyment of life in the modern age.

The Renaissance was also an emergent phenomenon of cultural, intellectual, and social transformations in Europe driven by emerging capitalism, and it was a combination of many factors that created it in its totality: interrelated social changes, scientific and geographic discoveries, a flood of inventions, a revolution in the arts, and the resulting new philosophy in science, which focused on the empirical exploration of nature, including the experimental verification of acquired knowledge. The Renaissance can be described as a movement from the exploitation of one's soul, with a focus on religious and eternal life, to the exploration of the entire world, with a focus on the enjoyment of life and its associated creativity. Exploitation is a deductive process. Exploration, on the other hand, is an inductive and abductive process, as it entails the generation or discovery of new ideas and their verification. What we find in the Renaissance is a transition from deduction to induction and abduction—with truly revolutionary consequences.

There is no single explanation as to why the Renaissance actually occurred. Many historians believe that the black death (bubonic plague) in the fourteenth century was the main reason for the Renaissance. It caused the rapid deaths of up to 50% of the population in parts of Europe and raised serious philosophical and religious questions about life in general, about human priorities, and, most importantly, about the role of the Roman Catholic Church. This inquiry eventually led to the rapid reduction of the church's influence. Noblemen were particularly exposed to the black death because of their travels and interactions with many people, and they died in disproportionately large numbers. The result was a concentration of wealth and capital in the hands of a very few surviving noblemen. Thus, a necessary condition for the emergence and growth of capitalism was created, and the transition from the feudal system to capitalism began. This process gradually changed social structures and redistributed wealth within large segments of European (and, subsequently, American) societies, with all its philosophical, social, and political consequences (Beazley 2003).

In his recent book, *1434—The Year a Magnificent Chinese Fleet Sailed to Italy and Ignited the Renaissance*, Gavin Menzies (2009) provides another hypothetical reason why the Renaissance began in Italy and why its pace was so rapid in the beginning. Menzies claims that in 1434, at the beginning of the Italian Renaissance, a large Chinese fleet visited Venice and Florence. The European fleets did not compare to the size of the Chinese fleet, which included hundreds of ships as well as tens of treasure ships, each approximately ten times larger than the ships used by Columbus 58 years later in his voyage to America (Figure 2.3). The fleet was sent to Europe by Emperor Zhu Zhanji as part of his strategy to create a global Chinese empire. The mission of the fleet was to demonstrate Chinese scientific, technological, and military superiority and to "instruct them (the Europeans) into deference and submission."

A generation before, Emperor Zhu Di, grandfather of Emperor Zhu Zhanji, had initiated the monumental work of creating an encyclopedia (the Yongle Dadian) containing knowledge accumulated by Chinese scientists over a 2000-year period and presented in 11,095 volumes. He employed over 3000 scholars on the project. A copy of this encyclopedia was carried by the fleet and shown in Italy. Menzies claims that many inventions in the encyclopedia were also copied in Florence and made available to numerous Italian scholars. With the landing of the Chinese fleet on European shores, a significant knowledge transfer from East to West occurred that ultimately led to an intersection of Chinese and

Figure 2.3 Chinese treasure ship. (With permission. Drawn by Michael Sikes.)

European bodies of knowledge and to the emergence of a new modern knowledge, which was the scientific foundation of the Renaissance. Menzies (2009) claims that when the pace of discoveries in the fifteenth century is considered, a significant increase can be observed after the year 1434, when the Chinese knowledge infusion took place. It is still an open research question for historians and ethnographers as to whether this increase can be entirely contributed to Chinese contact, although a probabilistic relationship between this knowledge and the increased pace of the Renaissance most likely existed. In addition, there is a good possibility that the Medici effect actually occurred on two levels—first on the global/international level with the intersection of Chinese and Italian bodies of knowledge (Menzies 2009) and then on the local Italian level as described by Johansson (2004). Menzies' discovery is significant for inventive engineering since it allows us to understand the emergence of transdisciplinary knowledge, which comes through integration, an integral part of our study.

The emergence of the Renaissance can also be explained strictly from an inventive engineering perspective. The sociocultural paradigm that was the Renaissance can be considered as a "system." As such, its evolution can be described by patterns of evolution (Clarke 2000, Zlotin and Zusman 2006), which provides an excellent explanation for the nature of expected change. Three types of relevant evolutionary patterns are provided here.

The *resources utilization* pattern states that a system evolves in such a way as to improve the utilization of resources. There is no question, in terms of the utilization of resources (capital), that capitalism represented a significant improvement with respect to the feudal system and that the concentration of capital as a result of the black death was necessary to initiate the evolution of the dominant sociocultural paradigm, in this case the Dark Ages, as it evolved into the Renaissance era.

The *increased system dynamics* pattern states that a system always evolves to increase its dynamics. The Dark Ages can be characterized by a very low mobility of people even within their own countries. That situation had to change, and therefore the Renaissance emerged as an age of significantly increased mobility both within individual countries and in all of Europe. Also, it was a period of geographic exploration of the world outside of Europe. These discoveries obviously contributed to the rapidly increased pace of the sociocultural paradigm of evolution.

The *increased complexity followed by simplification* pattern states that a system's evolution has periods of growing complexity followed by periods of simplification, or declining complexity. For example, in the Dark Ages, the number of rules describing socially accepted behavior imposed on the people was rapidly growing in terms of both numbers and complexity, and this had an increasingly harmful impact, particularly in science. This process had to be reversed for the next step of evolution to occur. When it eventually happened, this pattern led to a decreasing number of rules during the Renaissance.

2.3 RENAISSANCE MAN

A Renaissance man (defined here as a person of either gender) is a polymath or a person knowledgeable and successful in several unrelated domains—for example, in the fine arts as well as in mathematics. The emergence of polymaths was part of Renaissance culture. These individuals were celebrated and the concept inspired many scholars to excel in several domains. However, during the Renaissance, the available body of knowledge was very limited, so it was possible to acquire nearly encyclopedic knowledge in more than one domain. Today, it is impossible to do this, although at least a conceptual understanding of several domains is still feasible and even necessary to succeed in any field, especially science.

Three examples of Renaissance-era polymaths are Leon Battista Alberti (1404–1472), an architect, painter, poet, scientist, mathematician, linguist, and skilled horseman (Figure 2.4); Nicolaus Copernicus (1473–1543), an astronomer, mathematician, physician, jurist, Catholic cleric, economist, and military leader; and Galileo Galilei (1564–1642), a scientist, physicist, philosopher, and musician. Renaissance individuals exist today as well. Examples include entrepreneur Bill Gates, musician/astrophysicist Brian May, actor/poet/painter Viggo Mortensen, media personality Oprah Winfrey, and director of the National League for Democracy in Burma/Myanmar, Aung San Suu Kyi.

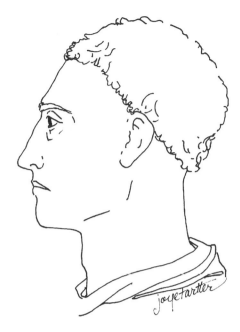

Figure 2.4 Leon Battista Alberti. (With permission. Drawn by Joy E. Tartter.)

A Renaissance man is a polymath, but that is only a part of what he or she is. A Renaissance man's interests, knowledge, and accomplishments reflect the ideas of the Renaissance. As such, this type of person is very confident in him- or herself. He or she believes in their own unlimited capacities for intellectual and physical development, and this belief leads to creative and world-changing results in many domains. In contrast, a person living in the Dark Ages considered him- or herself as an unimportant speck of dust in the universe and entirely controlled by a higher power. The balance of intellectual and physical development is a Renaissance-era idea that strongly contrasts with the Middle Ages' focus on religious doctrine. A Renaissance man has three other characteristics. He or she travels extensively, speaks several languages fluently, and has a good understanding of various cultures.

A Renaissance man is sometimes called a *generalist* because he or she has a holistic understanding of the world and is able to integrate knowledge from various domains in order to produce transdisciplinary knowledge. In this way, he or she is capable of advancing our understanding of the world and of contributing to the evolution of our civilization.

The best way to understand the concept of a Renaissance man is to examine Renaissance figures such as Leonardo da Vinci (Figure 2.5), a giant of the Italian Renaissance (Wrey 2005); and Krzysztof Arciszewski, a lesser-known figure of the Polish Renaissance, whose travels and achievements in

Figure 2.5 Leonardo da Vinci. (With permission. Drawn by Michael Sikes.)

many areas had an impact on history, engineering, and science in several countries, including Poland, the Netherlands, and Brazil.

Leonardo da Vinci (1452–1519) was among the most important Renaissance figures, and his impact on the evolution of European civilization cannot be overstated. Surprisingly, he had a very limited formal education and never studied at a university. This was his weakness, because in his early years, his formal knowledge was limited. It was also his strength, however, because it forced da Vinci to teach himself about the subjects he was interested in, thus avoiding the narrow understanding of the world and nature that still dominated at the universities of his time. His self-education was a process of discovering and experimentally verifying knowledge, which allowed him to learn objective truths about the world that later led to the development of a unique approach to science and engineering. This approach also eventually resulted in many one-of-a-kind scientific discoveries and inventions.

Da Vinci's more formal education began when he became an apprentice in the studio of the sculptor and painter Andrea del Verrocchio in Florence. The artist immediately recognized da Vinci's rare artistic talent and introduced him to his patron, Lorenzo "il Magnifico" de Medici. Soon after, da Vinci became a part of the Medici court and, in so doing, was exposed to a unique intersection of art and science known as the Medici effect (see Section 2.5).

In 1482, da Vinci was forced to leave Florence. He moved to Milan, where he worked under the patronage of Ludovico "the Moor" Sforza. In Milan, Leonardo produced several of his masterpieces, including *The Last Supper*. His paintings not only reflected his excellent understanding of human anatomy and nature but also represented a departure from the art of the Dark Ages. This shift between the art and philosophy of the Middle Ages and Renaissance art is evident when comparing a painting of St. Francis of Assisi by an unknown artist (Figure 2.6) to da Vinci's *Leda and the Swan* (Figure 2.7). Both paintings are focused on the human body. However, the first painting represents the ascetic male body of a saint, which is well hidden under a long monk's robe. In the second painting, a beautiful naked female body is depicted. The first painting expresses human suffering in the search for heaven, while the second is a manifestation of beauty, love, and

Figure 2.6 Saint Francis of Assisi. (With permission. Drawn by Michael Sikes.)

Figure 2.7 Leda and the Swan. (With permission. Drawn by Michael Sikes.)

eroticism. The first is inspired by religion, the second by a classical myth. The differences could not be greater.

While in Milan, da Vinci continued his studies of human and animal anatomy and also studied astronomy, botany, geology, and geography. He began studying flight principles and developed various civilian and military inventions. For example, he invented a tank (Figure 2.8) and, in the area of civil engineering, he invented a new type of bridge (Figure 2.9). When Sforza lost his power after the French invasion of Italy in 1500, da Vinci moved back to Florence. He continued his artistic career but also became a chief engineer working for Cesare Borgia, the commander of the papal armies, in 1502. In this capacity, da Vinci traveled extensively and became interested again in geography and cartography while making six maps of central Italy (Gelb 2004).

There is no easy way to characterize da Vinci. He was a figure who was truly larger than life and a man who had a strong impact on our civilization.

Figure 2.8 Da Vinci's tank. (With permission. Drawn by Michael Sikes.)

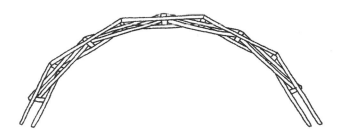

Figure 2.9 Da Vinci's bridge. (With permission. Drawn by Michael Sikes.)

Da Vinci was a talented artist—a painter and a sculptor. He made numerous contributions to the advancement of fine arts, including the introduction of his unique combination of realistic portrayals of humans and nature mixed with sophisticated symbolism, his bold use of perspective, and the development of new painting techniques. He was also a scientist who made discoveries in several areas including anatomy, hydraulics, fluid dynamics, and aerodynamics. He practiced engineering, including mechanical, civil, and structural engineering, and he was an inventor whose concepts, such as the helicopter, parachute, and tank, would be realized centuries later. Da Vinci has often been described as the archetypical Renaissance man. He was a polymath, but not just that. In the spirit of the Renaissance, he was always conscious of his body and health and considered his physical abilities and spiritual dimensions to be equally important as his intellectual power. Most importantly, da Vinci was a great scholar who developed a modern holistic/systems understanding of the world. From our perspective,

Leonardo's scientific methodology for studying nature is his most important contribution to humankind.

Krzysztof Arciszewski (1592–1656), depicted in Figure 2.10, was one of the leading figures of the Polish Renaissance, which occurred later there than in Italy. Arciszewski was well educated in the protestant schools in Poland (Śmigiel, Raków), Germany (the University of Frankfurt), and in the Netherlands. He served as a military and intelligence officer for the Lithuanian potentate Count Radziwiłł and the French Cardinal Richelieu. He traveled extensively all over Europe and participated in several wars. He also studied military engineering and artillery in the Netherlands and participated in a Dutch expedition to Brazil, where he took part in a Dutch war with Portugal and Spain over the control of Brazil. In regards to his engineering and military talents, Arciszewski was gradually promoted to the ranks of general and admiral of the Dutch fleet in Brazil and became the commander of all Dutch forces within that country. He was credited with the conquest of Brazil for the Netherlands and, among other honors, a special map of Brazil, prepared by the famous Dutch cartographer Blaeu, was dedicated to him (a copy of this map is owned by the Author and is shown in Figure 2.11). Arciszewski was also a scientist. During several trips to Brazil, he studied the customs of the local tribes (today extinct) and collected species of flora and fauna, which he brought back to the Netherlands. These specimens became the subject of research by biologists and botanists and subsequently led to the emergence of modern biology

Figure 2.10 Krzysztof Arciszewski. (With permission. Drawn by Michael Sikes.)

Figure 2.11 Seventeenth-century map dedicated to Arciszewski. (With permission. Drawn by Michael Sikes.)

in that country. When Arciszewski came back to Poland, he was given the post of general of the Royal Artillery and built a network of arsenals within Poland. He also created the Polish Royal Navy. While back in his home country, he participated in the Chmielnicki War in the Ukraine. Arciszewski was also a published poet and was interested in medicine. In 1643, he published a medical treatise called *Podagra Curate* about the treatment of podagra, a type of gout.

Arciszewski was one of the first formally educated military and civil engineers in Poland, and several military buildings and installations designed and constructed by him still exist in Poland and Brazil. For example, an arsenal designed by him in Warsaw, Poland, miraculously survived the Warsaw Rising of 1944 and presently houses a museum. Arciszewski was a military and intelligence officer, a poet, and a scientist interested in medicine and biology. In all these areas he achieved a high level of prominence. He could be considered a polymath. However, the combination of these achievements and the fact that he was truly a global man living and working in several countries earns him the title of Renaissance man as well.

Leonardo da Vinci and Krzysztof Arciszewski are figures of obviously different historical significance. However, there are striking similarities between their lives. Both men received a nontraditional education, which was the key to their accomplishments. Da Vinci learned during his apprenticeship with Verrocchio. It was a hands-on education driven by constant artistic and intellectual challenges and conducted under the spell of the Medici effect. Also, he was a self-educated man who widely used abduction and induction to acquire knowledge. In this way, he prepared himself to become a thinker and a creator of new ideas. Arciszewski studied in the protestant schools in northern Poland, which were much more socially and intellectually progressive than the traditional Catholic schools for noblemen of his time. Also, he studied in two different countries (the Netherlands and Germany) where education was provided in different cultural and social contexts. In this way, his education enabled him to develop a complex, multiperspective understanding of the world. This understanding is key to creativity and is simulated in Synectics (see Chapter 12). Like da Vinci, Arciszewski also spoke several languages, including Polish, Latin, German, and Dutch, as well as some French.

The second similarity between these two Renaissance men was that both had committed, or were accused of committing, crimes in their formative years and were forced to leave the environments in which they were raised and educated. Da Vinci had to leave Florence, while Arciszewski left Poland for many years. Both situations were extremely traumatic for these men and led them eventually to lifelong travels in periods of history where travel was not the norm. Their travels exposed them to various cultural environments and expanded their understanding of the world.

Finally, both men were never married for various personal reasons. Their entire focus was on work and on their never-ending travels and adventures, a situation partially forced by the circumstances of their times but also as a result of the choices they made. They became global men only loosely associated with specific geographic regions and with a good understanding of many cultures and peoples.

Looking at these men's lives, an interesting pattern emerges. These individuals can be considered living proof that true greatness can only be achieved by integrating knowledge from various domains and cultures and by balancing the contradictory, personality-based aspects of who they were with the professional components of their lives. Their successes can be directly attributed to their intuitive use of the theory of successful intelligence, which has as one of its foundations the ability to compensate for one's weaknesses through one's strengths. Obviously, there are lessons to be learned through da Vinci and Arciszewski for students like you who are looking for new ways to inspire themselves and become truly inventive engineers.

We can look at the Renaissance and the individual works that were created within that time period from a strictly engineering perspective as well. This may give us, as engineers, the insight with which to revitalize the profession in modern times. The dome of St. Peter's cathedral in Vatican City (Figure 2.12) is a good example of the creation of a novel structural system during that time. It was considered a miracle at the time it was designed and built. Five centuries later, the dome is still admired. The project was a result of a joint effort between artists, architects, and engineers, several of them polymaths. It is also an excellent example of idea integration from several domains. It was created by the Renaissance spirit and, even now, it is still inspiring generations of architects and structural engineers.

The dome was initially designed by the Italian architect Donato Bramante. The process of design and construction, however, took many years (1505–1590), and a number of Renaissance creators contributed to its completion, including Giuliano da Sangallo, Raphael, Peruzzi, Michelangelo, Fra Giocomo della Porta, and Fontana. St. Peter's was the largest and most complex dome structure of its time, with an internal diameter of 136.06 ft (41.47 m) and a total height of 448.06 ft (136.57 m) with respect to the floor of the basilica (Miller 2009). It has a complex ovoid shape and was designed as a two-surface brick shell structure integrated with a system of eight stone arches, very difficult to conceptualize and design even today.

The dome of the cathedral is a result of inventive engineering and the most creative thinking of the time. Artistic, architectural, structural, and construction knowledge was integrated in a sophisticated way, which produced results beyond comprehension for its contemporaries and which still offers modern civil and structural engineers a lesson in humility. There is no question that a structural engineer working alone today would not be able

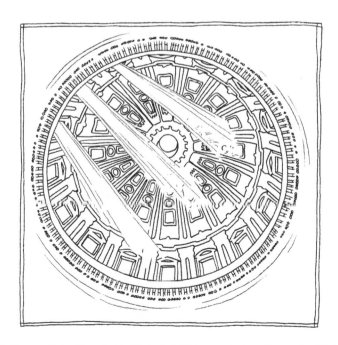

Figure 2.12 Dome, Saint Peter's Cathedral. (With permission. Drawn by Michael Sikes.)

to develop such a design. He or she would have a deep but narrow analytical knowledge limited mostly to the structural aspects of the project with very little, if any, understanding of the artistic and architectural components of the design. Indeed, the genius of the final design needed a long line of Renaissance men to add their input to Bramante's simple initial sketches.

All these individuals were true creators able to integrate knowledge from several domains and to produce a design that is absolutely unique in its novelty, sophistication, and complexity. Your objective as an aspiring inventive engineer is to learn the skills necessary to create comparable inventive designs. This requires a recreation of educational mechanisms, which will lead to a new generation of engineers similar to our Renaissance-era predecessors. It also requires discipline, inventiveness, and an open mind on the part of each inventive engineering student.

2.4 DA VINCI'S SEVEN PRINCIPLES

The essence of the Renaissance creative learning and thinking is captured by Leonardo da Vinci's "Seven Principles" as formulated by Michael Gelb (1998), a best-selling author and motivational speaker who specializes in creativity and innovation. Interestingly, Gelb is also a former professional juggler, and he uses the metaphor of juggling in order to inspire and motivate others to create life balance and work with the Seven Principles. He has authored several books on the subjects of creativity, performance, and success (Gelb 1998, 1999, 2004), the most well known of which is *How to Think Like Leonardo Da Vinci: Seven Steps to Genius Every Day*. In preparation for writing this book, Gelb spent months learning about da Vinci, poring over the dozens of notebooks da Vinci wrote in, and even visiting the places where da Vinci use to live and work. Gelb's Seven Principles are the keys to understanding this prolific engineer, painter, architect, and scientist as well as his genius in general. Most importantly, they provide an intellectual foundation for building a systematic approach to one's own life and work that can be called the *da Vincian approach*. As Gelb explains, da Vinci's thinking processes (which led to his numerous inventions and accomplishments in art and science) can be accessed by everyone.

2.4.1 Principle No. 1: Curiosità

In Italian, *curiosità* means "curiosity." However, as a da Vincian concept, this word has a much more complex meaning. In general, *curiosità* can be described as

> seeking the truth or an infinitely curious and open attitude to life and nature resulting in a never-ending learning process. (Gelb 1998)

The combination of intense curiosity and an open attitude of never-ending learning is the key to da Vinci's genius. *Curiosità* must be explained from several different perspectives in order to understand its transdisciplinary meaning.

First of all, there is the spiritual dimension to *curiosità*. This aspect is directly related to da Vinci's constant spiritual and scientific quest to understand the origins of life. The two seemingly contradictive concepts of spiritual and scientific growth describe best his passion for learning in general. While in Florence, da Vinci began his studies of human anatomy, including the anatomy of pregnant women and unborn babies. His drawing of a human embryo in a womb (Figure 2.13) is the first known anatomically correct rendition of such a subject. The drawing is a symbol of da Vinci's own scientific discoveries, but it is also ripe with spiritual and symbolic meaning because it shows the beginnings of human life.

The drawing also shows the moment when a child demonstrates his or her own *curiosità* in the womb. We are all born with unlimited and unrestricted *curiosità*. Unfortunately, over the course of our lives in the modern world and within the modern education system, *curiosità* gradually lessens. Very few people are able to maintain and cultivate it throughout their lives and, sadly, such people are often the subjects of ridicule. These individuals question established truths and confront authorities by asking difficult and often "strange" questions. Perhaps you recall an individual like this in school or in a job setting; perhaps you were or are one of these individuals.

Figure 2.13 Human embryo by da Vinci. (With permission. Drawn by Michael Sikes.)

By asking the tough questions, these individuals are actually engaging in the learning process and are actively expanding their knowledge about their lives and the world around them. If you want to be a successful inventive engineer in today's world, which requires out-of-the-box thinking and professional leadership skills, then you must preserve your own sense of *curiosità* and change your own perceptions regarding curiosity. Asking tough questions may make you stand out from the crowd and may lead to some raised eyebrows, but how else can you expand your knowledge about a given subject if you do not ask?

Developing a sense of *curiosità* is not easy for most at first, since our modern education system in general was designed to create workers and does not promote creative thinking for the most part. Fortunately, Gelb, and also psychologists, claim that *curiosità* is part of each person's inherent personality. They also claim that it can be a part of a body of knowledge and, as such, can be acquired using specific methods of knowledge acquisition (Gelb 1998). Although *curiosità* can be stunted by nonuse, it can also be restored or even expanded through proper training. The objective of such training, says Gelb, is to nurture a person's emotional intelligence (EQ) and to develop an individual's unique investigative style (Gelb 2004). Da Vinci's investigative style cannot be easily replicated today, but understanding its major components will help you to develop methods and tools of your own for cultivating *curiosità*, which will lead to a more successful career and a more enriching and authentic life.

The da Vincian investigative style can be considered as systematic process of knowledge acquisition and as such can be learned just like any other subject matter. In this process, a certain number of questions about a subject are generated beforehand while more questions are formulated as part of the process of inquiry. When the initial questions are gradually answered, these answers stimulate additional questions and lead to the building of a new understanding of the problem being investigated. The focus for this kind of exercise is not so much on looking for the right answers but on formulating the right questions.

There are several conditions that make the process of *curiosità* successful (Gelb 1998). First, the investigator must conduct formal studies. For example, if the subject at hand was a decision about which way to take your engineering career after graduation when you are a civil engineer, then you would first take some time to do an initial study of the current economic climate and options in general for beginning engineers as well as perhaps a more in-depth study of the particular fields of engineering that appeal to you. Then you must enhance these investigations with daily observations. In this case, this could be subscribing to or reading engineering-related or job-search-related publications as well as taking the time for personal reflection in order to become really solid in your choice of which way to go. Second, you must combine broad studies of many domains that are loosely related to your core issue with in-depth studies of your specific domain.

For the subject of career focusing, this means looking at all the periphery careers that connect to your career directly, such as domains of mechanical engineering, structural engineering, and transdisciplinary fields like bioengineering as well as more loosely related subject areas such as art, architecture, design, and software engineering. Learning about career-focusing techniques in general would fit into this category as well. Next, consider the problem from multiple perspectives and distances, literally and symbolically (Figure 2.9). Imagine yourself involved in your chosen career. What do you see yourself doing exactly? What skills are you using? What is your emotional state? What other factors are involved in your ideal life as you go forward with your dream career?

The above process must be balanced with periods of inactivity so that both the conscious and subconscious sides of the human brain will contribute to the final solution. The inquiry should also be infused with a "poetic language" and the use of metaphors that will stimulate human creativity and provide the necessary *sfumato* (see Section 2.4.4). As we give time to the logical investigation of solutions to our career problem, we must also let our right brain, or unconscious mind, simply "sit" with it for a while. Creative problems are often eventually solved by allowing all the information we have gathered to soak in to our reality.

Finally, throughout the entire process, results, points of inquiry, and "a-ha moments" should be regularly recorded in a notebook. Your notebook should be used to write down thoughts, ideas, impressions, and observations. The purpose of the notebook is not only to create a record but also, much more importantly, to create a foundation for knowledge integration and for the emergence of transdisciplinary knowledge. Carry your notebook with you at all times to jot down questions, inspirations, and ideas as they come to you.

In his books, Gelb (1999) provides detailed descriptions of various specific exercises that can be conducted to improve *curiosità*. One powerful exercise is called "Stream of Consciousness." This exercise requires selecting of one specific question that is important to you (about your work, your relationships, or your life in general or about a more academic subject) and then writing down your related thoughts, ideas, and answers regarding the subject without editing or censoring. The writing should be done without interruption for ten minutes straight. The results of this exercise are usually amazing, since it forces you to think with both hemispheres of your brain. The time limitation also forces you to disregard traditional analytical and deductive attitudes while using the entire brain's power. As mentioned, this exercise can also be used for more professional or academic questions. For example, if you are struggling with a particular step in a design project, take ten minutes and do this exercise. You may be surprised as to what you discover in the process.

Regular contemplation is also an important part of efforts to improve *curiosità*. In this case, Gelb (1998) recommends various exercises dealing with your personality as well as your observations of nature. Such exercises must be conducted in a quiet environment and they require time. However, if they are performed on a regular basis, they gradually lead to the desired attitude change. Other possibilities for improving *curiosità* include becoming involved in an ideal hobby, learning a new language, or working on expressing your authentic self.

As an engineering professor, I have used these exercises with my own students (and for my own subjects of inquiry) with great success. For example, while teaching the course Introduction to Design and Inventive Engineering, I routinely ask students to assess their own *curiosità* using the *curiosità* test from Gelb (1998). Their initial response is predictable, and most are surprised that engineers can even be asked such nonengineering questions. Soon after, however, frank answers are provided and a lively discussion usually ensues.

2.4.2 Principle No. 2: Dimostrazione

The Italian word *dimostrazione* literally means "presentation" or "demonstration." However, just as with *curiosità*, the da Vincian principle of *dimostrazione* means much more (Gelb 1998). It can be understood as an attitude of

- Experimentally verifying acquired knowledge
- Accepting both the analytical/abstract and practical nature of knowledge
- Accepting the value of both positive and negative results (mistakes) and learning from them
- Accepting social responsibility and serving society as a whole

The da Vincian and Renaissance-era concept of *dimostrazione* represents a fundamental departure from the purely abstract and faith-driven scholarship that was prevalent in the Middle Ages. During that time period, only consistency with religious doctrine was required to validate research results. *Dimostrazione* for da Vinci and for us today means direct feedback between research and the world and between processes of knowledge acquisition and verification.

Da Vinci also correctly recognized the significance of mistakes for research and inquiry. Mistakes (called negative examples in computer science) are critically important in the process of both human and automated (computer-generated) knowledge acquisition because both are equally necessary to acquire *decision rules*, which are the basis for formal knowledge and which represent the specific body of knowledge contained in any subject matter.

Finally, in regards to social responsibility, today's version of a Renaissance man, or a person who excels at a diversity of subject matters and has many talents and interests, is expected to be responsible with respect to society as a whole, not just to the Roman Catholic Church as it was in da Vinci's time and prior. In addition, such a person is expected to produce meaningful results and also to acquire knowledge that can be experimentally verified as being consistent with nature. Such knowledge should not only expand the existing body of knowledge but should have a pragmatic impact on society in terms of inventions created and ideas put forth afterwards. New improvements on bridge technology represent a great example of this in engineering. Inventions generated by the Renaissance man or woman should be beneficial to society and its evolution. In this way, a new pragmatic science is born as well.

Da Vinci's inventions represent the best reflection of *dimostrazione*. For example, his studies of nature and aerodynamics led to his invention of a parachute (Figure 2.14) and to specific parachute-related design recommendations that are still sound even in the context of modern science. In fact, Mr. Olivier Vietti-Teppa of Switzerland recently built a parachute inspired by da Vinci and actually used it to perform a perfect and complete 2,000-feet jump in Payerne, near Geneva, Switzerland (McManus 2008).

The Renaissance-era concept of apprenticeship is a good example of *dimostrazione* in action. Apprenticeship during the Renaissance period was a combination of fundamental studies and extensive hands-on training that resulted in experiential learning. This kind of training was very different from the educational opportunities of the Dark Ages and is also very different from the way we learn today. In apprentice-based education, a teacher

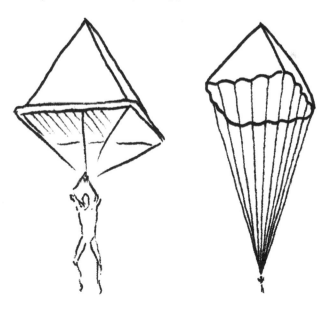

Figure 2.14 Da Vinci's parachute and its modern implementation. (With permission. Drawn by Michael Sikes.)

provides guidelines only to his students. Through these loose directions and the act of observation, students begin to learn the basic subject matter and also gain a sense of independence and confidence as they learn skills on their own. When da Vinci was a master, he always demanded that his students become independent thinkers, "inventores" (Gelb 2004) or *ideators*. Ideators are people who are capable of generating new ideas that may lead to a radical departure from the accepted theories and religious dogma of the day.

The ideal students of the Dark Age were passive followers. They were more like disciples who were only capable of interpreting the thoughts of their masters through deductive reasoning. Renaissance students, on the other hand, were experimentalists and creators. This was especially important as the experimental process began to become a part of actually scholarship. The contrast between the two could not have been greater. Unfortunately, today's way of educating students is often more similar to that of the Dark Ages than to that of the Renaissance. Very often, in the interest of time and practicality, modern students such as yourself are required to memorize facts and procedures without "wasting" time and resources on experiments, hands-on classroom activities, or experiential learning. In general, this kind of emphasis does not lead to successful engineers who can become leaders and the inventors of creative solutions to engineering problems.

Dimostrazione is an attitude, and as such it can be measured and improved in a systematic way. Gelb (2004) provides a reflective self-assessment test with eleven probing questions to begin to cultivate the attitude of *dimostrazione* in oneself. According to Gelb, taking this test is the first step in improving one's *dimostrazione*. The test can be done in many ways. For example, you can identify several of the most influential experiences of your life and describe them in short sentences. You can also reflect on how you use accumulated experience on an everyday basis. More difficult is the analysis of your core beliefs when considering several general and more esoteric areas such as human nature, politics, and the meaning of life. In each case, your responses should be written in your notebook. When you ask yourself these questions, consider the sources of your beliefs, their nature, and your personal emotional responses as well. This mental exercise should be followed by a reflection on the roles of media and family in the development of your beliefs. How have you come to believe what you currently believe? What or who has influenced you throughout your life? Obviously, the ultimate goal of this assessment is to develop your self-awareness about your own beliefs and their subjective nature. These inquires and your subsequent discoveries naturally lead to the development of a multiperspective understanding of the world and nature, which opens you up to the possibility of creative solutions to problems, both academically and in life.

Making mistakes is also part of learning. Acquiring knowledge happens inductively using both negative and positive examples; learning from mistakes when they occur can be considered a useful part of obtaining results in any inquiry. Having a positive attitude toward mistakes is difficult for anyone, but this attitude can be created, again, through the question–answer process. Conducting stream-of-consciousness writing (Gelb 1998) on questions such as "What would I do differently if I had no fear of making mistakes?" is a powerful exercise that can provide you with a different perspective on the mistakes you have made in the past or may make in the future.

Finally, creating affirmations for yourself that are believable to you can contribute to a growing sense of *dimostrazione* from within. For example, patience and resilience in the face of adversity are quite important. These attributes can be affirmed through writing and repeating sentences such as "I feel patient with myself" or "I never quit when problems arise."

2.4.3 Principle No. 3: Sensazione

In Italian, *sensazione* means "feelings," "sensitivity to feelings," or "experiencing a sensation." Da Vinci used this word to name a complex state that can be described in the following way (Gelb 1998):

- The development of all the senses
- Building both the rational/intellectual and emotional approaches to life, nature, and problems
- The integration of all abstract and physical inputs to create synesthesia
- Using synesthesia to acquire transdisciplinary knowledge or to create new ideas

Da Vinci strongly believed that every person is capable of being a complete human being—that is, a person with a transdisciplinary understanding of the world that reflects the perfect balance of art and science (as described in Principle No. 6, "Arte/Scienza"). He also believed that there should be a balance between emotions and intellect. To acquire transdisciplinary knowledge, a human needs to use the power of his/her entire brain, both left and right.

In the Middle Ages, prior to the Renaissance, education was mostly focused on analysis and on the development of the left (analytical) part of the brain at the expense of the right hemisphere, which deals with the human emotions, creativity, and sensory inputs. Unfortunately, this is similar to how students are taught today as well. If you want to acquire transdisciplinary knowledge and become an inventive engineer, developing the right hemisphere of your brain through focusing on all the senses and on the integration of the senses and thought is of utmost importance.

The ultimate goal with *sensazione* is to create the phenomenon of synesthesia. Synesthesia is a complete integration of various abstract inputs (thoughts coming from various domains) with our physical sensory inputs.

A human receives physical or sensory inputs from his or her five senses, including vision, hearing, smell, taste, and touch. The phenomenon of synesthesia can be considered as an intellectual and emotional state in which both the intellect and all five senses are activated and their interactions create a new state of awareness, leading to knowledge integration. Synesthesia requires a conscious activation of all senses, no doubt a difficult task for the average human. Fortunately, it is possible to learn how to become aware of all five senses and how to use them to create synesthesia. Da Vinci believed, and I agree, that this is the best way to create transdisciplinary knowledge, or new ideas, and is the key to creative intelligence.

Sensazione is an attitude that is difficult, but not impossible, to develop. When we are born, we are aware of all our senses and we use them all the time to learn about the world. We are open to all kinds of physical inputs coming from the surrounding environment. As we grow older, however, we are told not to touch certain things, not to smell certain things, and not to do certain things out in the world. Also, we gradually learn not to show our emotions. We learn how to control our emotions or, even worse, to focus only on the rational and nonemotional parts of ourselves and our lives. In this way, we learn how not to use all our senses and how to separate our emotions from our intellectual life. If we want to develop the attitude of *sensazione,* we need to rediscover our emotions and our senses. Interestingly, many successful and creative scientists in mathematics and engineering are people who are able to maintain a balanced attitude during their entire life, including keeping their emotional attitude alive and using their senses through various artistic activities, such as playing piano, painting, or dancing.

The attitude of *sensazione* can be gradually developed, but as is the case with the other da Vincian principles, it requires time and deliberate effort. The most difficult part of such training, but a part that is critical to inventive engineering, is to develop an ability to relate drawings and sketches to sounds, colors to textures, sounds to flavors, and so on. Such cross-referencing of sensory inputs is absolutely necessary to create true synesthesia and a full integration of sensory inputs. For example, we could say that as one listens to a composition by Debussy, it smells like the ocean; or that using belt truss systems in the structural systems of tall buildings is like seeing the color blue. Take a moment to reflect on the experience of sense integration and crossover in your own life. Was there ever a time that a sight appeared like a sound, or a texture conjured up a certain smell? That was *sensazione* at work in your life.

It is possible to measure your own level of *sensazione* so as to help you acquire or improve it. Gelb (1998, 2004) provides several self-assessment tests for the individual senses and also a self-assessment test of a person's ability to practice synesthesia. He also provides various mental exercises for the improvement of all the senses and their integration. Da Vinci believed in a clear hierarchy of senses from the perspective of synesthesia.

Vision was obviously considered the most important sense for him, closely followed by hearing, touch, smell, and taste. In this chapter, only vision will be discussed because of its importance to engineering. However, many details about the remaining senses and mental exercises to improve them can be found in Gelb's works on the subject (Gelb 1998, 1999, 2004).

The eye encompasses the beauty of the whole world.

This short statement by da Vinci clearly reveals his understanding of the supreme importance of vision. At the same time, it is an excellent example of his own *sensazione*. The use of metaphor and of emotionally charged words ("the beauty of the whole world") creates a complex rational and emotional description of the sense of vision and its significance. Da Vinci also claimed that looking is not enough, however. He wrote,

Do not look without seeing.

This statement means that receiving a visual sensory input is only part of the cognitive process of learning. *Saper vedere* ("knowing how to see") became an important part of the never-ending nature studies that da Vinci conducted throughout his life. Obviously, he clearly believed that seeing was a part of learning. Such an understanding is entirely consistent with modern concepts of vision. For example, the use of sketching is considered vital for problem solving in business (Roam 2008). Vision is also a part of the human-reasoning methodology called *visual thinking* (Arnheim 1969) and is even thought of as part of human intelligence as a whole (as in the concept of *visual intelligence* introduced by Hoffman 1998). Can you think of an experience, either within the classroom or in your life in general, when you suddenly "saw" a subject from a different or larger perspective, where maybe before you could only see the parts? That is an example of the importance of "vision."

There are at least two ways to improve your vision and your ability to "see," in the larger context. Learning to draw by hand is one way. Learning how to sketch serves at least two purposes. First, drawing teaches you how to present and represent complex concepts quickly and in a natural way. Second, hand drawings are never perfect, but their fuzziness is desirable in accordance to Principle No. 4 (*sfumato*) because they allow the imagination and a sense of possibility to emerge. Technically perfect, computer-generated drawings always seem to be complete in and of themselves, implying that the work is over. Such drawings, however, are produced using a limited number of objects that are available in a given computer tool. This obviously restricts the imagination, forcing you to think only in terms of a specific domain. On the other hand, the unfinished nature of hand drawings creates a proper framework for further thoughts and for the continuation of work. The feeling of imperfection and work in progress associated with hand drawings is also inspiring and provokes further ideas. As a side note, the desire to evoke a feeling of inspiration is one of the reasons why true visual artists were hired to prepare illustrations for this book. When sketching, you can easily produce your own objects, which expands your search for ideas in general. Did you know that hand drawing was widely taught as part of civil engineering education until only about 30 years ago, before it was replaced by a mechanistic approach based on the use of various computer tools? Gelb (1998) provides a short "Beginners' Da Vinci Drawing Course" in his book.

There are many strictly mental exercises to improve one's vision and ability to see as well. For example, in the exercise "Describe a Sunrise or Sunset" (Gelb 1998), a person watches a sunrise or sunset in a quiet place. First, however, before the actual watching takes place, the person is asked to relax his or her body and mind through several deep, full breaths followed by extended exhalations. After the event, the person describes the experience in as much detail as possible in his or her notebook. In another mental exercise called "Create Your Own Internal Masterpiece Theater" (Gelb 1998), the participant is asked to select a famous painting, for example da Vinci's *Mona Lisa*. Next, a quality reproduction of the painting is hung on a wall and studied for at least five minutes every day for a week. Then, as the individual is preparing for sleep, he or she is requested to recreate the painting in his or her mind's eye with all its details. In addition to shapes and color, appropriate sounds and smells can be added. Such a visualization should be performed for several days, and each time its results and comments should be recorded in a notebook. When the exercise is over, the person can reflect on what they wrote about the experience and, if possible, compare their observations with others doing the same exercise. A growing sophistication of images and impressions can usually be observed by individuals doing the exercise over time.

Da Vinci believed that painting was the highest form of art. This was a reflection of his belief in the supreme importance of vision, but also because paintings were the best way to present *sensazione* in a way acceptable to a diversity of individuals during his time. His famous painting *St. John the Baptist* (Figure 2.15) is an excellent example. For engineers, the flow of robes in the painting is perfectly consistent with fluid dynamics (and was inspired by da Vinci's studies in this area). Also, the entire composition seems to be balanced, bringing a unique kind of joy to many structural engineers. At the same time, nonengineers can perceive this balance as a picture of spiritual harmony and peace. There are no sharp edges in the painting; it is a reflection of the complexities of life and love as well as the complexities of engineering theories and models. The colors are warm and inviting. A sensitive person practicing *sensazione* can easily generate many interpretations of the painting, as it is rich intellectually as well as thought-provoking and highly sensory.

Figure 2.15 *Saint John the Baptist* by da Vinci. (With permission. Drawn by Michael Sikes.)

2.4.4 Principle No. 4: Sfumato

Da Vinci was a master of complexity. He always managed to present complex and sometimes not fully understood notions in simple terms and often in a way that was purposefully contradictory. In modern times, this is done in Synectics as well (see Chapter 8). Synectics is a problem-solving methodology that stimulates thought processes of which the subject may be unaware. It was developed in the 1960s and is now widely used. *Sfumato* is also a term used by visual artists to describe particular color and technical attributes and techniques where a blurring of defining lines occurs within a particular image. Such concepts, when combined, create a delicate balance between various meanings reflecting the complexities of the world. The principle of *sfumato* addresses all these related and important issues.

In Italian, *sfumato* literally means "going up in smoke," "smoked," "turn to mist," or, in the case of colors, "soft" or "mellow." As an attitude, *sfumato* may be interpreted as (Gelb 1998)

- A willingness to accept and understand the world in its infinite complexity
- Keeping an open mind in the face of the unknown and uncertainty
- A willingness to embrace contradictions and paradoxes
- Acquiring, accepting, and using ambiguous knowledge

The attitude of *sfumato* represents a paradigmatic shift from the scholarly attitudes promoted in the Middle Ages. During those times, scholars were supposed to discover the truth in the form of crisp, or binary, results with a clearly distinguished boundary between good and bad; that is, between God and the devil. *Sfumato* represents a much more sophisticated attitude. It is like the shift from binary thinking to multivalue logic, which led to the development of the theories of fuzzy (Zadeh 1965, Ross 1995) and rough sets (Pawlak 1991) in engineering. These theories offer a much more realistic and complex understanding of the world than traditional binary logic. Their introduction in the 1980s and 1990s led to revolutionary changes in many areas of engineering and science.

In the same way, *sfumato* represents the end of black-and-white thinking about the world in general. It teaches us that a complex concept (e.g., the concept of "a man") cannot be reduced to a single point (i.e., a single example of "a man"). These complexities must be described by using

Figure 2.16 Old Man and a Boy by da Vinci. (With permission. Drawn by Michael Sikes.)

the entire spectrum of concepts contained within two extreme opposites. Unfortunately, engineering formulas are usually presented in terms of the mechanics of materials with very limited discussion of the assumptions behind them and their often controversial nature. These discussions must be had, however, if true understanding is to occur about the very nature of engineering itself and if inventive engineering is to be developed that can produce truly revolutionary inventions.

In more abstract and overarching terms and adhering to our example of the concept of "a man," Figure 2.16 provides a good example of a spectrum that describes a human life in its extremes. Supposedly, both figures represent da Vinci. On one side, he is in his youth, and on the other side, he is an old man. Looking closely at the lines in the drawing, you can see that there is no sharp boundary between figures, only a continuity representing gradual changes in the subject's life and its constant evolution. When such understanding of complex concepts is accepted and when the extreme and opposite ends of such concepts are taken into consideration, the determination of where the spectrum ends on either end can be qualitatively determined by each individual. This is a critical step in any learning process.

An example of the importance of looking at extremes in complex concepts can be seen by looking at da Vinci's life and work in particular. Da Vinci was always engaged in a search for beauty, and he obviously studied beautiful people. However, he was equally interested in ugly or grotesquely disfigured people. Intuitively, da Vinci discovered the fundamental assumption of learning: to acquire knowledge, we need both positive and negative examples. They are equally important.

Sfumato also means that to learn about a complex concept, contradictions represented by paradoxes must be accepted and considered as a legitimate part of the acquired knowledge. Such understanding is surprisingly modern. Only in the 1940s did Altshuller (1969) propose discovering contradictions (and using them) as a part of inventive problem solving. The use of paradoxes is also promoted in Synectics. In the case of Synectics, the formulation of a problem as a paradox is expected. It is intended to initiate a complex cognitive process, which is still not fully understood but has proven to be powerfully effective.

In addition, *sfumato* means that a Renaissance person can willingly accept the frequent ambiguity of his or her observations and purposefully incorporate this ambiguity into his or her body of knowledge. Such an attitude is partially a reflection of the mysticism that was prevalent in the Middle Ages and at the beginning of the Renaissance. However, it also represents an attitude of openness to uncertainty in general and is consistent with the modern understanding of cognitive processes. In the case of knowledge discovery and inventive design, such processes can be lengthy and have subsequent periods of conscious and subconscious activity. To produce novel ideas, all kinds of input are desired in order to use the entire power of the human brain, both the analytical left hemisphere and the creative right hemisphere. The latter, that is, the right side of the brain, particularly benefits from ambiguous knowledge as it has the potential for stimulating all kinds of associations. These associations are critical for inventive problem solving and are also a key part of Synectics.

Figure 2.16 shows St. John the Baptist in his famous pose with a raised finger. The message of this painting is subject to many interpretations. Nothing is clear about the painting. Therefore, it is much more stimulating intellectually and emotionally than a painting with a clear message communicated only on the rational level. The understanding of complex concepts in the world as proposed in *sfumato* is acceptable today, and at least some within the engineering community have accepted such approximate theories as the theories of fuzzy (Ross 1995) or rough sets (Pawlak 1991). Modern scholars are much more open to ambiguity than ever before. Six centuries ago, however, this attitude was truly revolutionary.

2.4.5 Principle No. 5: Arte/Scienza

The concept of *arte/scienza* ("art/science") means that a Renaissance person should be a *whole-brain thinker*. He/she should build an understanding of the world by using two entirely different but complementary perspectives with roots in art and science, respectively. These two perspectives should be balanced since both are necessary but neither is sufficient. In more specific terms, this attitude can be described as (Gelb 1998)

- Always maintaining a balance between art and science
- Using both artistic and engineering knowledge
- Using both emotional and rational approaches
- Using both imagination and logic
- Using both the right and left brain hemispheres
- Seeing both the whole picture and the details

The common assumption about engineers is that they are left-brained. Such individuals have a profile that can be described as analytical, with logical/deductive reasoning being the preferred type of thinking. This assumption is generally correct. In my opinion, however, the practice of reinforcing this existing brain-dominance profile in engineering education is damaging to the potential that exists in the profession. By maintaining this image, we significantly reduce the chances of amazing engineering innovations being brought into being by future engineering leaders such as yourself. Engineers who consciously employ whole-brain thinking, on the other hand, possess the potential for truly inventive engineering practices, which can benefit not only the profession but the entire world.

It is not easy to develop and practice whole-brain thinking when one is not used to thinking in this way. Gelb (1998) provides a self-assessment test to determine your level of arte/scienza as well as a number of exercises that develop this type of thinking.

Mind mapping is a powerful way to cultivate whole-brain thinking, and some believe that it is the key to whole-brain thinking. The concept of a mind map goes back to da Vinci's practices. However, it is also supported by modern research in cognitive psychology (Buzan and Buzan 1994). Surprisingly, this concept has also been reborn in computer science as ontology, and its use in engineering is considered to be state of the art (Gruber 1993). Within engineering, the use of mind mapping and ontologies is limited to the building of complex, multimedia, knowledge-based systems for training and decision making (Oguejiofor et al. 2004). As you will see, however, the practice can in fact be used for every aspect of your academic and personal life.

A mind map is a visual representation of a body of knowledge related to a specific central concept. It is designed to show all related concepts and their relationships in a nonlinear way, with the main concept centrally positioned and secondary concepts surrounding it. A mind map should be produced using pictures, words, colors, and differentiated line thicknesses and fonts, but there are no formal rules on how to actually create one. With practice, however, your artistic richness and sophistication usually gradually improves. A mind map should be drawn by hand if possible so as to increase its spontaneous nature and to maximize the involvement of the right brain hemisphere. Over the last several years, however, many computer tools for mind mapping have been introduced. For more details and examples, see Section 5.3.

If you are interested in thinking outside the traditional linear and hierarchical knowledge structures that are predominant in engineering and in becoming a truly inventive, creative, and successful engineer, then I suggest you use mind mapping as a tool for problem solving and idea generation whenever you can. The use of pictures and colors requires right-brained thinking and imagination, which leads to a truly cognitive "revolution" and to novel, rewarding results.

The principle of arte/scienza is very important, and mind mapping is the fastest way to bridge this gap within yourself. It represents a significant departure from how you might be used to thinking about academic and life problems, but the rewards for your learning and life in doing so will be well worth it.

2.4.6 Principle No. 6: Corporalita

The Italian word *corporalita* in English means "corporality," or "the state of being in physical or in bodily form rather than in spiritual form" (MSN Encarta). Da Vinci's attitude of *corporalita* has a much more complex and interesting meaning, however. This attitude is described by Gelb (1998) as *sana in corpora sano*—a sound mind in a sound body. In more specific terms, it can be described as (Gelb 1998)

- Recognizing the fact that a human being has mind, spirit, and body and that they are wholly integrated
- Seeking a balance between intellectual and physical development
- Recognizing the importance of the human body
- Recognizing the fact that a fully integrated, balanced, and optimally functioning man or woman must have a healthy body
- Seeking health through the holistic approach in which a human body is considered a system
- Recognizing the fact that a human body is like a microcosm and that its energy field and the complex interactions within it exist among various parts

The Dark Ages was an era of ascetics. During that time, the human body and its needs were considered unimportant. It was believed that the body provided only a temporary shelter for the spirit and, as such, did not need to be respected. In fact, sometimes the physical body was abused in order to demonstrate the purely spiritual nature of human existence. The Renaissance changed this way of thinking. The human body was again important, as it was during Greek and Roman times. This new way of thinking stated that the body not only supported human existence on Earth but was also a powerful instrument for the enjoyment of life. Most importantly, during the Renaissance, the body was thought of as an integrated system that included the mind and spirit as well. This system was a necessary component for serving both God and society.

When da Vinci was an apprentice in the workshop of Andrea del Verrocchio in Florence, he became interested in human anatomy. First, he wanted to learn the "divine proportions" or the "Golden Rule" describing the ideal proportions of the human body and its ultimate beauty. Next, he became intrigued by the incredible complexities of the body itself. Several historians even claim that in 1472, da Vinci joined the Company of St. Luke—a guild of apothecaries, doctors, and artists—in order to have access to hospitals and to continue his anatomical studies. As a result of these studies, da Vinci developed a unique understanding of a human being as an integrated system. Da Vinci's *Vitruvian Man* (Figure 2.17) became a symbol of his holistic/systems understanding of the unity of mind, spirit, and body.

Da Vinci intuitively used the concept of fractals to make the claim that "the human body, and the energetic system that surrounds it, is a miniature replica of a larger, universal system" (Gelb 2004). This fascinating view of the body was immediately rejected by his contemporaries in medicine. Only today, several centuries later, is da Vinci's holistic understanding of the human body and its energy field finally accepted in the area of energy healing and used to heal people through various practices such as Reiki.

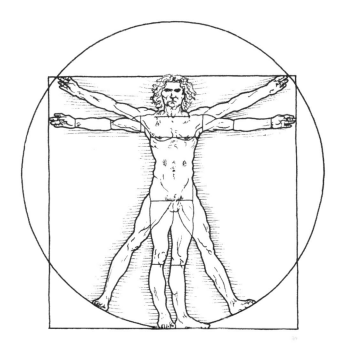

Figure 2.17 Vitruvian Man by da Vinci. (With permission. Drawn by Michael Sikes.)

Da Vinci believed that a human being should carefully maintain a balance between his or her intellectual and physical development in order to realize his or her full potential. This line of thinking was also a reflection of the Renaissance-era belief that a genius must be physically superior to ordinary people. Da Vinci was a skilled equestrian, and many other Renaissance leaders were also outstanding sportsmen. Da Vinci developed a surprisingly modern physical development system, which combined three major elements: a healthy diet, a complex training program, and body awareness.

Da Vinci practiced these tenets in his own life. He was a vegetarian. He drank lots of water and very little wine. He ate meals that included fiber-rich vegetables, grains, beans, and fruits, all prepared with olive oil. The best example of such a diet is his famous minestrone soup. He also recommended that eating should be a moment of enjoyment with full focus on the food. One should find pleasure in every bite while contemplating, at the same time, the nature and origins of all the ingredients and their unique taste.

Da Vinci's physical training system had three distinctive components, including aerobic conditioning, strength training, and flexibility exercises, each with a different goal and all carefully balanced for maximum benefits. Aerobic conditioning was intended to improve awareness, emotional stability, and stamina. Also, it was supposed to prevent arteriosclerosis and premature aging (a health theory he proposed as a result of his anatomical studies). Strength training was used to burn excess fat, strengthen muscles and, again, prevent osteoporosis. Finally, flexibility exercises were intended to prevent injuries and to improve the circulatory system.

The development of body awareness was a uniquely da Vincian idea, and it should be considered within the context of his holistic understanding of the human body, which always balanced the purely physical and purely spiritual aspects of life. Becoming aware of one's body means having a clear picture of it, understanding how it operates, and understanding how individual parts are balanced in physical, functional, and spiritual terms. Again, such awareness can be developed through various physical and mental exercises described in Gelb's book (Gelb 2004). For example, in one such exercise, a person is to witness his or her naked body in a mirror and to prepare a drawing. Next, the entire body map is to be explored to find places where the head balances on the neck, where shoulder joints are located, and so on. Ultimately, body awareness is intended to help a human being utilize his or her body in the most optimal way. This leads to improved performance in all activities.

On the surface, this principle may seem unrelated to inventive engineering. I believe, however, that *corporalita* is particularly important for inventive engineers. We need to maintain a balance between body and mind for mental stamina as well as to attain a relatively high level of physical fitness in order to work for many hours, sometimes days and nights, on our inventions.

2.4.7 Principle No. 7: *Connessione*

Connessione means "connection" in Italian, but like da Vinci's other principles, this concept signifies much more. It can be interpreted as (Gelb 1998)

- Recognition of the interconnectedness of all phenomena in nature and life
- Understanding the world as a single system with all its elements connected by direct and indirect feedbacks
- Understanding that the world is a complex and chaotic system
- Understanding that knowledge is created within a nonlinear system

Da Vinci studied human and animal anatomy. This inquiry led him to the understanding that there are many similarities between human and animal organs. It also led him to realize that all living systems could be viewed within a larger framework. From da Vinci's vantage point, differences between mammals are insignificant and, in fact, a unified picture of nature can and does emerge. This concept was rediscovered several centuries later in the scientific community by systems scientists. In fact, only recently, in the second half of the twentieth century, has the science of the holistic understanding of the world called *cybernetics* emerged. Cybernetics has gradually led to the development of *systems analysis*, which is based on the principle of wholeness in its approach to the analysis of all systems, built and natural, real and abstract, small and large. *Connessione* may be interpreted as a "systems" view of the world.

In relation to this, during the last 20 years, chaos theory (Alligood et al. 2000, Arciszewski et al. 2003) has been reluctantly accepted by the scientific community as well. Chaos theory takes systems theory a step further in recognizing that complex systems may become unstable and may display chaotic behavior. As part of this behavior, even small quantitative changes in certain critical parameters may produce significant, qualitative changes in the behavior of the entire system. This may include a trajectory change in the case of planetary systems, for example, or a buckling of structural members under compression. Da Vinci understood such behavior in the sixteenth century. He wrote, "The earth is moved from its position by the weight of a tiny bird resting on it."

Da Vinci left thousands of pages of his personal notes covering all areas of science and art. However, he never attempted to organize them nor to categorize individual items. Da Vinci strongly believed that all his thoughts were interconnected and equally important. As such, they too should be seen in their "systems" context. This understanding was surprisingly modern. Da Vinci was a highly creative person who used various pieces of knowledge from many domains to create new concepts. This is similar to creating an ontology, or way of thinking and studying reality, in which many concepts are interrelated and exist without a hierarchical structure. Da Vinci had a lifelong practice of combining and connecting disparate elements to form new patterns. This process was not much different than the process of morphological analysis proposed by astrophysicist Fritz Zwicky (1969) in the late 1940s. In this method, a problem is divided into a number of subproblems that are considered entirely independently. For each subproblem, various solutions are found and presented in a *morphological table*. A solution to the entire problem is found as a randomly produced combination of solutions to the individual subproblems. For example, when da Vinci was a young boy, he created his first dragon (Figure 2.18) and described the process of drawing it: "make dragon looking naturally—take for its head that of a mastiff or setter, for its eyes those of a cat." By following his own step-by-step instructions for adding relatable images together, young da Vinci was able to create something quite fantastical and new.

The attitude of *connessione* can be quantitatively assessed using tests developed by Gelb (2004). One very personal and life-focused example is a mental exercise called "Family Dynamics." This exercise involves answering questions about family and about the functions of its individual members. In addition, there are questions about how the family has changed over time and about the patterns of evolution behind these

Figure 2.18 Dragon by da Vinci. (With permission. Drawn by Michael Sikes.)

changes. Gelb (1998) believes that understanding family dynamics also improves one's understanding of one's own personality and life goals. In this particular exercise, Gelb recommends drawing a diagram representing your family as a system. This drawing may subsequently lead to metaphorical questions about the family as seen as a single human body. Some questions include: Who in the family is the head? Who is the heart? Who would be at the limbs? Who would be standing on solid ground as the feet? For the engineering profession, this same exercise can be extremely helpful when analyzing the dynamics of a specific company (one that you may be considering working or interning for) or it can be used when evaluating engineering issues themselves, such as looking at the relationship between the parts of a vehicular system and determining their interrelating importance.

Connessione in terms of creativity may be developed through other inventive exercises as well. When seen through the lens of *connessione*, the key to human creativity is assumed to be the ability to see unfamiliar combinations and connections. For example, if you were asked to invent a new concept of a floodwall that combines knowledge from both foundation and naval engineering, how would you go about doing so? What interrelated factors would you have to consider? What materials would you use, and how would you assemble them in a way that promotes interrelational harmony and functionality? Using the da Vincian concept of *connessione* (as well as tools such as mind mapping) may help.

Another worthy general exercise that is simple and also fun involves contemplating the origins of all components of a meal. Once a meal is visualized, prepare a drawing or mind map representing it and how each individual ingredient contributes to the whole of the meal. Of course, this contemplation exercise can also be conducted using engineering subjects. Consider, for example, the structural components of a large-span suspended bridge. How do all of the individual parts relate to create a unique whole? What would happen if you left out or removed one part? How would this effect the entire structure? How would adding a new component change the whole?

Perhaps one of the most important, indeed life-changing, exercises for general life evaluation is called the "Time Line River of Life" (Gelb 1998). This is also a great exercise to do when you are at a crossroads in life, such as just after graduation and before entering the job market. The goal of this exercise is to help you develop a bigger picture of your life in order to be a better human being, to live a meaningful life, and to increase your chances of success in all aspects of life. The exercise requires the use of metaphorical thinking as well as the commitment to devoting at least one week to it.

For this exercise, first consider life as a river flowing from the mountains to the sea. As the river flows to its ultimate destination, it encounters all kinds of dams, levees, rapids, and so on. For the next seven days, focus on just one aspect of your time line river of life per day. When you focus on these aspects, imagine them as part of your river:

- Day 1: Sketch the big picture of your dreams (a mind map): What does your river look like in general?
- Day 2: Explore your goals: As you move along your river, what do you see up ahead?
- Day 3: Clarify your core values: Imagine a challenge coming your way upstream. What do you do about it?
- Day 4: Contemplate your purpose: Why are you in the river in the first place? Where are you going? What are you moving away from?
- Day 5: Assess current reality: What does the water you are moving on look like? Is it flowing fast or slow? What do you look like? What are you wearing? Are you sailing or swimming? What does your craft look like? Are others with you? How are you feeling, and what are you thinking about?
- Day 6: Look for connections: What do you see around you? Are there fish or wildlife? What do you see on the bank of the river? What is the weather like? Can you see any connections or synchronicities (interesting, interrelated coincidences) happening as you move down the river?
- Day 7: Strategize for change: Imagine yourself getting closer and closer to your intended goal. Is this goal changing or staying the same? How do you feel about this goal? Are you excited, happy, or anticipating what will happen next? Or are negative emotions coming up for you? If you need to change your goal, that is OK. Part of this process is to determine if our current goals are still valid based on changes that have occurred in our lives and those that continue to occur.

Do this exercise and all the exercises presented in this section using mind maps. The insights that can be gained by looking at issues and problems with a sense of *connessione* can be quite dramatic and eye opening as well as extremely insightful for your life and work.

2.5 THE MEDICI EFFECT

Frans Johansson (2004), in his book *The Medici Effect*, recently proposed an intriguing concept that speaks to creativity and the generation of novel ideas. He describes the specific environment that eventually created the intellectual foundation for the Renaissance as part of a transformational paradigm he calls the "The Medici Effect."

The Medici family lived in Florence, Italy. Their most well-known family member was Lorenzo Medici, or Lorenzo il Magnifico. In the fifteenth century, the Medici family had accumulated a lot of wealth through banking and was considered one of the richest and most influential families in Europe. They also became great supporters and sponsors of the arts and of science. They pursued their philanthropic projects for a combination of reasons, including their desire to increase the prestige of their family, the need to expand their power base, and also because of a sense of guilt for their wealth and prestige. During the time the family was thriving, the practice of banking, and particularly lending money with interest, was considered sinful and an obstacle to salvation. Apparently, the Medici family was concerned about their collective salvation and tried to balance their banking practices with positive contributions to society.

Many leading Italian artists and scientists of the day became members of the Medici court and lived together there at the invitation of the family in conditions of relative comfort. There, these individuals had time to focus on their work but were also forced by circumstances to interact with each other. Together, over time, they created a small community. This was a rare occurrence, since the Middle Ages, or the Dark Ages, was a time when interactions between individuals of different disciplines was not the norm. All members of the Medici court talked, debated, and even argued about their individual works. In the process, they began to break down the barriers between disciplines and began to generate all kinds of ideas, often randomly and in the moment. These ideas were then analyzed from various perspectives. Many of them were rejected, but those that were accepted often led to new understandings of each discipline and to the gradual emergence of transdisciplinary knowledge. This knowledge represented a new understanding of the arts and sciences. More importantly, it led to a heightened understanding of the world in general. Ultimately, the knowledge and ties between the individuals in the Medici court became the very foundation for the Renaissance.

Johansson (2004) also proposes the concept of *intersection* as it relates to the development of Renaissance-era thinking. Intersection is a result of the Medici effect and can be described as "the time and place-specific integration of knowledge with components coming from various disciplines, cultures, and personalities." When a new idea is generated within a given discipline, it usually happens along an existing line of evolution and therefore can be considered "directional" (Johansson 2004). When an intersection occurs, however, that new idea represents a true paradigm change, or the beginning of a new line of evolution. Such an idea can be called an *intersectional idea* because of the unique mechanisms behind its development and because it often has a great impact on the evolution of its domain. A great example of intersection is the public introduction of the Internet as the World Wide Web in the 1980s. Not only did this introduction change its domain (computer science and technology), it also created a global paradigm shift that, as we know, was to affect every aspect of modern life.

In Chapter 14, "Bio-inspiration in Designing," we will discuss intersection in the context of biology and engineering and how using knowledge from biology in conceptual design requires the integration of knowledge from these two domains but at the same time how this process changes our understanding and leads to inventions.

From a social perspective, intersection is a phenomenon that naturally results when people of various backgrounds, professions, and cultures are put together and encouraged to interact. Over time, such interactions lead to dialog and an exchange of ideas and concepts. They also lead to the development of new ideas reflecting integrated thinking with roots in several separate disciplines (Figure 2.19).

It takes time to create intersection. Intersection is a knowledge acquisition process and can occur when professionals from two or more domains work together on a problem and along the way discover that the same concept often has entirely different meanings in various disciplines. Both learn the *situatedness* of these concepts (Gero 2007); that is, how the concepts change depending on their domain-related context. This process is similar to Synectics, in which we intentionally change the meaning of various concepts using contexts from many domains that are randomly selected. In the case of Synectics, changing the context is done purposefully and in a semisystematic way, and this has the result of *making the strange familiar*. The results of Synectics also include *making the familiar strange* and making excursions. Excursions are created by selecting words (concepts) from outside the problem domain and trying to interpret them in the context of a new domain. Such a process obviously expands our body of knowledge through the introduction of new concepts with roots in various domains that are not necessarily related to our area of expertise.

Intersection can be considered on the level of a community—as was the case at the Medici court in the fifteenth century—or as a phenomenon that simply occurs between two people. It can also be considered on the level of

Figure 2.19 Intersection. (With permission. Drawn by Michael Sikes.)

a single person who randomly generates ideas using his or her knowledge from various disciplines and situations, which are informed by different cultural experiences. Intersection that occurs within a single person is also closely related to Synectics.

There are at least three factors that are driving the process of intersection in the world today. They are globalization, the evolution of science in the direction of transdisciplinary knowledge, and the information technology revolution (Johansson 2004). Globalization encourages, and indeed often forces, people of various cultures and languages to work together and to combine, merge, and create new ideas from diverse backgrounds and perspectives. In terms of scientific evolution, there is growing interest in the integration of existing scientific disciplines such as genetics and computer science in order to develop a transdisciplinary, or integrative, understanding of the world. This understanding is absolutely necessary when dealing with complex, adaptive systems and has, in fact, created new scientific disciplines such as cybernetics, the systems sciences, chaos theory investigations in physics, and evolutionary design in engineering. Finally, rapid progress in information technology (IT), particularly in the area of computing, has created new opportunities to simulate complex phenomena and acquire advanced knowledge. This ultimately has led to the intersection of disciplines and the creation of transdisciplinary knowledge as well. We have more advanced intellectual resources today than ever before. If forward-thinking scientists and artists are able to initiate another period of qualitative changes resulting in a new wave of modern inventions, then the next Renaissance would be much more powerful than the last one.

Chapter 3

Lessons from modern science

3.1 POLITICAL SCIENCE

3.1.1 Creative community and the creative class

In 2002, urban theorist Richard Florida (2002) published the groundbreaking book *The Rise of the Creative Class*. In it, Florida describes a new class of professionals he calls the "high bohemians" and makes the claim that in metropolitan areas with a high concentration of artists, musicians, gay and lesbian individuals, and high-tech workers, economic development reaches a higher level than in other areas. Many concepts proposed in Florida's book are relevant to our central concept of inventive engineering. These concepts will be used in the following pages as well as in Chapter 4 in order to explain the importance of creating a successful environment for inventive engineering education and learning. According to Florida, modern American society has three major classes. They are manufacturing, the services, and the creative classes. He defines creativity as the production of nonmaterial goods, that is, knowledge. Therefore, all people involved in the production of knowledge are creative and are members of the creative class. Florida estimates that approximately a century ago, roughly 10% of the US population belonged to the creative class, while today this number is about 30%. The growth of the creative class obviously occurred at the expense of the manufacturing class and reflects the shifting distribution of people within our society. Today, the creative class is credited with approximately 50% of all wages and salary income. The importance of the creative class is constantly growing, as this class is shaping the future of our society in the wake of globalization and the outsourcing of manufacturing jobs and routine services.

There are two groups within the creative class, states Florida. They are the "super-creative class" and the "creative professionals." The first group is involved in knowledge creation, the second in knowledge processing. Scientists, engineers, artists, and political leaders are classified as members of the super-creative class. The creative professionals include all workers in the knowledge-intensive industries and high-tech and financial services, as well as legal and health professionals.

Interestingly, Florida classifies engineers in the same category as artists. He correctly recognizes the fact that the key to engineering is not routine work (knowledge processing) but the development of new ideas (knowledge creation), which is potentially a basis for inventions and the evolution of engineering innovations that drive the changes in our society.

Florida discovered that members of the creative class are not uniformly distributed throughout the United States. They are similar to the nineteenth-century Bohemians in that they tend to live in several *creative regions*, such as the Raleigh–Durham area in North Carolina, the San Francisco area in California, and the Washington–Baltimore area, among others.

A creative region, or a creative-class center, can be described as a geographic area with a high concentration of members of the creative class, or a region where knowledge creation and processing is the dominant part of the economy. Similarly, regions where manufacturing is the dominant part of the local economy have a high concentration of manufacturing-class members. Good examples of manufacturing regions are Pittsburg and Detroit. Finally, a service region attracts a large number of members of the service class. The best example of this is Las Vegas, Nevada.

To understand more about how this new class structure that Florida proposes works, let us compare two areas: Washington DC and Las Vegas. In 1999 (Florida 2002), members of the creative class in the Washington DC area constituted 38.4% of employment positions, with members of the super-creative class at 15%. Members of the service class were at 43.8% in that area. The same set of numbers for Las Vegas was 18.5%, 4.9%, and 58%, respectively. There is no question that Washington DC is a magnet for members of the creative class, including scientists. Whenever a new position is announced at George Mason University, where the Author used to teach, the vacancy usually attracts a large number of applicants, sometimes numbering in the hundreds. Many of these applicants come from much more established universities that are located in manufacturing or service regions. The candidates simply want to move to a creative region, perhaps subconsciously knowing that such a move will have a positive impact on their own creative and professional activities. Similar to what takes place with the Medici effect, members of the creative class naturally gravitate toward other creative individuals for inspiration and the sharing of ideas.

From a public policy perspective, creative regions are much more than merely geographic areas; they are creative communities. A creative community can be interpreted as a community wherein knowledge is created and processed. A description of such a community obviously covers any community producing inventions and providing engineering education as well. Therefore, the concept of a creative community and its descriptive features are important for us also in our quest to create a successful environment that is perfect for the inventors and for the education of inventive engineers.

As a result of many years of research, Florida has identified four distinctive features of a creative community. They include (1) a physical place, (2) diversity, (3) authenticity, and (4) identity.

Surprisingly, in the era of globalization and virtual worlds, a physical place is still important for members of a creative community. "[A physical place] facilitates the matching of creative people to economic opportunities," says Florida (2002). Creative production requires frequent intersections between individuals, and these still occur mostly through social interactions. For such interactions to occur, a complex urban environment has to be created with places for mingling and gathering such as bookstores, coffee or pastry shops (my favorite), small restaurants, and farmers' markets. In other words, creative people need a third place to enjoy life in addition to the primary places of home and work. The availability of third places is very important for the members of the creative class, and it is often the decisive factor when they are considering job offers in several locations.

The town of Fairfax, where George Mason University is located, is a good example of a place that is changing and expanding in order to attract creative people. Recently, the Fairfax City Council made a bold decision and invested tens of millions of dollars in the revitalization of its small downtown. In the process, old but typical buildings were erased, the post office and library were moved, and an entire complex of new "old-style" buildings was constructed. These buildings offer a lot of space for various shops and restaurants, both chains and local. The place has already attracted many such shops, including an excellent Italian ice-cream parlor and a very good wine-tasting venue, which also offers light meals. Students and local professionals alike continue to come to downtown Fairfax in order to discover its charm every year, including my family.

Diversity is also an important feature of a creative community. In this context, diversity covers a broad spectrum and includes ethnic and racial diversity, age differentiation, various sexual orientations, and even alternative dressing and adornment styles, such as body piercings and tattoos. Interestingly, Fairfax recently allowed the location of a tattoo shop in the downtown area in order to increase the diversity of people coming to Fairfax. As a matter of note, the owner of this shop, Michael Sikes (Figure 3.1), a wonderful artist himself, has created all the illustrations for my previous book. Diversity, as it is described here, means a diversity of ideas, styles of thinking, backgrounds, and personalities. All the dimensions of diversity are necessary to produce intersectional events and should be considered in creating a successful environment for any genre of creative education, including inventive engineering.

Creative people need to be surrounded by an environment that is authentic for the most part. In an urban environment, this means one that contains historical buildings as well as rivers, streams, and natural landscapes. Typical shopping malls, or even strip malls with chain stores and chain

Figure 3.1 Mike Sikes. (With permission. Drawn by Michael Sikes.)

Figure 3.2 New Fairfax downtown. (With permission. Drawn by Michael Sikes.)

restaurants, do not attract creative people. Such people need a human dimension for their physical environment in order to maintain face-to-face relationships and in order to create (Figure 3.2). Authenticity of place is increased if it offers original forms of art and music as well, which allow creative people to relate emotionally to a given place.

Finally, creative people need to identify themselves with their place. Their community must contain features that can be used to describe its own inherent uniqueness as well as that of its inhabitants.

All the features of a creative community are important for truly creative activities to occur. Imagine if you were in such an environment as you pursue your inventive work. Would you feel and be more creative, inventive, and even confident if you were surrounded by creative people and environments? Perhaps you live in such a place right now. If so, think about how your creative environment enhances your overall sense of well-being, ability to focus, and creative inspiration.

3.2 PSYCHOLOGY

3.2.1 The theory of successful intelligence

What does it mean to be successful? Does one have to be rich or famous to achieve true success? Maybe only those who rise to the top of their field can truly be called successful. How do *you* measure success in yourself and in others?

According to Dr. Robert J. Sternberg (1985, 1996), a cognitive psychologist who studied human intelligence in the 1980s and 1990s, there is no absolute way to measure success. Instead, his theory of successful intelligence (also known as the triarchic theory of intelligence) states that success is defined by each individual in relation to the sociocultural context in which he or she lives as well as the person's unique desires. Sternberg says that successful people are also able to achieve their goals by adapting to, shaping, and selecting environments to work in that will facilitate their success. He states that these individuals come out on top by simply focusing on their strengths and compensating for their weaknesses as best they can. According to Sternberg, there is no absolute measure for success.

In Sternberg's world, everybody has a chance to be successful. His theory of successful intelligence is based on three main assumptions:

- Successful intelligence can be learned (anyone can be successful).
- Successful intelligence is based on a combination of three other intelligences, or abilities—analytical, practical, and creative—and each of these intelligences can be learned independently of each other.
- Successful intelligence is dynamic—how an individual achieves success may change over their lifetime.

3.2.1.1 The three intelligences

When we say someone is "intelligent," we usually equate this intelligence with "book smarts" or being able to achieve academically. According to the theory of successful intelligence, however, there are in fact three intelligences. A truly intelligent person maintains a balance between all three.

First of all, let us take a look at the kind of intelligence that most people equate with the word. Analytical intelligence is what most traditional intelligence quotient (IQ) tests measure when determining intelligence levels. Analytical intelligence is the ability to solve analytical problems. It is also the ability to use deductive reasoning and to think critically. Need to figure out the dimensions of a building and the best materials to use for its design and structure? Use your skills of analytical intelligence to determine what you already know about the subject, fill in the blanks with the steps that make the most sense, and you are on your way to develop a design that is safe and on-spec. What about an assignment that asks you to compare and contrast differing opinions in a court case? Critical thinking, or analytical intelligence, will help you flush out each side and write a clear and compelling essay. Analytical intelligence is acquired through a combination of rote learning (memorization) and deductive skills. It is emphasized to the exclusion of other forms of intelligences in modern classrooms, including most engineering classrooms.

Practical intelligence is the ability to solve simple problems in everyday experience. It is mostly learned through rote learning or trial and error as we mature. Babies and children begin to acquire practical intelligence the minute they are born. "If I put my finger in this fire, it will hurt. That will not feel good, so I think I will keep my fingers out of the fire!" says the three-year-old to himself in his uniquely three-year-old way. Practical intelligence also allows us to respond to a changing environment, either by shaping it, selecting it, or adapting to it. Practical intelligence, for example, propelled those living in the Dust Bowl of the United States during the 1920s to leave their home environment in search of a healthier place to live. This was an example of our ability to use practical intelligence in order to select our environment. An example of adaptation would be a person who learns how to use a computer when their job or the economy demands it. Practical intelligence is mostly acquired through rote learning, memorization of facts, and heuristics or experience-based knowledge.

Finally, creative intelligence is the ability to solve problems and meet challenges through means other than rote response, deductive reasoning, or relying on past experience. Creatively intelligent people can rely on rote learning and deductive reasoning, but they can also rely on abductive reasoning and imagination. A creatively intelligent person working on a creative solution to a problem allows for the possibility of an outcome to grow out of that which is other than past experience or a set formula. The use of creative intelligence produces novel ideas and inventions. Every major technological advancement in our modern era has come about at least in part through the use of creative intelligence, including in the field of engineering.

3.2.1.2 Successful versus conventional intelligence

Did you know that IQ is only a small factor in determining both academic success and income levels? According to a study done by cognitive psychologist Ulric Neisser and colleagues (Neisser et al. 1996), IQ accounted for only 25% of the variance in grades in the university students they studied. Another study done by Sternberg and others (Bowles et al. 2002; Duckworth et al. 2007; Sternberg 1997) took a look at the various factors that led to income differences among working adults. Their findings were similar to Neisser. Despite the widespread belief that analytical intelligence (measured by IQ tests) leads to higher income levels, IQ only accounted for one-sixth of the variance in incomes in the subjects they studied.

Conventional theories that were developed in the past and that are still used today describe human intelligence solely in terms of its ability to analyze and solve abstract problems. Furthermore, an individual's IQ, as measured by IQ tests and other similar means, assumes that intelligence is "static" and cannot be changed through education, experience, or interaction with a given environment. These theories, however, are not connected to what is happening in the real world because they do not account for actual human beings and their actions in constantly changing realities.

It is interesting to note that Sternberg's lifelong study into the subjects of intelligence, success, and test taking was sparked by a severe case of test anxiety he himself suffered from when he was young. As he advanced in his own education, he had a hunch that the conventional tests he was being subjected to, such as the IQ test, were not adequate measurements of his knowledge and abilities. This conclusion caused him to create his own intelligence tests later on in his career, including the Sternberg Test of Mental Agility. It also led him to eventually create the triarchic theory of intelligence. Current research, such as that being conducted by Neisser, Sternberg, and others, call into question the assumed "static" nature of human intelligence. They say that instead of being static, human intelligence is, in fact, in a state of constant flux, change, and growth; we could say that it is an adaptive system.

Take a moment to see the relevance of this statement in your own life. Go back in time to when you were 15 years old. How much did you know about the field of engineering then compared to how much you know about it now? Reflect on how you were able to learn the things that you now know. Was it all through rote memorization or relying on past knowledge? Or was creative intelligence involved in some way? Now fast-forward into the future. Is it safe to assume that you will know more about engineering 10 years from now, especially after you have gained some practical experience working directly in the field and solving the unique engineering problems that may come your way? Can you predict from where you are now how you may gain engineering knowledge, especially after you are out of college?

3.2.1.3 Bloom's taxonomy versus Sternberg's intelligences

Benjamin Samuel Bloom (1913–1999) was an American educational psychologist who contributed greatly to modern educational thinking. In particular, he was studying the concepts of "talent," "mastery-learning" and how individuals achieve "greatness." He is most known for his creation of Bloom's taxonomy (Bloom et al. 1956), which is a way of classifying educational objectives and goals. Bloom's taxonomy remains a foundational rubric for educators today. It distinguishes six levels of achievement: knowledge, comprehension, application, analysis, synthesis, and evaluation.

A fundamental difference between Bloom's approach to successful learning and that of Sternberg is that Bloom's taxonomy says that all abilities are hierarchically dependent on one another. It also says that one must sequentially learn these abilities, starting with the ability to retrieve knowledge (mainly through the memorization of facts). The top of the hierarchical pyramid for Bloom is the ability to evaluate solutions. In contrast, Sternberg's three intelligences—analytical, practical, and creative—can be acquired by an individual independently of each other; no hierarchy exists among the three. According to Sternberg, a genius inventor with many patents to his name may have a high level of creative intelligence but may seriously lack in practical and analytical intelligence. Likewise, an accountant may have a high level of analytical intelligence about how to save people money and about specific tax laws, but he may also be operating with a relatively low level of creative intelligence. For Bloom, abilities and intelligences are dependent on each other and are hierarchical. For Sternberg, they are independent of one another and are not hierarchical.

What does Bloom have to say about creative intelligence specifically? His taxonomy does hint at the concept, but its importance in the overall scheme of things is minimal. In general, creative intelligence for Bloom is just one of the many sequential steps one most go through on one's journey toward achievement. However, without the time-tested ability to think out outside of the box and think creatively about solutions to problems, especially in engineering, many students may be missing out on opportunities for jobs and may be relegated to static secondary positions within the engineering field. Are challenge, creativity, growth, and perhaps even better pay important to you as you begin your engineering education and eventually your engineering career? If so, then creative intelligence is a must-have for you.

According to Sternberg and his theory of successful intelligence, no one can be successful without creative intelligence and, more specifically, without a balance of the three intelligences: analytical, practical, and creative. Engineering education, and in fact the American educational system as a whole, has never provided a complete balance of all three abilities. When the master–apprentice paradigm was in place in the Middle Ages and into the Renaissance, only creative and practical abilities were taught. In the twentieth century, with its emphasis on science and industry, the analytical component was substituted for creativity. According to the theory of successful intelligence, in order to truly succeed, educational experiences must include a complete balance of analytical, practical, and creative abilities.

3.2.1.4 The importance of significance

In engineering, one must also consider the added component of success significance. In order to be successful, an engineering creation may be unique, beautiful, and potentially useful. However, if it is not *significant* as well, it will likely not contribute to society as a whole and to the advancement of engineering in particular.

The significance of engineering inventions for society requires that an engineering solution fills a need or addresses a gap in a way that other solutions do not and also in a way that is superior to existing solutions in terms of simplicity, cost, durability, beauty, or in another way. In order to develop a significant solution, an engineer must be able to combine a high degree of comprehension of engineering principles with a broad knowledge of existing solutions and a high degree of understanding of the challenges and issues surrounding the solution domain. The solution could also employ a novel approach to creating it. An example of a significant engineering solution could be the creation of a new type of beam–column connections in steel skeleton structures that are subject to blast. The domain may be structural engineering in this case, but this significant invention also addresses the important social need for the safety of such structures.

3.2.2 Thinking styles

Everybody has a different style of thinking. You must recognize what yours is in order to have the optimal learning experience in inventive engineering. A style is not an intellectual ability but rather it describes, and often predicts and prescribes, how a given individual *uses* his or her abilities. In terms of knowledge, a thinking style can be described as a collection of methodical decision rules and heuristics that are used by a given person to decide how to conduct various processes, particularly those involving other people. In engineering, it is common for the same result to be produced using a variety of methods. In human interactions, however, very often the nature of an interactive style is equally important as its final result. The right process builds friendships and even communities, while the wrong one destroys them.

Styles can be defined using five symbolic attributes (Sternberg 1997): functions, forms, levels, scope, and leanings. A total of 96 basic styles can be distinguished (see Table 3.1. for a sampling), each defined by a single combination of values, one for each attribute. This list of styles is still incomplete. There is an infinite number of thinking styles that each person

Table 3.1 Attributes of thinking styles

Attributes	Attribute values			
Functions	Legislative	Executive	Judicial	
Forms	Monarchic	Hierarchic	Oligarchic	Anarchic
Levels	Global	Local		
Scope	Internal	External		
Leanings	Liberal	Conservative		

could possess, considering the fact that many people may combine several basic styles.

Sternberg also speaks of thinking styles. He says that a successful human operates like a democratic society or like a balanced system (Sternberg 1997). As such, he or she has three major functions: the generation of ideas (legislative), the implementation of ideas (executive), and judging if the results are consistent with the initial ideas (judicial). Each function is best fulfilled if the appropriate style is employed. Generation of ideas (concepts) for a design project in engineering (conceptual designing), for example, is best performed using a legislative style of thinking, which mostly utilizes creative intelligence. The implementation of ideas, such as the actual detailed designing of an engineering product using the design concepts created, requires an executive style, which mostly utilizes deductive thinking and analytical intelligence but also involves using practical intelligence. Finally, judging results, that is, evaluating the ways in which the implemented design is a success and the ways it needs to be improved, requires a judicial style, which, of course, relies heavily on analytical intelligence (Figure 3.3).

Each human uses a combination of all three styles in their life, but usually one style is dominant in a given individual. Your style of thinking can also reflect the nature of a given job as well, since the best results are usually produced when the style reflects the demands of a task in addition to the human using the preferred style. Which style of thinking allows you to operate best? Can you think of a situation or job where you had to use a specific style?

Sternberg (1997) also distinguishes at least four types of thinking styles: monarchic, hierarchic, oligarchic, and anarchic. If you lean toward a monarchic style in your thinking, you are entirely focused on a single issue or goal. You are single minded and consider the entire world in the context of your particular issue (Figure 3.4). If you have a dominantly hierarchic style, on the other hand, you are very well organized and can focus on many goals, usually hierarchically arranged in the form of long lists that are systematically pursued (Figure 3.5). The oligarchic person usually has several competing goals of perceived comparable importance going on at once, some of which may even seem contradictory (Figure 3.6). Finally, a person with an anarchic style has neither goals nor priorities. His or

Figure 3.3 Thinking styles. (Prepared by Michael Sikes based on the Author's sketches.)

Figure 3.4 Monarchic-style thinking. (Prepared by Michael Sikes based on the Author's sketches.)

her selection of goals and activities appear to be entirely spontaneous and unpredictable (Figure 3.7).

Sternberg also defines two levels of thinking: global and local (Sternberg 1997). People with global thinking styles tend to be focused on the big picture and on a holistic understanding of the world. By contrast, local-thinking individuals focus on details. Obviously, there are also people who can change their level of thinking depending on the circumstances. Are you normally detail oriented in your thinking or do you most often look at the big picture in any given situation?

In terms of scope of thinking (Sternberg 1997), an internal style is used primarily by introverts who are mostly, if not entirely, focused on their own reality and do not pay much attention to the thoughts and activities of others. By contrast, people with an external scope are always focused on other people and mostly get their ideas and motivations from interactions with others. Such people are called "extroverts."

Finally, *leanings*, according to Sternberg, deal with the level of risk averseness in an individual (Sternberg 1997). People with liberal leanings are not afraid of risks and novelty and often seek them out. They thrive in conditions of uncertainty and change. People with conservative leanings, on the other hand, avoid taking risks as much as possible. Leanings are not related to the political orientations of people. There are many political liberals

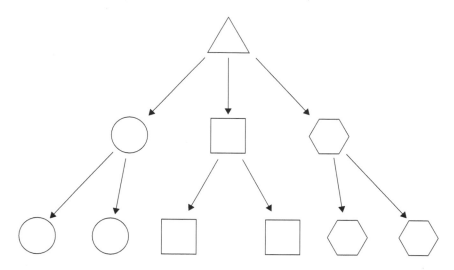

Figure 3.5 Hierarchic-style thinking. (With permission. Drawn by Michael Sikes.)

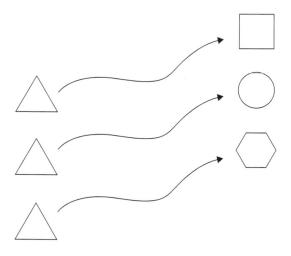

Figure 3.6 Oligarchic-style thinking. (With permission. Drawn by Michael Sikes.)

afraid of change (e.g., trade unionists) and many conservative individuals looking for change (e.g., Senator Mark Rubio).

The importance of leanings, or thinking styles, in engineering in general is still poorly understood and utilized. Many students, perhaps yourself, may have experienced frustration when dealing with instructors who simply have a different thinking style than you do. For example, if you are inherently a global thinker and you are working with an instructor who

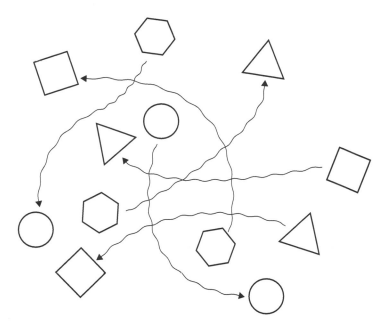

Figure 3.7 Anarchic-style thinking. (With permission. Drawn by Michael Sikes.)

is a local thinker, you may be forced to focus only on details and on the analytical aspects of the knowledge being taught, such as facts and specific analytical models. If you resist or become visibly frustrated, the instructor may not understand your resistance. It is at this point where knowledge of thinking styles on both parts can lead to better interactions, and thus a better learning experience overall. Reflect for a minute about your own thinking style or leaning. Are you a global thinker or a local thinker in general? If you are a global thinker, can you think of one or two reasons why thinking locally might be to your advantage in some situations? If you are a local thinker, can you think of particular situations where taking in the big picture might give you a needed perspective?

3.2.3 Positive psychology

The Author's ultimate goal is to inspire and facilitate the transformation of readers from engineering students to successful engineers and inventors. Such a transformation obviously requires the acquisition of a specific body of knowledge, but this knowledge is simply not enough. The transformation has two dimensions: intellectual and psychological. In this section, we will develop a good understanding how modern psychology may help us to prepare for this most important transformation in our life and show us how to conduct it.

Marcel Proust (1871–1922), a French philosopher and writer, once wrote that "The real act of discovery is not finding new lands but seeing with new eyes." This is the essence of your transformation, finding new eyes, changing your perspective and in the process changing your attitude to life and work and thus greatly expanding your opportunities. We will address this life-changing issue, and most likely it is the most important part of this book.

Over the last century or so, the new science of psychology has emerged. Its initial and still dominant focus has been on the negative, or pathological, states of the human mind. Only recently, during the last 30 years, a new focus has begun to emerge: a focus on the positive, or optimal, psychological states. Appropriately, this new kind of psychology has been called *positive psychology*. Learning positive psychology is a fascinating experience, but, much more importantly, it will ultimately lead to practicing positive attitudes, and that makes developing creative ideas easier and definitely contributes to one's life and professional success.

Two main pioneers of positive psychology are Diener (1984, 2009), who proposed the *science of well-being*, and Argule (1987), who developed "The Psychology of Happiness." Both published their initial findings in the 1980s and initiated a global research movement to verify and expand their work. Today, there is a group of young American psychologists whose entire focus is on positive psychology. They have already made significant contributions to our understanding how positive psychological states emerge, how they are maintained, and how they impact all kinds of human activities, particularly acts of creation, leadership, and general well-being. This group includes, for example, Kim Cameron (2008), Marci Shimoff (2008), Barbara L. Fredrickson (2009), Sonja Lyubomirsky (2013), Kashdan and Ciarrochi (2013), and Shane Lopez (2013). One of the leading research centers in the area of positive psychology is the Center for the Advancement in Well-Being at George Mason University (your Author is a senior scholar at this center). Three leading positive psychology scholars are associated with this center, including Tojo Thatchenkery and Carol Metzker (2006), the "father" of *appreciative intelligence*; Todd Kashdan, a Renaissance scholar in this area (2009, 2013); and Nance Lucas. Lucas is the brain and heart of the entire positive psychology movement at George Mason University and the leader of that school's transformation into the first well-being university, where the well-being of students and faculty is becoming equally important as learning and research.

Probably the most interesting and potentially useful construct coming from the field of positive psychology is that of appreciative intelligence. Recent research has discovered that this kind of intelligence is closely associated with highly successful entrepreneurs and inventors, if not with all successful people. Therefore, learning about it and how to acquire it seems natural for all future inventive engineers. The concept of appreciative intelligence has been recently proposed by Tojo Thatchenkery and Carol Metzker (2006) and can be described thus:

Appreciative Intelligence is an ability to appreciate all positive possibilities in any situation.

This ability can be understood as a result of the integration of three specialized abilities, including

1. An ability to reframe, consciously or subconsciously, a given situation leading to a desired outcome
2. An ability to see and appreciate all positive aspects of a given situation, including all opportunities
3. An ability to see a process leading from a present given situation to a future desirable situation

An ability to reframe a given situation means being able to perceive, interpret, or see this situation from an entirely different perspective: from the perspective of a different person, from that of a different scenario, or from a different physical perspective. Human learning about the world is always situated; that is, it is done in the context of a given situation or of the knowledge of the person doing the learning. Reframing is then a psychological process in which a person intentionally changes the context of a given situation to develop a better understanding of it.

The classical and casual example of reframing is that of a glass half empty or half full. The actual situation is the same in both cases, but viewers' perspectives can be entirely different (Figure 3.8).

Reframing is particularly important when dealing with complex engineering or decision-making systems. For example, a designer of a tall building should be able to see the structural system from at least three perspectives: those of a structural engineer, of a fireman, and of a construction manager. A structural engineer will see the structure from the perspective of a flow of internal forces. The same structure will be seen by a fireman from the perspective of a gradually evolving fire, which has just begun on the 27th floor and is spreading. Finally, for a construction manager, the same structure will be seen as nothing else but a flowchart specifying delivery and installation times for the individual columns, beams, and elements of bracing during the construction process.

A successful military leader on a battlefield should be able to see the situation from his perspective (as it is expected) but also from the perspective of his opponent. Only these two entirely different views will give him a full understanding of the situation and will allow him to make appropriate decisions. Similarly, the same military operation may be conducted during an attack or during defensive maneuvering, only the context scenario is different; it may be a subtle difference, but it may decide who wins or loses.

Thatchenkery and Metzker (2006) have provided an excellent engineering example showing the power of reframing. In 1990, the Hubble Space Telescope was launched. Initially, this event was celebrated as the greatest

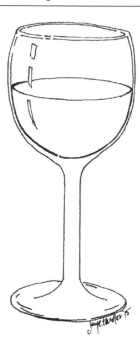

Figure 3.8 Glass half-full. (With permission. Drawn by Joy E. Tartter.)

of NASA's achievements. Unfortunately, it was very soon discovered that the telescope was producing fuzzy pictures because of misaligned mirrors. Suddenly, the greatest NASA success was perceived as its biggest multi-billion-dollar failure. Every engineer involved in the program was literally in shock and simply unable to address the problem; everybody but Charlie Pellerin, who earlier oversaw the launch of the telescope. When everybody was thinking that the Hubble Telescope Program ended with a failure, Pellerin reframed the situation and convinced his NASA colleagues that launching the telescope was only a stage in the Hubble program, not its end; and, moreover, that the situation had produced specific engineering data that could be used to improve the situation. With this new understanding, a solution was found and the astronauts made the necessary repairs. The Hubble Space Telescope is still working, contributing to space exploration, and its pictures have already led to several significant space discoveries.

Ultimately, seeing a situation from various physical perspectives is relatively easy for engineers who are trained to produce and read technical drawings showing engineering systems in Cartesian projections.

The second specialized ability is an ability to see and appreciate all positive aspects of a given situation and to see all emerging opportunities. It may be interpreted as an ability to activate on demand an attitude of seeing only what is good in a given situation, no matter what the initial perception of this situation is, or no matter how bad the situation seems to be.

It is a psychological mechanism of filtering all available information and selecting only the pieces of positive information. It can be also explained in the context of knowledge acquisition and learning from both positive and negative examples when both kinds of examples are necessary and equally important.

For example, the collapse of a small bridge without any loss of human life or injuries may be considered as a positive event, revealing various deficiencies in this particular bridge design and preventing their occurrence in a much bigger bridge of the same type that is just about to be designed. In fact, the progress in engineering is always driven by new ideas successfully (and safely) implemented but also by disasters that show which ideas are simply wrong or require more development. Therefore, inventive engineers should always feel that there are no bad situations but only sources of positive or negative feedback, both equally important for an ever-learning inventive engineer.

Inventors are sometimes called *opportunity seekers*, and this ability is critical for their survival and growth. Usually after their first invention, after solving a problem given to them, inventors are in a *seeking mode*, looking for more problems waiting for inventive solutions. In fact, the so-called bad situations are most attractive for inventors; such situations may need inventive solutions, and therefore they may create the best opportunities for inventors.

In the case of the third specialized ability, we refer to an ability to connect the present situation with a future and desired outcome through a feasible process. Thatchenkery and Metzker (2006) have called this ability "Seeing the mighty oak in the acorn" (Figure 3.9), and that poetically describes the essence of this ability. In fact, this ability is like seeing a famous professor in a student failing a graduate course or seeing a successful engineer in a boy breaking the furniture. In more specific engineering terms, this ability will allow us to see a future and safe structure in a destroyed frame, which has just undergone the process of a plastic collapse; during the collapse process, the location of the first and most important plastic hinge has been revealed, and that will allow us to reinforce the structure at this location and thus significantly increase its load-carrying capacity.

Thatchenkery and Metzker (2006) have formulated four requirements for practicing appreciative intelligence, which provide another perspective from which to understand this construct in its entire complexity. They include

1. Persistence
2. Conviction that actions matter
3. Tolerance for uncertainty
4. Irrepressible resilience

One of NASA's leading engineers and inventors once said, "Success is just having the fortitude to keep going." This captures well the idea of

Figure 3.9 Acorn and oak tree.

persistence behind appreciative intelligence. From the psychological point of view, persistence has two dimensions: the behavioral and the cognitive. The first one is understood as a sequence of interrelated activities that are required over a period of time and are sufficient to reach the stated goal. Cognitive persistence identifies a kind of thinking that focuses on the goal all the time, even after the actual activities have been terminated.

Conviction that actions matter is an emotional driver of persistence; it creates motivation. An individual practicing persistence must be motivated to continue, and that comes from the conviction that his or her actions matter and will make a positive impact on a community, on society, or on the profession. The term *self-efficacy* well describes this state of mind, dominated by the strong belief that the individual can be proactive, in control of his or her life and, most importantly, capable of achieving any goal as a result of his or her actions.

The next requirement, tolerance for uncertainty, does not require any additional explanation for future inventive engineers who have already learned about da Vinci's principle of *sfumato*, which addresses the issue of feeling comfortable while working under conditions of uncertainty. Finally, the last requirement, irrepressible resilience, means an exceptional ability to survive periods of crisis, periods of criticism, troubles, resistance, and so on.

Some people are seemingly born with appreciative intelligence and practice it effortlessly, but the majority of people have very little of it. Fortunately, it is possible to acquire or improve appreciative intelligence, as it is with successful intelligence. In fact, its "parents" (Thatchenkery and Metzker 2006) claim that "Identifying, developing, and enhancing appreciative intelligence in yourself or other individuals, and applying it for personal or organizational success, can lead to great advantage and reward." They propose a four-stage process called the "conscious competence model of learning." It consists of four stages:

1. Unconscious incompetence
2. Conscious incompetence
3. Conscious competence
4. Unconscious competence

In stage No. 1, a given person is simply unaware that he or she is not a good driver, but after several fender-bender accidents in a month, this realization gradually emerges. The person is now in stage No. 2, conscious incompetence. He or she begins thinking about their incompetence and how to address it. First comes an assessment of their driving skills and next the decision to take a three-day course in high-performance driving in order to improve these skills through practicing with professional drivers, former car racers. During the course, the learner is consciously building his or her competence and therefore is now in stage No. 3, conscious competence. The learner participates in various exercises. For example, he or she drives with an instructor on a wet surface in a car with eight wheels. The instructor raises the front or rear additional wheels, simulating a change in the car's behavior from oversteering to understeering, and the student learns the correct reactions; the same correct reactions are repeated again and again for several hours. In the process, the learner's brain becomes properly "hardwired" and at the end of the course he or she can unconsciously correctly control the car. There is no question that the learner is now in stage No. 4, unconscious competence, and is finally ready to drive safely on the public roads.

An incompetent driver is dangerous, but a person with a very limited appreciative intelligence may also cause of lot of harm to himself and to others, even physical harm, particularly if the person is a military officer or a policewoman. You cannot be a successful inventive engineer without constantly improving your appreciative intelligence. It is a process similar to becoming a competent driver.

The process begins with assessing your personal appreciative intelligence and creating your personal appreciative intelligence through answering a number of probing questions provided by Thatchenkery and Metzker (2006). There are three major lines of activities leading to improving somebody's appreciative intelligence.

In the first case, undesired behaviors are directly changed. We usually behave using specific behavioral patterns associated with various situations. For example, when we read a technical book, we never take notes, but we can change this pattern and start taking notes. At first, changing the pattern will be difficult, but after several conscious attempts to take notes you will be surprised to discover that you began reading and that you subconsciously reached for your notebook or iPad.

The second approach is more demanding but also more effective in the long run. We change our thought processes directly. We will use a phenomenon known as *neural Darwinism*. During the learning process, our brain creates new synaptic connections between neurons—new neural networks reflecting our acquired knowledge in the form of complex patterns (they may represent decision rules, pictures, shapes, smells, etc.). Connections used frequently are kept or even reinforced, and the least used connections are destroyed. Because of that mechanism, the more we use a given mental process, the stronger it becomes. If we intentionally work on the development of feeling optimistic or feeling appreciative, neural connections related to these feeling will be gradually strengthened. It is like a mental workout and building the neural "muscles."

In the third case, we change our mind by changing our actions. We utilize the phenomenon that the brain does not distinguish between actual signals from the body caused by true emotions and identical *fake* signals created only to cheat the brain. Thatchenkery and Metzker (2006) provide the excellent example of a forced smile, which tricks the brain into thinking that we are happy, and even when we stop smiling we may still feel the happiness created by our fake smile.

There are many tools developed by positive psychologists that may be used by inventive engineers to improve their appreciative intelligence. A good selection of such tools is provided by Thatchenkery and Metzker (2006), but other tools can be also found in Shimoff (2008) or in Kashdan (2009).

Interestingly, when writing on appreciative intelligence, the Author received a call from his patent attorney informing him that his patent application had been rejected and that an interview with the patent examiner probably would be necessary. The entire "bad" situation was discussed, but at the end of the conversation, the Author began unexpectedly talking about licensing his invention, an unusual idea in the context of the situation. At this time, he realized that he was already learning from his book or that he had at least some appreciative intelligence, a nice thought indeed.

3.2.4 Well-being and inventive engineers

The notion of well-being is foreign for the majority of engineers, but it should not be so for inventive engineers. Reaching a state of well-being is the ultimate goal of all people, even if it is not explicitly articulated by them. Also, there is already a proven positive relationship between well-being and human creativity. For these two powerful reasons, inventive engineers should be familiar with the present state of the art in the area of well-being and should also understand how to benefit from the most recent advances in this area.

Well-being may be understood as a phenomenon of having a good life, that is, a life that is meaningful and desired and brings everyday happiness (Diener 2009). Many people would describe such a life as a life of high quality. Obviously, well-being is a subjective notion, and psychologists usually talk about *subjective well-being*. We have already discussed the notion of success, (see the preceding section on the theory of successful intelligence), which is also subjective. Both notions can be only understood in the context of perceptions of a given person living at a given time in a specific location and in a given political, social, and cultural environment. Both notions are based on the same assumption that each person is different and each person is living in his or her own world (at least as it is perceived by each person).

For example, let us consider twin Igbo sisters born in the southern part of Nigeria where the Igbo tribe mostly resides (the Author used to teach there at the University of Nigeria in Nsukka). One sister is still living in a small village in a rural area, while her twin sister is living in New York City. The first sister believes that she has a good life because she has access to potable water and can buy cheap subsidized rice for her family. The second sister, living in New York City, believes that she has also a good life because she has a high-level managerial position with the J.P. Morgan Chase Bank and lives in a penthouse in a high-rise building on Fifth Avenue. As we may see, a "good life" or "well-being" are subjective notions driven by human perceptions, which in turn are based on human emotions. Again, we discover that human emotions are important, particularly for inventive engineers.

There are several entirely different perspectives on well-being, and they need to be known to understand well-being in its complexity.

First, in economics, well-being is associated with a given social group or with society as a whole and is quantitatively measured as the material quality of life, which can be understood as the state of infrastructure, the gross national product for a country or for a given region, the state of the environment, and so on. Second, in sociology, life happiness is related to such external conditions as income and social status. In psychology, well-being is a matter of attitudes, personalities, perceptions, and successfulness. Finally, in metaphysics, well-being is associated with an individual's happiness as driven by his or her ability to achieve a balance between material and spiritual life.

All these perspectives are meaningful because only their combination allows us to grasp such a difficult notion as well-being. The last two perspectives are probably the most important from our point of view, because they explain the importance of well-being for human creativity. It is obviously not a coincidence that Google is paying a lot of attention to the well-being of its employees, not only through providing excellent salaries but also an inviting working environment to make them happy while working.

Another good example is George Mason University (GMU) and its Center for the Advancement of Well-Being. Recently, the president of the university announced that GMU would become the first major university whose mission is (among others) to create a well-being environment for learning and research; that is, all major academic initiatives will be also considered from the perspective of making an impact on the well-being of the entire academic community. What will actually happen still remains to be seen. There is no question, however, that the term "well-being" is used more often with each passing month and usually in the context of creativity. That means a changing perception of well-being from a luxury phenomenon that only very rich people can afford to think about to nearly a requirement of a modern work environment.

The message for all inventive engineers is simple: always be concerned about your well-being and do your best to improve it. In practical terms, this may mean making fundamental changes in your life to improve your level of happiness. Also, it may mean making very easy changes in your environment and your everyday life, which can be surprisingly effective. For example, you may bring a piece of art or some living plants into your office, or begin regular meditation. Meditation in particular may significantly improve the well-being of sensitive people, and inventive engineers are usually such people.

Chapter 4

Basic concepts

4.1 INTRODUCTION

Learning about *engineering designing* requires first acquiring several basic concepts, which will be used by us to define more complex concepts and to develop our understanding of this fascinating domain. Our approach to engineering designing will be based on a combination of systems and knowledge approaches. That means that we will be using thinking based on systems analysis and that we will understand engineering designing as comprising various activities associated with knowledge—its acquisition, transformation, and utilization. Therefore, we need to understand first several concepts that are directly related to our investigative approach. We will also discuss such concepts as *transdisciplinarity*, *emergent behavior*, *knowledge convergence*, *synthesis*, and *synesthesia* and will explain how these relate to inventive engineering. These ideas are not well known in the field of engineering and therefore present a new framework with which to look at the subject of our interest.

4.2 SYSTEM

There are many definitions of a *system* (e.g., in Rechtin 1991), but let us use a relatively simple definition introduced several years ago by Mark Maier and Eberhard Rechtin (2000):

> A system is a set of different elements so connected or related as to perform a unique function not performable by the elements alone.

A system's elements are understood as separate entities, each with a specific function that is usually different from the function of the entire system. Let us use a *glass box model* to present a general model of a system. It has interconnected elements and relationships between these elements, which are called *feedbacks*. Our system is active in a given environment and communicates with its environment through input and output. *Input* represents

77

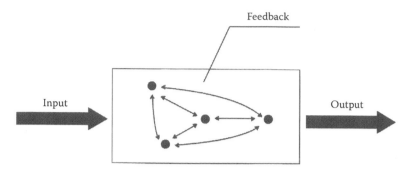

Figure 4.1 System.

whatever comes in from its environment, while *output* represents whatever leaves the system and affects its environment. In other words, the feedback between a system and its environment is represented by the system's input and output. When we develop a model of a system, we need to identify its input and output and its structure, that is, all the elements and their feedbacks (Figure 4.1).

For example, when a model of our system represents a skeleton structural system, its input may include gravity, wind, and earthquake forces. Its output is the pressure transferred from the structure through the foundation to the ground, as shown in Figure 4.2.

From a different perspective, we may say that input represents the action or impact of the environment on a given system operating in this environment, while the system reacts through its output; we have here an *action-reaction model*. Another interpretation is that a system transforms its input into output, and this *transformation model* is particularly important for us, as we will see in Section 4.6, where we discuss various engineering designing models.

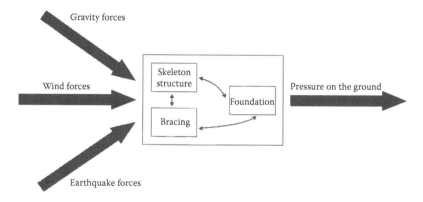

Figure 4.2 Systems model of a structural system.

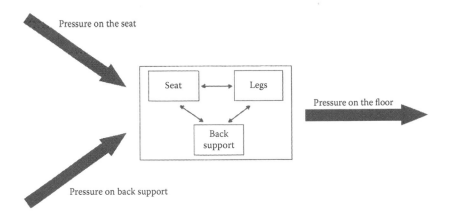

Figure 4.3 Chair and its subsystems.

Our definition of a system describes all kinds of systems, including real or built systems, for example a car or a streetcar; and abstract systems like an analytical model of a fuselage in an airplane. When elements of a system are complex and each element is composed of a number of elements or sub-elements, we often call them *subsystems* and their functions *subfunctions*. For example, when a chair is considered as a system, it can be said to have three main subsystems: a seat, a back support, and legs. The entire system provides support for a person sitting on it, and it is its main function. The seat provides support for buttocks, the back support provides support for the back, and the legs transfer loads from the seat and the back support to the floor; these are subfunctions, which are obviously different from the system's main function, that is, to provide support for a sitting person (Figure 4.3).

Sometimes, it is convenient to call the environment of our system a *supersystem* because it is a system having a dominant impact on our system. In this case, we have a trio: a supersystem, a system, and subsystems. For example, a plane is a system. However, when we consider a wing as a system, its components will become subsystems, and the entire plane will become a supersystem, as is shown in Figure 4.4.

A system is much more than a collection of its elements. For example, a railway truss bridge can be considered as a system. Its main function is to provide continuity of a railway line across a river or a valley. Our bridge—our system—has many elements. It is composed of many subsystems, for example abutments, trusses, decking, bracing, railway, and hand-rails. In Figure 4.5, only the superstructure is shown with its wind bracing, sway bracing, deck, and so on. When all these subsystems are brought to a storage facility by the river crossing, they are still not a bridge and not a system, since they do not provide the desired function; that is, they do not provide continuity of the railway line.

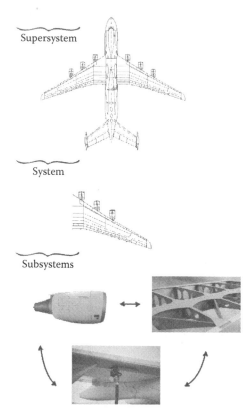

Figure 4.4 Supersystem, system, and subsystems.

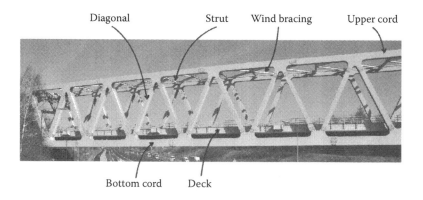

Figure 4.5 Railway truss bridge and its major components.

Several years ago, my systems engineering students reminded me that we should distinguish between systems and *objects*. We have already defined a system. An object (real or abstract) is an entity with only a single component that provides a single simple function. We could say in this context that an element of a system may be considered as an object.

For example, when a truss is considered, it could be considered as a system, and a single member, a diagonal, could be called an element of our system. At the same time, however, this diagonal could be also considered as an object when it is the focus of our attention and is considered outside the truss. Such a member could be treated as an object because its function is simple: to transfer an axial force. Such a function cannot be easily subdivided into subfunctions, but it is still possible.

We must remember, however, that concepts of a system and of an object are relative. A wing of an airplane may be considered a subsystem when the entire plane is considered as a system, but it can also be seen as a supersystem for its subsystems. Also, each system is a part of a larger system. A house is a part of a subdivision, a subdivision is a part of a town, a town is a part of a metropolitan area, and so on. We must decide what part of our reality will be considered as a system and what in any given case will become a supersystem, a system, and subsystems.

We will now consider *engineering systems* and *complex engineering systems*. Engineering systems will be understood as systems created using exclusively engineering knowledge and operating in strictly engineering environments. No other kind of knowledge is necessary, from example from history, to design or to use them. For example, foundation systems could be designed using only knowledge from the areas of foundation engineering, soil mechanics, structural mechanics, and so on. Unquestionably, we do not need any nonengineering knowledge to deal with foundation systems or with other strictly engineering systems. In the case of such systems, we are usually able to formulate optimality criteria (e.g., cost or weight) and to use these criteria to find the single optimal design.

Complex engineering systems will be understood as systems created using engineering and nonengineering knowledge and operating in environments that cannot be fully understood in purely engineering terms. Such systems are becoming more and more important for society. Therefore, engineers must know how to deal with their design and utilization. In fact, many engineering systems of the past would be considered as complex engineering systems today. For example, in the past, bridge designing was considered the exclusive focus of bridge engineers. Today, designing bridges requires considering political and social factors, and even their esthetics. This expanded understanding of bridges naturally leads to much more complex design situations in which a design of minimum weight (optimal from the strictly engineering perspective) may often not be the right design if its environmental or social impact has not been considered.

4.3 DATA, INFORMATION, AND KNOWLEDGE

4.3.1 Data and information

When we conduct an experiment and record our readings, we acquire *data*. Therefore, data can be defined as facts collected together for the future use. After we conduct several experiments and perform a statistical analysis of their results, we can present these results together with data in an organized form, creating *information*. Therefore, information can be defined as data organized in a meaningful way.

4.3.2 Knowledge

When we study information and discover an unknown relationship between two subsystems in our system or a relationship between two variables describing it, we acquire *knowledge*. In fact, the concept of knowledge is a little more complex than the concept of information. Therefore, we will introduce first a general definition provided by the *Oxford Dictionary*:

> Knowledge is a system of facts, information, and skills acquired by a person through experience or education; the theoretical or practical understanding of a subject.

In our study of inventive engineering, we will understand knowledge in more pragmatic terms. In our case, a subject will be a specific engineering domain, and we will introduce the concept of *domain knowledge*. Our definition of domain knowledge will use the terms *decision rules* and *heuristics*. We will be using these terms quite often and we need to understand them well. Therefore, we will discuss them first. These terms can be interpreted in many ways because they are both relative; that is, they have various meanings depending on the source of information. Generally, the difference between them is in the level of abstraction. Both can be interpreted in the design context as a *call for action*. In the case of decision rules, this call may be very specific, as the example below shows, while heuristics are usually more general and recommend attitudes rather than specific actions (Arciszewski and Rossman 1992).

Decision rules are usually in the form:

> If conditions A, B, and C are satisfied, then take action D.

For example

> If you design a railway bridge (condition A) with span 100 m or more (condition B), then design your bridge as a steel bridge (action D).

A heuristic can be understood as an abstraction of experience (Maier and Rechtin 2000). A heuristic is usually presented in a natural language and is much less specific then a decision rule. For example

Be proactive.

In engineering, decision rules and heuristics are acquired in different ways. Decision rules are a product of *deduction*, a reasoning process in which available knowledge is used. Therefore they are consistent with the state of the art in a given domain. Heuristics usually represent experience or the knowledge accumulated by engineers over a long period of time as a result of working on various design cases. Heuristics are a product of *abduction*, a reasoning process in which available knowledge and observations or examples are used to generate new hypotheses or heuristics.

Usually, a collection of decision rules is supposed to be internally consistent, it may be considered as a knowledge system, and the individual rules are expected to be correct. On the other hand, heuristics may be contradictory, not always valid, and useful only for very few specific situations.

In computer science a single term, decision rule, is used. In the area called automated knowledge acquisition, or machine learning, the focus is on an abductive process of automated (by a computer program) learning of decision rules from examples. So acquired decision rules may not be consistent with the state of the art in a given domain and may be valid only in the context of a collection of examples used to produce them. The quality of decision rules is determined by considering their *predictive power* or their *performance*, that is, their ability to predict examples, which have not been used for acquiring these decision rules. Predictive power is measured using various *empirical error rates*, which measure the probability that a given collection of decision rules fails, that is, does not correctly predict an unknown example. Values of empirical error rates are determined through computer experimentation. A detailed discussion of these terms is available, for example, in Arciszewski et al. (1992).

Machine learning may produce an entire spectrum of decision rules in terms of their predictive power. At one end of spectrum we have *strong*, or quasi-deterministic, rules with error rates close to zero, which very rarely fail and are almost always valid. At the other end, we have very *weak*, or probabilistic, decision rules with error rates close to one, which usually fail and are very rarely valid, and then only for very few special situations. In the area of machine learning, such rules are sometimes called heuristics. Unfortunately, there is no clear division between decision rules and heuristics, and our judgment must be used to distinguish between them.

4.3.2.1 Domain knowledge

Domain knowledge is a verifiable and consistent system of facts, models, decision rules and heuristics associated with a specific domain.

Domain knowledge represents our understanding of a given domain and is considered objective since it can be repeatedly verified through experimentation and it does not change depending on its user, although its interpretation may change. All forms of knowledge are useful in engineering designing, but decision rules and heuristics are particularly important since they may guide us through the process and may be used in knowledge-based computer tools. Unfortunately, they are usually very difficult to acquire.

4.3.2.2 Background knowledge

When we are talking about the engineering designing process, we usually are not interested in the entire body of knowledge associated with our domain, but we want to acquire and use *background knowledge*. By this term we mean a part of domain knowledge that we believe is directly, or at least indirectly, related to our design process. Therefore, background knowledge can be defined thus:

> "Background knowledge" is a part of domain knowledge, which is, up to our best understanding, directly or indirectly useful for conducting our designing process.

The concept of background knowledge (BK) is relative. What our BK is very often changes during the design process as we develop a better understanding of a design problem, acquire additional knowledge, consult experts, and so on.

4.4 TRANSDISCIPLINARITY

Transdisciplinary knowledge (Sage 2000) happens when new concepts (and the understanding behind them) emerge from the transformation, restructuring, and integration of knowledge from multiple fields of study. One interesting example of this is ongoing and is a heart patient who is also a structural engineer decided to work with medical doctors and vascular researchers to create a new form of arterial stint that would be more flexible and less prone to breakage. This new medical technology is created as a collaborative effort between a structural/mechanical engineer and medical scientist. In this example, the transdisciplinary knowledge that is created is medical engineering.

Transdisciplinary knowledge provides a holistic understanding of a brand new domain (or field of study) through the integration of several existing domains. The best example of this is the unique body of transdisciplinary knowledge that is called *bio-inspiration*. This knowledge results from the integration of concepts coming from structural engineering, biology, and psychology (Arciszewski and Cornell 2006), and it is discussed in Chapter 14, "Bio-inspiration." Transdisciplinary knowledge should be distinguished from fusion, however, which leads to "knowledge soup" (Koza

2003). In fusion, knowledge from several domains is combined but not integrated.

Inventive engineers welcome the integration of different concepts from several domains into transdisciplinary knowledge. Engineering, mathematics, physics, chemistry, and computer science as well as cognitive psychology, history, political science, and the fine arts can all be integrated to create unique and new fields of knowledge. Of course, these are only some examples of the hundreds of ways knowledge can be meaningfully combined and integrated.

4.5 EMERGENT BEHAVIOR

The concept of emergent behavior has been the subject of studies by philosophers, biologists, engineers, and computer scientists since the late nineteenth century (Arciszewski et al. 1995; Blitz 1992; Bedau and Humphreys 2008; Corning 2002; Goldstein 1999; Holland 1998; Morowitz 2002; Wolfram 2002). The concept is used to better understand complex systems. *In general, it is a* behavior of a system that does not depend on its individual parts so much as their relationship to one another—relationships that are very complex and nearly impossible to identify and fully understand. In the case of your course of study in inventive engineering, it can be seen in the context of the relationship between factors that leads to the overall driving force for your success.

In metaphorical terms, an emergent behavior is like a perfect storm. Such storms can never be predicted. They are complex and rare occurrences and are unique outcomes of several basic and interrelated phenomena, such as wind speed, humidity, solar output, and water temperature, all happening simultaneously. In addition, one can never really predict what will happen as a result of the storm. The emergent behavior of a storm could be a new shoreline or waterway, or it could be the displacement of an entire population. Whatever the case, an entirely new quality, or reality, is formed.

An emergent behavior is always a product of a qualitative or revolutionary change driven by a unique combination of individual contributing phenomena. It can be described by the following three factors:

- It usually leads to the acquisition of new concepts, which expands the body of knowledge within the domain in which it occurs.
- It is unpredictable and unexplainable and therefore cannot be derived simply from the known phenomena that contribute to it.
- It is irreducible and cannot be presented as a simple resultant of its components.

If we want to recreate an emerging behavior, a holistic understanding of the contributing phenomena is necessary. In our storm example, we would need to know the exact wind speed, temperature, solar output, and so on.

The proper combination of phenomena must also be known in order to create a specific situation in which the desired behavior will occur (i.e., emerge); that is, we would need to know at exactly what time and in what way all the elements converged to create the upheaval that was the storm. In the sciences, an emergent behavior can be recreated with a fair amount of certainty. For example, it is well known that if a magnet is heated gradually, it suddenly loses its magnetism at a specific temperature. In many cases, such as with a storm, there is much less certainty. Even if we systematically put all of the known factors together again and in exactly the same order, this does not guarantee that the result would be exactly the same, simply because some factors may not be known.

An example of emergent behavior in a social setting would be the phenomenon of less face-to-face communication between individuals as a result of texting, cell phone use, email, and social media. The end result of less face-to-face interaction was created by the *relationship* between these factors more so than the individual factors themselves. In the case of inventive engineering, the emergent behavior of your success in this field (and what that will look like specifically) will depend on the relationship between many factors, including motivation, your understanding of inventive engineering study habits, discipline, physical and mental health, and many others.

4.6 KNOWLEDGE CONVERGENCE

To continue with the last example, the driving force for success can also be understood in the context of knowledge convergence. This is a process that also involves the acquisition of knowledge from several domains. When a specific critical stage of this process is reached, or when the necessary and sufficient body of knowledge is acquired, knowledge integration and the acquisition of transdisciplinary knowledge is initiated. Knowledge convergence is an emergent phenomenon and can be compared to many other natural phenomena that humans have learned from in the past, the accumulated knowledge of which has created the intellectual foundation for our civilization. For example, the transdisciplinarity of bioengineering is a knowledge convergence that emerged from cross-disciplinary studies that grew out of the sharing of knowledge around the globe via the Internet. It was also born out of successes and failures within separate fields of study (such as biology, engineering, and physics), which led to individual advances that eventually converged. As this example shows, newly acquired transdisciplinary knowledge contains newly acquired concepts that are only partially explainable in the context of the domains contributing to it. Indeed, when these concepts converge, they are also the result of it.

In regards to inventive engineering education, transdisciplinary knowledge can be considered a convergence of knowledge from such domains as education, cognitive psychology, history, sociology, and engineering, and it can be viewed as a new quality independent from the domains contributing to its emergence.

4.7 SYNTHESIS

In engineering, *synthesis* is usually an abductive process in which existing knowledge is used to produce new hypotheses, new ideas, or new concepts. In the context of engineering designing and creativity, synthesis is a process that occurs during the conceptual design stage, since synthesis covers a large class of processes leading to the creation and generation of design concepts. Russian engineer, inventor, scientist, and writer Genrich Altsculler (1984), the creator of the Theory of Inventive Problem Solving (TRIZ), distinguishes at least five types of synthesis that may occur in the conceptual design stage of a project. The five types and examples of each are listed below (Arciszewski et al. 1995).

- *Selection:* An example would be catalogue design (selection of a design concept from a list of several concepts previously developed) of columns in an industrial building.
- *Modification:* An example would be when two existing car concepts from a given domain (a sports car and a utility vehicle) are combined to produce the concept for a sports utility vehicle (SUV).
- *Innovation:* An example would be when the concept is a product of combining concepts from two different domains: An airbag for a car is produced using concepts of a safety car pillow (crash engineering) and of a balloon (aerospace engineering).
- *Invention:* An example would be when the concept of ceramic disc brakes is developed by combining the concept of disc brakes with concepts coming from the new technology of high-strength ceramic materials.
- *Discovery:* An example would be the concept of an x-ray machine developed in the 1890s using the discovery of x-rays 20 years prior.

The concept of synthesis is also used by Bloom et al. (1956) in his *Taxonomy of Educational Objectives*. Synthesis in engineering education has been recognized as the key to design creativity (Arciszewski 2009; Arciszewski and Rebolj 2008). Unfortunately, however, the existing available learning environments in the majority of engineering programs almost exclusively focus on the quantitative or numerical aspects of design, not on creativity and synthesis in design. There is still a dominant belief within these departments that synthesis cannot be taught in a systematic way, and that it must be acquired as a by-product of a gradually growing design practice begun once one has left academia. The Author believes that this conclusion must be abolished and that opportunities for systematically learning design creativity must be implemented in higher education in order for inventive engineering students like you to have true success in the increasingly demanding profession of engineering.

4.8 SYNESTHESIA

Synesthesia is a concept that describes a relatively rare phenomenon known in cognitive psychology, wherein the stimulation of one cognitive sense causes a reaction in a second sense or in several senses (Robertson and Sagiv 2005, Ward 2008, Harrison 2001). For example, synesthesia occurs when, upon hearing the number seven, a person sees the color green, or the number seven is simply perceived as green. In a more complex form of synesthesia, casual engineering components such as trusses are personified. For someone who is experiencing synesthesia, a truss can be seen as an "adventurous" truss or a "beautiful" truss. Synesthesia is mostly a naturally emergent phenomenon. It may be hereditary, or it may occur as a result of various medical conditions such as a stroke or an epileptic seizure. It can also be induced through the use of psychedelic drugs or can come from placing oneself in a highly creative space through creative practice.

Synesthesia has been the subject of research in psychology for more than 100 years. Current synesthesia research has focused on the phenomenon as it relates to human creativity. Synesthetic experiences have an impact on human behavior, and many artists in the areas of visual art, music, or theater report experiencing synesthesia in their creative processes. In addition, a heuristic method called Synectics uses *personal metaphors* with the goal of purposefully creating a synesthetic state in order to expand the available body of knowledge for sense connection and minimize the impact of the vector of psychological inertia (see Chapter 1). In Synectics, one's ability to produce personal metaphors is encouraged as a means for becoming more creative and more able to produce novel ideas. There are even exercises and computer tools available today that can aid in developing this skill, for example Mind Gymnasium in the MindLink software (MindLink Inc.).

The Author understands synesthesia in an even more complicated way in the context of Leonardo da Vinci's work on engineering creativity. Da Vinci's process can be described as the emergence of the human ability to generate novel ideas when a unique combination of conditions takes place, that is, when all da Vinci's principles are active and the creator reaches a rare state of mind entirely focused on a problem. In this process, each condition is necessary but none is sufficient alone. Synesthesia requires the activation of all five senses and the integration of knowledge from several domains. It can also be considered as a process leading to the emergence of transdisciplinary knowledge (Sage 2000), which is the foundation of human creativity.

We can use a variation of the storm analogy to explain better the concept of synesthesia. Synesthesia for engineers is also like the perfect storm. We know that the perfect natural storm is a negative and random event entirely controlled by nature. Synesthesia, however, would be a positive and desired convergence of phenomena, of which at least some of the conditions can be created in a systematic way (see Chapter 2, Section on da Vinci's seven principles).

If inventive engineering education can be considered a complex adaptive system (Arciszewski and Russell 2013), then synesthesia would be its emergent property. Even if we know the conditions stimulating learning human creativity, we cannot guarantee the results. However, we can assume that if we do know the conditions that stimulate creative learning and synesthesia, then more than likely synesthesia will take place. Have you ever experienced synesthesia in regard to your engineering studies or in some other aspect of your life? If so, can you recall the experience?

4.9 ATTRIBUTES: DESIGN DESCRIPTORS

We will be using the term *attributes* when dealing with engineering systems. An attribute is a descriptor associated with a specific feature of an engineering system, for example with the diameter of a pipe but also with the type of cross section of this pipe. Therefore, we divide attributes into two classes:

- Numerical attributes
- Symbolic attributes

Numerical attributes describe quantitative and measurable features of an engineering system, for example its dimensions. In the case of the pipe mentioned earlier, we could use a numerical attribute "diameter of a pipe." It might have, for instance, a value $D = 12''$. Good examples of numerical attributes are length, depth, weight, or speed.

Symbolic attributes are a little more difficult to define and comprehend. In traditional engineering education, we talk almost exclusively about numerical attributes, and students are quite familiar with them. The concept of symbolic attributes is new. Symbolic attributes describe qualitative, abstract, or conceptual features of an engineering system, and it is for this reason that they are so important in inventive engineering.

Symbolic attributes can be divided into *nominal* and *structured attributes*. Symbolic nominal attributes take values from an unordered set of values. For example, when we consider the symbolic nominal attribute "color," it may have the values white, yellow, black, and so on, and we cannot easily order them using a simple criterion. When we consider the symbolic nominal attribute "temperature," it may have the values medium, very low, high, low, and so on. These values can be easily ordered or structured and presented in a structured form as a clear sequence of values: very low, low, medium, high. We can identify many symbolic attributes that are quite important in designing. For example, in designing we use such symbolic attributes like the shape of a member with such values as straight or curved, or a type of connection with such values as fixed or flexible.

We need to know that all numerical attributes could be converted into symbolic attributes. For example, the numerical attribute "diameter of a pipe" could be converted into a symbolic structured attribute with such values as very small, small, medium, large, and so on. Such conversions are often justified when we do not want to be flooded with technical details and need to develop the big picture of a given designing situation.

Symbolic attributes can be converted into numerical attributes, but it is much more difficult and not always possible. Earlier, we introduced temperature as a symbolic attribute. This one could be easily converted into a numerical attribute with numerical values representing a certain range of variation. However, a symbolic attribute "type of a cross section" with values "Box," "I Section," "C Section," and so on, cannot be converted into a numerical attribute.

4.10 DESIGN REPRESENTATION SPACE

We could say that engineering knowledge is available everywhere; it is in the textbooks and in previous designs, experienced designers have it, and even you, as a student engineer, have already acquired a large body of engineering knowledge. However, when we want to conduct an engineering designing process (defined in the next section) using various formal methods and computer tools, we need engineering knowledge in a specific form, and this is called *design representation space* or simply *design space*.

> "Design representation space" is an organized collection of attributes and their feasible values, which is necessary and sufficient to describe all known designs and has a potential for finding many new and unknown designs. It represents the State of the Art of knowledge in the problem domain.

We could say that a design representation space (DRS) is an equivalent of our background knowledge (BK), but in this case our BK is presented in a formal and systematic way, which is useful for both the human designers and for various computer tools. A DRS has two major parts: the symbolic and the numerical. The symbolic part contains all the symbolic attributes and their feasible values, while the numerical part contains all the numerical attributes and their specific feasible values or their feasible ranges of variation.

DRS can be presented as a table with rows representing individual attributes and their values. Such a form is useful for design purposes and is consistent with one of the most popular inventive designing methods, which is

Table 4.1 Design representation space for a class of small objects

Attributes	Attribute values			
A1: Material	Steel	Concrete	Wood	Plastic
A2: Shape	Cube	Cuboid	Cone	Other
A3: Homogeneity	Solid	Hollow	—	—
A4: Height	10″	15″	20″	Other

called the *morphological analysis* and uses a *morphological table*; it is in fact a DRS.

For example, we would like to represent our knowledge about a class of small objects of various shapes. In this case, we could use, for example, only three symbolic attributes and one numerical attribute. Our first symbolic attribute is

A1: Material and it defines the kind of material used with four feasible values: steel, concrete, wood, and plastic. The second attribute *A2: Shape* is also symbolic. It describes the shape of our object with the three feasible values of cube, cuboid, and cone. The last symbolic attribute, *A3: Homogeneity*, determines if our object is solid or hollow. The attribute *A4: Height* is a numerical attribute with three specific numerical values. Obviously, more specific values could be used for the individual attributes, depending on our needs (Table 4.1).

The following example represents a specific combination of four values of our attributes:

A1: Material	Plastic
A2: Shape	Cuboid
A3: Homogeneity	Solid
A4: Height	15″

This describes a single solid plastic cuboid with a height of 15″. It is a specific cuboid. However, consider what happens if only one value in this combination of values of our attributes is changed, and we assume that

A4 : Height = other

This combination describes an entire huge class of solid plastic cuboids with various heights.

Using attributes and their values to identify objects and systems is very powerful and is absolutely necessary when developing computer programs for engineering design. Also, an attribute-based approach is used by patent offices in many countries. In this case, each patent claim represents a different feature of an invention—a different symbolic attribute of a given invention.

4.11 DESIGN REQUIREMENTS AND CONSTRAINTS

Design requirements specify features of our object or system that are required in the case of our specific design. They may be stated in both symbolic and numerical terms. For example, a design requirement may state that our truss should have sufficient stiffness. A symbolic attribute will be used, and this requirement presented as

R1: Stiffness = Sufficient

This requirement may be also much more specific and may be presented using a numerical attribute and its specific numerical value:

R1 : Stiffness = Maximum Deflection $\leq 16''$

In more general and computational terms, a design requirement can be defined thus:

"Design requirement" is a combination of attributes and their values that must occur in a design.

Design constraints identify features of our object or system that must be avoided in our design. They may also be stated using both symbolic and numerical attributes. For example, a constraint may state that concrete may not be used as a material for our bridge. In this case, obviously only a symbolic attribute may be used and the constraint will be presented as

C1 : Material = Not Concrete

A numerical attribute will be used in a constraint specifying that steel with a specific yield point (yield strength) cannot be used. For example,

C1 : Yield Point of Steel = not 250 MPa

A design constraint may be defined in computational terms thus:

"Design constraint" is a combination of attributes and their values that must not occur in a design.

Differentiating between design requirements and constraints makes a lot of sense from the perspective of a human designer who wants to know and understand what is expected from him or her in terms of the desired and undesired features of his or her future design. From the computational point of view, however, this distinction is not important and only complicates calculations. Since it is possible to formulate design requirements as constraints, as shown below, sometimes we use only constraints. In fact, there is a class of design methods called *constraint search*, in which only constraints are used, but they are functionally equivalent to both requirements and constraints as we defined them earlier (Hajdo and Arciszewski 1991).

For example, the requirement discussed earlier,

R1 : Stiffness = Maximum Deflection $\leq 16''$

may be presented as C1

Stiffness = Maximum Deflection not $\geq 16''$.

4.12 ENGINEERING DESIGN

Engineering design is a product of an activity called the *engineering designing process*, which is discussed in Section 4.3. In general, it may be described thus:

Engineering design is a description of a future engineering object, or a system, and

1. It contains all the organized information (a body of knowledge) necessary and sufficient to build, maintain, and eventually to demolish this object or system to recover materials.
2. It has two components, including a design concept and a detailed design.

A design concept represents the qualitative or abstract part of an engineering design. It can be defined thus:

"A design concept" is an abstract part of an engineering design describing a future engineering object, or system, in terms of symbolic attributes. In computational terms it is a sequence of symbolic attributes and a feasible combination of attributes and their values.

Examples of design concepts may include an arch bridge or a cable-stayed bridge for a bridge (Figure 4.6), a prestressed concrete I beam for a large-span floor structure in an industrial building, or a touring bike or a racing bike for a bike (Figure 4.7).

When we design a roof structure and consider various design concepts, a steel truss may be one of design concepts analyzed. This design concept may be identified by only three symbolic attributes:

A1: "Material" with feasible values "steel," "concrete," and "wood"
A2: "Member shape" with feasible values "straight" and "curved"
A3: "Connection type" with feasible values "rigid" and "pinned"

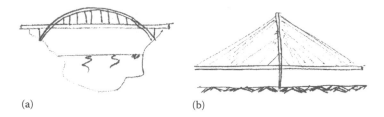

Figure 4.6 (a) Arch bridge and (b) cable-stayed bridge.

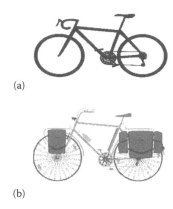

(a)

(b)

Figure 4.7 (a) Racing bike and (b) touring bike.

These attributes and their values are shown in a conceptual design representation space below. A combination of values describing a truss is shown in the table as bolded words. Obviously, a steel truss is a steel structure with straight members that are connected by hinges; that is, a steel truss is described by a specific combination A1: "Material" = "Steel," A2: "Member Shape" = "Straight," and A3: "Connection Type" = "Hinged" (Table 4.2).

A *detailed design* is a quantitative or numerical part of an engineering design. It can be defined thus:

> "Detailed design" is a numerical part of an engineering design describing a future engineering object, or a system, in terms of numerical attributes. In computational terms it is a sequence of numerical attributes and a feasible combination of their values.

Table 4.2 Conceptual design representation space

Attributes	Values			
A1: Material	**Steel**	Concrete	Plastic	Wood
A2: Member shape	**Straight**	Curved		
A3: Connection type	Rigid	**Hinged**		

Table 4.3 Design concept and detailed design comparison

Feature	Design concept	Detailed design
Specificity	Abstract	Numerical
General nature	Qualitative	Quantitative
Describing attributes	Symbolic	Numerical
Product of	Conceptual designing	Detailed designing
Impact	Novelty	Optimality
Scientific foundation	Inventive engineering	Analysis and optimization

A detailed design specifies values of all numerical attributes, such as the specific dimensions of the individual members of a structural system, the number of members, material yield points, and so on. How to develop detailed designs when the design concept is known is the subject of numerous textbooks on machine design, steel design, and chemical design.

A simple comparison of the terms "design concept" and "detailed design" is provided in Table 4.3.

Sometimes the term *design specification*, or simply *specification*, is used. Unfortunately, this term has two meanings. One is associated with the designing process product and is equivalent to the term just introduced, *engineering design*. The second one is related to the beginning of the designing process and its input. In this case, design specification is defined thus:

Design specification

- Specifies the desired features of the future engineering object, or system, including features of its behavior
- Provides specific design requirements and constraints, which must be specified
- Presents the background knowledge available in a given design situation

This definition can be interpreted as a combination of what has to be achieved (desired features and formal requirements), what has to be avoided (constraints), and what is known (background knowledge).

4.13 PROBLEM, DESIGNING PROBLEM, WICKED PROBLEM

The term *problem* is usually used to describe a situation that requires an improvement or a solution. This definition is too narrow for inventive engineers interested not only in solving existing problems but also, and much more importantly, in seeking opportunities and creating life-changing inventions. Therefore, our understanding of the related term *designing problem* will be thus:

> A designing problem is a description of a designing situation which requires understanding and improvements. It is a body of knowledge necessary and sufficient to find a solution improving the situation and/or leading to the identification of new opportunities potentially resulting in patentable inventions.

This is a holistic definition of a designing problem. Solving designing problems may be fundamentally different from problem solving as it is known in decision science: systematic, deterministic, algorithmic, and ready for automation. There were attempts to consider engineering designing as exclusively problem solving in the context of decision science (Hazelrigg 2012), but they were rejected by the design research community and by practitioners as too narrow and limiting.

There are all kinds of problems, from simple problems that can be easily solved in no time to very complex problems that seem nearly impossible to solve. Even definitions of such problems are not entirely clear, change with time, and usually are the subject of never-ending discussions. Usually it is impossible to develop more than one solution, and it is obvious that this solution is most likely not optimal in terms of money, time, or other formal quantitative or numerical optimization criteria. We all know of such problems, for example, global warming, the well-being of a nation, or social justice. Such problems are formally called *wicked problems*. This term was introduced by Rittel in 1972 (Rittel 1972).

Formally, a wicked problem has 10 unique features (Rittel and Webber 1973):

1. It has no definite formulation.
2. It has no stopping rules.
3. Solutions are not true-false but better or worse.
4. There is no immediate solution and no ultimate test of a solution.
5. Every solution to a wicked problem is a "one-shot operation" because there is no opportunity to learn by trial and error; every attempt counts significantly.
6. There is no enumerable (or exhaustively describable) set of potential solutions, nor is there a well-described set of permissible operations that may be incorporated into the plan.
7. It is essentially unique.
8. It can be considered to be a symptom of another wicked problem.
9. Its causes can be explained in numerous ways. The choice of explanation determines the nature of the problem's resolution.
10. The problem solver has no right to be wrong.

According to the one of the British pioneers of design research, the late Sydney Gregory, all designing problems can be roughly divided into "routine problems" and "inventive problems" (Gregory, private communication with the Author). A routine problem can be solved using well-known design concepts and a deterministic analytical procedure. Inventive problems cannot be solved using only analytical methods and require the use of heuristic methods involving abduction.

The huge majority of designing problems are routine. Very few designing problems are inventive, but when they occur they are usually unavoidable, very important for stakeholders, and critical for progress in engineering. In Chapter 5.6, "TRIZ," we will present a formal classification of inventive problems. We should bear in mind, however, that some inventive problems may be considered "wicked." Wicked problems are basically outside the scope of inventive engineering, but this science creates at least an opportunity to tackle them.

Chapter 5

Science of inventive engineering

5.1 INTRODUCTION

We will define the term *engineering science* as a body of knowledge that is necessary and sufficient to understand, design, and use a class of engineering systems. For example, *structural engineering* is an engineering science. It is a body of knowledge necessary and sufficient to understand, design, build, and maintain a class of engineering systems called *structural systems* like roof structures, foundations, tunnels, or bridges.

We will assume that *inventive engineering* is a body of knowledge that is necessary and sufficient to understand, develop, and use a class of inventive designing processes and various related computer tools. It is a transdisciplinary science with roots in engineering, cognitive psychology, systems engineering, political science, history, and so on. Knowledge from all these separate domains has been acquired and integrated in the context of inventive designing, creating a unique understanding of this activity and a foundation for learning about it.

The Author proposed the name inventive engineering only recently, when, in the 1990s, he began teaching a new course at George Mason University on Design and Inventive Engineering, and Inventive Engineering was taught as a separate and distinct part of the course. However, the term *design engineering* was introduced in the United States in the 1990s when the Defense Advanced Research Programs Administration (DARPA) created a research program called Design Engineering. Earlier, in the 1980s, the National Science Foundation (NSF) introduced a new research program called Engineering Design, which initiated and stimulated design research and created the initial momentum in the United States. We should add, however, that in Europe at this time, design research had been active for more than 20 years.

As we will immediately discover, the history of inventive engineering was created by polymaths, people who were talented and extremely accomplished in several separate domains. As a matter of fact, only such people are able to acquire knowledge from different domains, integrate it, and subsequently create a new quality or a new understanding in the form of transdisciplinary knowledge. Such knowledge is absolutely necessary to achieve a real breakthrough in science or engineering and to become truly creative—to become an inventor.

The history of human engineering creativity, and thus of inventive engineering, began when people discovered fire, invented the first simple tools, and began dreaming about a better life. This dream led to a long stream of inventions and in no time, in historical terms, people were walking on the moon, talking on cell phones, and still dreaming about a better world and more wonderful life-changing inventions. However, the written history of inventive engineering goes back "only" about 2000 years.

Marcus Vitruvius Pollio was a Roman architect, civil and military engineer, and writer. He was a polymath of his time. Probably around the year 25 BC, he published a monumental treatise (10 volumes or books) on architecture and civil engineering, called *De Architectura* ("On Architecture," published as *Ten Books on Architecture*). The treatise is surprisingly modern; it presents architecture and civil engineering from many perspectives, including the social and environmental, creating in this way a fascinating interdisciplinary understanding of this domain. The first volume is focused on town planning and on designing in architecture and civil engineering in general. One of Vitruvius's heuristics was that "The ideal building has three elements: it is sturdy, useful, and beautiful" (Wikipedia, seen on May 15, 2015), and this sounds quite modern. We could say that Vitruvius wrote the first known text on designing methodology. Since he also addressed the development of various design concepts, and this process may be creative, his work can be considered as the first publication in the area of inventive engineering. The science of inventive engineering was born.

Heng Zang (Figure 5.1) lived in China (AD 78–139). He was a true polymath, being successful in several fields, including astronomy, mathematics, and geography. Zang was also an artist and a poet. Most importantly, he was an inventor who developed several important inventions. His best-known invention was a seismometer, a device for roughly finding the location of an earthquake. Zang not only wrote poems and rhapsodies but also reports for the emperor on various subjects, including his inventions. In this way, he became one of the first scholars in Asia writing about inventions and commenting on their development (Wikipedia, accessed on May 15, 2015).

In Europe in the fifteenth century and at the beginning of the Renaissance, Leon Battista Alberti (Figure 2.4, Chapter 2) was one of the key scholars and creators of his time. He was also a polymath: an architect, artist, poet, philosopher, and even a cryptographer. From our perspective, his most important publication was the book on *De Pictura* ("On Painting"), a treatise in Latin on painting and drawing, in which the fundamental principles of perspective were formulated. The book directly contributed to design methodology and inventive engineering. Even more importantly, it inspired the creativity of generations of architects and builders, who used these principles to design beautiful buildings like St. Peter's Cathedral in Vatican City with its absolutely gorgeous

Figure 5.1 Hang Zang, a Chinese scholar and inventor. (With permission. Drawn by Joy E. Tartter.)

dome; and the entire urban system, with buildings, streets, monuments, trees, and so on, which would be impossible to design without the use of perspective.

Leonardo da Vinci (Figure 2.5) was a true giant of the Renaissance. He lived in the fifteenth century and made many fundamental contributions to art and science, which changed not only his society but also had a significant impact on Western civilization. He was also an inventor, developing inventions that were so far reaching that they were implemented only centuries later, like a submarine, a parachute, and a tank. He also believed in the power of knowledge and regularly recorded all his findings in his diaries. These diaries also contain descriptions of his inventions, and most importantly from our perspective, they contain his comments about the development of these inventions. In other words, da Vinci shared with us his methodological experience and therefore can be considered as the first European inventive engineer, capable not only of developing inventions but also of understanding how this could be done.

Figure 5.2 Rene Descartes, creator of the universal method. (With permission. Drawn by Joy E. Tartter.)

In the seventeenth century, a French polymath, philosopher, mathematician, and writer, René Descartes and Lafleur (1960) (Figure 5.2), published a treatise titled "The Discourse on the Method." Four centuries later, it still is one of the most influential publications in philosophy and science. In the second part of the treatise, Descartes presented a method that became known as the *method*, the *scientific method*, or the *universal method*. It is the first attempt to formulate a universal, domain independent method of scientific inquiry. It is also a method applicable to inventive designing, and some of its rules, for example the *rule of division*, became incorporated in various inventive designing methods. Therefore, Descartes should be recognized as one of key figures in the history of inventive engineering.

The modern history of design research began in Great Britain in the early 1960s. The initial focus was on the design methods (methodics) in architecture, but it soon became truly interdisciplinary. The British "design revolution" was strongly stimulated by the Design Research Society, established in 1966 (your Author was a member of this society in the 1980s and in 1986 gave a talk during a special meeting in Coventry) and by the journal *Design Studies*, established in 1979. Chris Jones, Bruce Archer, Nigel Cross, Sydney Gregory, and Thomas Maver: these are several of the British design research pioneers who started the "design revolution" and were able to create a global movement, one that is still growing and increasing its impact on design practice.

Gradually, during the 1970s, design research spread to Germany (Koller, Johannes Müller), Switzerland (Hermann Holliger-Uebersax, Vladimir Hubka), the Netherlands (Harry van den Kroonenberg), and Poland (Wojciech Gasparski, Andrzej Goralski). In the early 1980s, design research was initiated by John Gero in Australia at the University of Sydney. More information about the history of design research may be found in Bayazit (2004).

The history of inventive engineering will never be complete without mentioning Herbert Simon. He was an American scholar and a modern polymath with interests and accomplishments in political science, economy, sociology, and psychology. His work was integrative in nature and resulted in transdisciplinary knowledge, offering a new understanding of science. In 1969, he published a book on *The Sciences of the Artificial* (Simon 1969). The book became an immediate intellectual foundation for the science called *artificial intelligence* and provided a breakthrough understanding of engineering designing as an abstract reasoning activity that is domain independent and focused on finding designs through searching the design space. Such an understanding is still widely accepted and used for research purposes and is implemented in various computer tools for conceptual designing, for example, in Amadeus (Hajdo and Arciszewski 1991), Inventor (Murawski et al. 2000), and Emergent Designer (Kicinger 2004).

Today, inventive engineering is emerging as a separate science, and its importance has been recognized by scholars and instructors in several countries. For example, inventive engineering is taught under various names in the United States (George Mason University, Stanford University, the Massachusetts Institute of Technology [MIT], Carnegie Mellon University), in Taiwan (The National University of Kaohsiung), in China (Tsinghua University), and in Poland (Wroclaw University of Technology, Kielce University of Technology). There is a simple common-sense explanation for the recent emergence of inventive engineering. If you want to drive a car, you need to take several driving lessons, but if you want to become a professional race driver, in addition to the basic driving lessons you need to take several advanced high-performance and racing courses and do a lot of driving and racing under the supervision of experienced racers. By contrast, if you want to become a designer, you take a number of analytical courses and courses *about* structures, which teach you very little about *how* to design structures. Unfortunately, there is huge gap in engineering education, and very few academic programs teach students how to become inventors.

5.2 ENGINEERING DESIGNING: FIVE DEFINITIONS

5.2.1 Introduction

We are all proud engineers. We have created a foundation for our civilization to exist, and we are its custodians. Moreover, we shape the future.

Our imaginative and creative products not only serve our societies but also create new opportunities and inspire people. Railways and bridges in the nineteenth century and tall buildings in the last century are excellent examples of engineering systems that had a tremendous impact on the societies that produced them. They have changed the way people think and created new opportunities for them to explore.

Today, the Internet and smart communication devices are tools that not only make our lives easier and allow us to work much more effectively than before. They have also created a new reality of instant communication in which virtual communities are formed. Traditional borders and communication barriers no longer exist, and people exchange information and immediately acquire knowledge in the process, knowledge that changes their understanding of the world and which will ultimately change their lives. This new paradigm has a tremendous positive impact on engineering communities. For example, the American Society of Civil Engineers Global Center of Excellence in Computing (established by your humble Author) provides teaching materials on computing to instructors on 6 continents and in 24 countries. We should bear in mind that only several years ago, this would have been simply impossible. This paradigm also changes political landscapes in many countries and allows people living in nondemocratic societies to learn about democracy and to organize themselves and rise, as the recent Arab Spring so clearly recently demonstrated.

As we may see, the evolution of our societies is driven by progress in engineering and by the resulting stream of new engineering systems that are changing our reality and shaping our future. All these new systems are products of engineering designing, which is the most important engineering activity considering its direct and long-term impact. Therefore, we need to develop a strong understanding of what is meant by this process.

No single and widely accepted understanding of engineering designing exists. In fact, such a single definition would not serve us well or would be sufficient for us, so we will learn several definitions that deal with various perspectives on engineering designing. We will introduce and discuss five definitions from a holistic definition (proposed by the Author) and a cybernetic definition (proposed by Johannes Müller [1970]) to our working two-stage definition, which will serve us well. In all cases, we will assume that engineering designing is a knowledge-based and knowledge-intensive process. We will also assume that in this process knowledge is acquired, transformed, and utilized. In such a case, reasoning and designing processes are viewed as knowledge processing activities; that is, they can be understood in the context of knowledge. Therefore, they are rational activities that can be understood, learned, and conducted repeatedly and on demand and include inventive designing aimed at producing novel and patentable solutions, the area of our main focus.

5.2.2 Definition No. 1: Holistic

When we consider engineering designing as a process occurring in a social environment, obviously its ultimate objective is to satisfy existing or future needs of a society for new, or modified, engineering objects or systems. Therefore, we assume that the entire process begins when these needs are realized, and it ends when a specific design, or a class of designs, is developed. Such an understanding of engineering designing may be summarized in a definition called *holistic*, because it has a nonengineering or social context.

> Engineering designing is a process beginning when a need for a new, or modified, engineering object, or a system, is realized and it ends when its final feasible product, an engineering design, or a class of designs, is developed.

5.2.3 Definition No. 2: Cybernetic

In the late 1960s, a German design scholar and a systems engineer, Johannes Müller (also the Author's friend), developed a design science that he called *systematic heuristics* (Müller 1970; Arciszewski 1978). He also proposed a simple *cybernetic definition* of problem solving, which has been adopted by the Author to introduce our second definition of engineering designing (Figure 5.3).

> Engineering Designing Process is a system transforming a design problem into a design or a class of designs. Probability of this transformation P(T) determines the nature of the process. P(T) = 1 for deterministic processes and P(T) < 1 for probabilistic, or heuristic processes.

In the case of this cybernetic definition, a design problem, or specifically *problem formulation*, is considered as *input* while the design is considered as the *output* of our designing system. The term *transformation* describes all complex design activities that occur during the entire process. In the case of routine design processes, the probability of this transformation P(T) = 1. It means that there is no doubt that a transformation will take place and a feasible design will be produced. Such situation may be called

Figure 5.3 Designing: cybernetic model.

"deterministic," since we know that the transformation will take place. For example, if we want to design a retaining wall 8′ or 10′ tall by a highway, we simply know that we will be able to do it and it is only matter of time before we produce our design. However, if we are asked to design a retaining wall 600′ tall in the mountains to support loose rocks, we know that our challenge is impossible. In this case, the probability of a transformation $P(T)$ is 0. These are two extreme design cases representing opposite ends of the spectrum of variation of $P(T)$. There is also a huge class of design situations when the probability of transformation $P(T)$ is definitely <1.0 but >0; such situations may be called "probabilistic" because we have only probabilistic relationships between the input and output. In all such cases, we are not sure if a feasible design can be developed, but we know that if new designs are developed, they will contribute to the progress in engineering and they will have potential to become patentable inventions. Therefore, all such cases are very important for society and engineers and are the subject of our interest.

The cybernetic definition does not specify the nature of the transformation, and this is its huge advantage. As we will learn, there are many ways to conduct this transformation ("there are many ways to skin a cat," as Sydney Gregory, one of the leaders of the British design revolution, used to say), and they are the focus of inventive engineering.

5.2.4 Definition No. 3: Multistage

In accordance to the cybernetic definition, the engineering designing process is a single-stage process. Such understanding can be considered as a "big picture" view and can be also associated with our holistic definition. It is not sufficient, however, for our studies, and we will introduce three other definitions of engineering designing. First, we will introduce a definition proposed by the Author (Arciszewski 1977a,b). At this time, the Author was studying cybernetics and was strongly influenced by systems thinking.

As a result of this thinking, the engineering designing process was assumed to be a multivariant designing process, which is a complex multistage process with several lines of activities. Each line of activities represents designing activities associated with a different variant of the final design. Some of these lines may need to be terminated before leading to the final designs, as happened with the line "N" in Figure 5.4. That means that they were based on ultimately infeasible ideas, which could not be eliminated earlier because of the lack of an appropriate knowledge necessary to evaluate their feasibility.

The entire process begins when a design problem is identified and formulated and problem formulation is developed, that is, we know the nature of our design problem, the design requirements and constraints, and the background knowledge, as discussed in Section 4.11. The analysis of problem formulation inspires the emergence of several initial concepts for various designs. These initial design concepts may not represent complete design concepts but

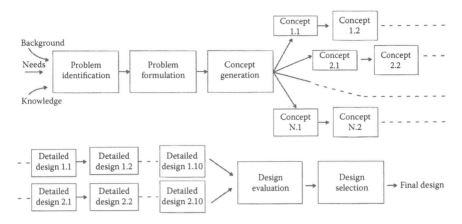

Figure 5.4 Multistage model.

may ultimately evolve into them. For example, when a bridge is considered, the initial concept may be related only to the abutment or to the deck design.

As we remember, the design (a description of a future engineering object or system) can be understood as organized information about a future engineering system or as a body of knowledge about it. Therefore, we can assume that at the beginning of the designing process, nothing is known about the future design (and it is an obvious simplification), and that at the end of this process, 100% of the necessary knowledge has been acquired (probably also a simplification). Therefore, the entire process of designing can be considered as one of gradually acquiring and transforming knowledge about the future engineering system. This process is conducted using models—very simple conceptual models at the beginning and extremely complex analytical and optimization models at the later stages of the designing process. With each subsequent designing stage, a more complex model is used and the number of attributes used in it also grows. In the early stages of the process, only symbolic attributes are used, but after design concepts are developed, the focus shifts to numerical attributes and their number also grows with each stage.

The described process is a process of knowledge acquisition and transformations, which occur at the individual stages. Thus, it is a learning process. When such an understanding of engineering designing was proposed in 1977, it was understood exclusively in the context of human designing and was considered esoteric with no immediate practical use. However, about 20 years later, it was used by the Author and his research team to develop several computer programs for the automated evolutionary designing of structural systems in tall buildings (Murawski et al. 2000) as a part of NASA-sponsored research (Kicinger et al. 2005a–c). This kind of understanding of engineering designing can be summarized in our "Multistage Definition," provided below.

Engineering Designing Process is a learning process in which knowledge about a future engineering object, or a system, is gradually acquired through a multistage process with several lines of models, each line corresponding to a different variant of the design. In each line, subsequent models are of growing complexity and they contain a growing body of knowledge.

5.2.5 Definition No. 5: Five-stage

In the 1970s and 1980s, the *five-stage definition* of the engineering designing process was widely accepted as a reflection of common designing practice at this time. From this perspective, the objective of the first stage is to develop a class of several design concepts. Therefore, this stage is called *conceptual designing*. Concepts developed in this stage are used next to develop *preliminary designs* in the second stage, called appropriately *preliminary designing*. Preliminary designs are simplified designs, which contain only selected information about future engineering systems. They are produced through a process of analysis and dimensioning in which simplified analytical models are used. The objective is to determine values of only key numerical attributes describing the designs. Although such design descriptions are incomplete for actual use, they contain enough information to compare various preliminary designs in order to select the best one so as to further develop it into the final design during the third and the last stage, called *detailed designing*. During this stage, the selected preliminary design is completely analyzed; all its dimensions and other features are thoroughly analyzed and optimized. As a result of such activities, the final detailed design is produced, which, together with the developed earlier design concept, constitutes the final design (Figure 5.5).

In the first stage, only symbolic attributes are used to identify design concepts. In the second stage, a limited number of numerical attributes is used to conduct simplified calculations. Finally, in the last stage, all meaningful numerical attributes are used.

The popularity of the five-stage definition was a reflection of the fact that in the 1970s and even in the 1980s the use of computers was quite expensive, not to mention very time consuming in terms of the input

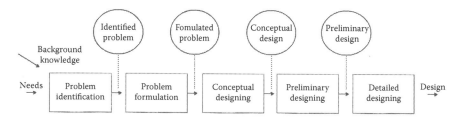

Figure 5.5 Five-stage model.

data preparations and actual computation. The cost of calculations was proportional to the complexity of the used models, the number of attributes considered, and so on. In fact, the Author worked in the early 1970s on a Ph.D. dissertation that was partially focused on the development of simplified analytical models of complex systems of wind bracings in skeleton structures of tall buildings. The motivation for this research was to develop simplified models for the manual preliminary analysis in order to prepare more accurate data for the computer calculations and thereby reduce the cost of these calculations through the elimination of many expensive computer runs.

The described process is summarized by the following definition:

Engineering designing process is a five-stage sequential process with stages called "Problem Identification," "Problem Formulation," "Conceptual Designing," "Preliminary Designing," and "Detailed Designing."

Today, the cost of computer calculations is so much lower with respect to the cost of manual calculations that there is no justification for using preliminary models and following the five-stage process. Therefore, today the dominant understanding of the engineering designing process is that of a four-stage process.

5.2.6 Definition No. 5: Four-stage

The first two stages are the same as in the previous definition, that is, problem identification and problem formulation. The objective of the third stage is to develop a design concept or a class of design concepts. These concepts are used in the next stage to develop a class of detailed designs, and this is done using traditional engineering methods of analysis, dimensioning, and optimization. When the final designs are known, they are formally evaluated and compared, and the best one is selected for optimization. In the process of conceptual designing, only symbolic attributes are used, while during detailed designing, only numerical attributes are used. This process is shown in Figure 5.6 and defined thus:

Engineering designing process is a four-stage sequential process with stages called "Problem Identification," "Problem Formulation," "Conceptual Designing," and "Detailed Designing."

5.3 PROBLEM IDENTIFICATION

Our approach to designing is knowledge based. That means that we understand the entire process as a process of acquiring knowledge about

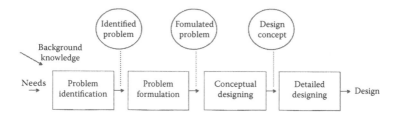

Figure 5.6 Four-stage model.

the future engineering system to be designed. This process of knowledge acquisition starts when designing begins, that is, at the stage we call *designing problem identification* or simply *problem identification*. The objective of this stage is to develop a good understanding of a given designing situation and as a result of that to determine specific design goals, which must satisfy all imposed constraints and requirements.

Specifically, we must acquire knowledge about

- The needs to be addressed by the future engineering system
- What is known about the available resources; human, physical, and intellectual (available knowledge)
- What the specific design goals are
- What the design constraints and requirements are

This body of acquired knowledge, which we call problem identification, will be analyzed and reformulated into a specific format in the next stage, called *designing problem formulation* or simply *problem formulation*. As we will discover later, each inventive designing method is based on a different understanding of inventive designing and requires a different formulation of a given designing problem.

When working on the problem identification, any technology-specific terms should be avoided at all costs. Such terms have the power to put our problem in a certain context and thus to significantly limit our focus to only a small part of the relevant knowledge. Consider the following example:

1. Provide a beam supporting distributed loading
2. Provide support for distributed loading

In the first case, a design concept is suggested in the provided statement (a beam), while in the second case the designer needs to develop his or her own design concept, and that creates the potential for novel designs.

Usually, a clear technical specification for a future engineering system is provided. Sometimes, however, the needs are presented in a descriptive form, on purpose or by necessity, and may be interpreted in many ways. It is not necessarily a bad situation, because it allows us to develop a broad understanding of needs, and that may ultimately lead to novel designs. In fact, in the case of inventive designing, needs should be presented in abstract terms so as to avoid unnecessarily and prematurely specific problem identification.

For example, the same designing situation may be presented in two ways:

1. Towns A and B located on the opposite banks of the river C need a steel truss bridge.
2. Towns A and B located on the opposite banks of the river C need a transportation system connecting them.

There is a big difference between these two situations. In the first case, it is clear that the intent is to limit the solution space only to steel truss bridges and, even more importantly, that no other means of transportation should be considered. In this case, only various design concepts for steel truss bridges should be developed and the best one used to prepare the final design. In such a designing situation, a traditional engineer specializing in steel truss bridges most likely would do a good job.

The second situation is entirely different—much more interesting and challenging with huge potential for innovation. Not only bridges, and particularly not only steel truss bridges, should be considered. Other means of transportation should be considered, including tunnels or ferries, for example. Even in the case of bridges, our solution space to be considered would be much bigger. The concept of a bridge includes steel bridges (including steel truss bridges) but also concrete bridges, prestressed concrete bridges, suspended bridges, arch bridges, cable-stayed bridges, and so on. In this second situation, a traditional engineer may not have sufficient knowledge and, more importantly, may not know how to acquire the necessary knowledge and how to develop various transportation concepts beyond steel truss bridges.

This example shows the importance of the proper problem identification but also explains why in *inventive engineering* our understanding of a given designing situation must be truly holistic. It will not hurt us if we know more than necessary, but the results may be catastrophic when we know less than necessary and we are missing a part of the available knowledge, which is potentially critical for solving our problem. Also, we need to know much more than in the case of routine designing, which is deterministic; all necessary knowledge is available, and we know that a design will be produced. Unfortunately, this is not the case when we invent, even when using formal inventive designing methods.

5.3.1 Systems approach

Using the systems approach is usually the best way to develop an initial and broad understanding of a designing situation. In this case, our first priority is to prepare a systems model of this situation (Figure 5.7).

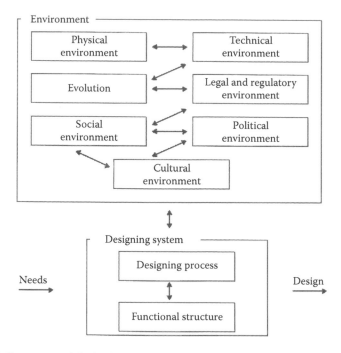

Figure 5.7 Systems model of a designing situation.

In our model, the input represents the needs to be satisfied and the available resources, while the output represents the final design. It is a model of a designing system with two main subsystems representing our designing process and a functional model of the future engineering system to be designed, or more specifically its initial functional structure. Developing such a structure and relationships among its elements (feedbacks) is appropriately called *structuring*. Many experts, particularly systems engineers, claim that structuring is the most important but also most difficult part of engineering designing. Structuring is so important because the abstract functional structure is usually reflected in the final actual physical structure of the system. In other words, we create various elements of our actual system to perform specific functions identified in the abstract functional structure.

Each system is unbounded by definition. Therefore, what constitutes its environment and this environment's boundaries are always subjectively determined by a systems analyst or by a designer, as in our case. These subjective decisions are extremely important and have a critical impact on the nature and novelty of the final results. In the case of inventive designing, the environment should be considered very carefully and its boundaries assumed *generously* to avoid at all costs any omission of important information, which might have impact on the design. At least six interrelated elements of the environment should be considered, including

1. Physical environment
2. Technical environment
3. Process of the long-term evolution of the engineering system to be designed
4. Legal and regulatory environment
5. Social environment
6. Cultural environment
7. Political environment

These elements may be considered as static, which means they are analyzed as they exist at a given time; or as dynamic, assuming that they have been gradually evolving with time over the last 20–40 years. The second option is the right one, since it gives us a much better understanding of our designing situation in the context of a gradually evolving environment. Also, it may be the first step in the direction of predicting the future evolution of the engineering system being designed.

Determination of the *physical environment* in which our engineering system will operate is the first step in the process of learning about our system's environment. If we design a new type of a bicycle, we need to know if our bicycle is intended for use in a flat urban environment or in the mountains. Obviously, we need to know much more than that. For example, we also need to know if our bicycle will be used in the hot and dry desert climate of Niger or in the tropical forest climate, hot and humid, in Amazonia in Brazil.

The technical environment represents the state of the art in the domain of our engineering system, which is the subject of our designing process. However, it is only a bare minimum and is grossly insufficient in inventive designing. Such designing usually requires knowledge also from the related domains and even knowledge from apparently unrelated domains. When the problem domain is mechanical engineering, the related domain may be structural engineering, and a seemingly unrelated domain may be biology. Such an expansion of our focus may seem to be counterintuitive, but we need to understand that knowledge from outside our problem's domain is important, and its acquisition may be the key to solving an inventive problem.

We will discuss *patterns of evolution of engineering systems* in Chapter 9, "TRIZ." For now, we need to know only that all engineering systems gradually evolve with time and that their evolution is driven by the universal rules of evolution of engineering systems, called *patterns of evolution*, which are domain independent; that is, they are the same for all engineering systems, from watches to airplanes. Patterns of evolution are known and can be applied to the evolution of the system being designed, providing a deep understanding of its evolution and allowing us to predict changes that might occur in our system with respect to its predecessors.

Each engineering system is designed in a specific legal system and must conform to all regulatory requirements, which are changing all the time, usually becoming more stringent. There are safety requirements, environmental protection requirements, and so on. It is necessary to identify all specific regulations and their requirements relevant to our design and to determine which regulatory requirements must be addressed during the conceptual designing and which may be considered and satisfied later.

On the surface, engineering designing is a purely technical activity in which only engineering knowledge is exclusively used. Also, it is believed that this activity is society independent; that is, the process of designing a bicycle is identical in the United States and in Iran. However, such an understanding is simply grossly inadequate for inventive engineers, who want to create inventions and change the world. In fact, engineering designing is a situated activity (Gero 2007); that is, it is an activity conducted in a social environment that has a great impact on it. This impact must be understood and the resulting knowledge must become a part of the social environment in which our designing system operates. In the case of our bicycle designing example, the social environment has at least two kinds of impact. First, the designing team in the United States may be much more diversified than that in Iran in terms of gender, religion, or political views. Next, the product in Iran must accommodate various needs specific to female riders who wear very long and heavy traditional robes; such needs simply do not exist in the United States.

The *cultural environment* may be interpreted as the culture surrounding our designing process. Usually, the term *culture* is narrowly understood as all the intellectual and artistic products of a given society. All books, paintings, sculptures, music, and so on create and identify culture. In a broader sense, culture is also defined by customs and the system of values and beliefs of a society or a group of people. For inventive engineers, this broader understanding is more useful. Therefore, they should acquire knowledge not only about the intellectual and artistic accomplishments of a given society but also about its customs, beliefs, and values. Understanding literature and the fine arts is necessary to know the context of various needs gradually emerging in a given society, while knowing customs, beliefs, and values is necessary in order to interact with people and to avoid conflicts. For example, working with a Swiss designer creates a punctuality challenge; even being a minute late may be considered by him or her as being offensive or simply unprofessional. On the other hand, time in Black Africa is rarely considered literally; usually it is at best only one of many flexible reference points when planning future activities.

The political situation sometimes has a tremendous impact, both positive and negative, on engineering designing and particularly on inventive designing, but this impact is often unpredictable. For example, the political climate in Germany in the 1930s and 1940s had such a huge and negative impact on German chemists, many of them Jews, that very little progress in chemistry was made, although German chemists were considered the best in the world. On the other hand, the enthusiastic support of the US government for sustainable energy and wind power had a strong positive impact on the designers of wind turbines who see their jobs as secure and consider their activities as making positive contributions to the future of the United States.

5.3.2 Mind mapping

Problem identification is about acquiring and organizing knowledge about a given designing situation, including learning about needs, available resources, design goals, and design constraints and requirements. From this perspective, *mind mapping* is a powerful method for all inventive engineers.

The term mind mapping has many meanings. It is usually understood as a kind of visual thinking in which depictions of various concepts are shown together with their names and links showing how these objects are interrelated. What is most important is that creating a mind map requires *whole-brain* thinking, which engages all parts of the human brain. Such thinking is exponentially more powerful, particularly when creativity is concerned, than exclusively deductive thinking, associated with the left hemisphere; or even inductive and abductive thinking, considered the domain of the right hemisphere. (Although the clear division of functions between the brain's hemispheres has not been entirely supported by recent neuropsychological research, it is used here to explain in simple terms the concept of whole-brain thinking.)

There is also a fundamental difference between the operations of a traditional computer and of a human brain. A computer executes a linear program based on a "step-by-step" deduction and conducts the retrieval of data. A human brain, when it is in the whole-brain mode, thinks nonlinearly or *radiantly*. This is a term introduced by Buzan and Buzan (1994), which perfectly describes the nature of human thinking during a mind-mapping process. It considers simultaneously many lines of reasoning; it not only uses data retrieval from memory and deduction (as a computer program) but also uses induction and abduction. In this way, the human brain not only uses all available resources but also acquires knowledge through induction and generates new ideas through abduction.

The concept of whole-brain thinking can be also explained in the context of the theory of successful intelligence, discussed in Section 3.2. A computer program may exhibit both practical intelligence (retrieval of data and decision rules) and analytical intelligence while conducting complex deductive analytical processes. A human being obviously possesses these two types of intelligence, but in addition they may also possess creative intelligence and will use it while in whole-brain mode.

For all these reasons, mind mapping can be described as a process stimulating and facilitating whole-brain thinking. It leads to the acquisition, organization, and formal representation of knowledge, all of which are crucial for problem identification. It can be defined thus:

"Mind mapping" is a form of human visual thinking. Its purpose is to acquire and organize knowledge about a given domain while stimulating human creative processes and provide a knowledge foundation for solving inventive problems.

The name mind mapping was introduced only recently by Tony Buzan (2010), but the practice of mind mapping began thousands of years ago. In 1940, an 18-year-old boy called Marcel Ravidat discovered the entrance to a cave located near the village of Montignac in southwestern France. The discovered cave was in fact a part of a large cave complex, called later the *Lascaux Caves*. These caves contained realistic drawings of various animals and groups of animals (Figure 5.8) that lived in the area during the Upper Paleolithic era, about 17,000 years ago (seen on Wikipedia May 14, 2015). These paintings of groups of animals may be considered the first examples of mind mapping. They are best known as one of first examples of complex human art, but they are much more than that. They can be also considered as the first attempt to acquire and preserve knowledge about the environment in which the Paleolithic people lived. We could speculate that the paintings were produced to create images of animals for young hunters learning about their prey, showing the relative size of the individual animals and their interactions. Most likely, paintings of single animals were produced first, and only later, when understanding had developed gradually, more complex paintings with various animals and herds of animals were done, reflecting Paleolithic people's growing understanding of the environment around them.

Figure 5.8 Lascaux Caves: a painting of a group of animals. (With permission. Drawn by Joy E. Tartter.)

In ancient times, the Greek philosopher Porphyry of Tyre, who lived in the third century BC, used sketches to show how his ideas were developed and interrelated, and this may be considered a form of learning about a domain and organizing ideas that is mind mapping. Also, Leonardo da Vinci used mind maps, (Gelb 1998; Arciszewski 2009) mostly to record his ideas but also to discover how various ideas were interrelated or how they could eventually be interrelated. As discussed in Section 3.2, these discovered relationships are often the key to knowledge integration. Therefore, they may lead to transdisciplinary knowledge, a foundation of so many inventions.

In modern times, in the late 1960s, Ross Quillian introduced the abstract concept of "semantic networks" (Quillian 1968). This was done in the context of human semantic memory and is defined thus:

> A semantic network is a system of interrelated concepts (notions) that provide objective meaning of these concepts and their relationship with other related concepts.

Later, in the 1990s, computer science research on the abstract concept of a semantic network led to the concept of *ontology*, which was defined in the computational context by Tom Gruber (1993):

> An ontology is a specification of a conceptualization.

On the surface, these two definitions are similar; both describe a knowledge representation scheme. However, the difference is in the context. The first one describes a theoretical, abstract concept, while the second one, proposed 25 years later, describes the same concept as used in a computer system. Today, ontologies (based on the concept of a semantic network) are widely used in information technology but also in engineering to represent and store in computers complex knowledge systems, which can be continually updated and which can contain definitions of various concepts, their relationships, pictures, and even audio and video materials (Ogujejofor et al. 2004).

Mind mapping is based on seven basic assumptions:

1. Knowledge, which is known as *intellectual property* in the business world, is the key to innovation.
2. Knowledge has to be acquired and organized before the process of inventive designing begins, which ultimately leads to inventions.
3. More is better No. 1.
   The more knowledge is acquired, the better chance that the knowledge necessary to develop inventive design concepts has been acquired.
4. More is better No. 2.

The more knowledge diversity there is, the better chance that knowledge from outside the problem domain is acquired, which is necessary for the development of inventive design concepts.

5. Knowledge should not only be acquired from the problem domain and from related domain but also from unrelated domains, particularly when highly novel design concepts are sought.

6. Complex, differentiated knowledge cannot be acquired using exclusively linear or deductive thinking. It must be a result of radiant thinking, that is, thinking that is not only deductive but allows us to use the power of the entire human brain. (Such thinking is also called whole-brain thinking.)

7. Radiant thinking is a kind of visual thinking reflecting the way the human brain acquires knowledge associated with a specific concept, which in the case of mind mapping is a design problem for which inventive design concepts are to be found.

The process of mind mapping has been well described (Buzan and Buzan 1994), and our description is mainly based on this description but with several modifications and heuristics added by the Author. Its main steps are as follows:

5.3.2.1 Step 1: Preparation

Discover the enjoyment of drawing. If you are serious about mind mapping and you have time, take several drawing lessons simply so as to feel comfortable sketching and coloring your sketches, no matter how simple and unprofessional they look. Your objective is not to create art but to acquire knowledge.

Find a large table in a quiet location and play soft music in the background.

Prepare several large sheets of drawing paper, the bigger the better.

Find various pencils, crayons, color pencils, and one or two erasers.

Spend an hour or two warming up your brain while drawing images of dog houses, houses, cats, and dogs.

5.3.2.2 Step 2: Creation of the central image

Images activate your entire brain and are much more powerful in inventive designing than words. Therefore, try to maximize the use of images and minimize the use of words in your mind mapping. The *central image* should reflect your ultimate goal as expressed in the way unique to you; it may be a simple abstract sketch or and an elaborate drawing. It may be in black and white only, but preferably it should be in many colors. The artistic quality of your drawing does not matter; however, its spontaneity does. Feel like a child truly enjoying drawing even if her drawings look terrible to adults. Remember, engaging in the process is much more important than the artistic beauty or engineering perfection of the results.

The central image should be placed in the center of your large sheet of paper, in exactly the same way that the concept behind this image will be placed at the center of your brain.

5.3.2.3 Step 3: Creation of the first primary branch

Draw the first *primary branch*, that is, the thick line radiating from the central image. This branch should be thick and imposing and should represent the most important association relating to the concept behind the central image. In fact, Buzan (2010) recommends using two lines connected at the tip and filled with a color. Both lines should flow smoothly as dendrites in the brain to be attractive to your eye while creating the impression of a certain level of *sfumato,* or ambiguity, which will stimulate your brain and will activate your creative talents. On the top of your new primary branch (or below, if you prefer), write a single word or key word that will best describe knowledge associated with this branch and will be called a *basic ordering idea.* A single word is so much more powerful than a sentence or a phrase. Also, finding this single word will force your brain to start thinking in abstract terms, immediately expanding your knowledge space.

5.3.2.4 Steps 4–9: Creation of remaining central branches

Your mind map should not have more than seven central branches and seven central images. This limitation is related to the natural limitation of human short-term memory, which on average is able to handle up to seven pieces of information.

Repeat the activities of Step 3 and use different colors for the individual central branches. It is natural to move clockwise around the central image, starting from the first central branch positioned on the right side of the central image. However, if you prefer to move counterclockwise, simply do it.

You might also use differentiated shapes for your *primary branches* so as to make your drawing more interesting and "artistic." More important, however, is the psychological finding that such shapes will make it easier for your brain to visualize and memorize your drawing, both of these activities being very important for starting the subconscious activities of your brain, which may significantly contribute to your final success.

5.3.2.5 Step 10: Creation of the second, third, and higher level branches

This is a process of building associations and creating structures that look like networks of dendrites in the human brain. This activity begins with the first primary branch and subsequently continues with the remaining primary branches.

When you consider the first primary branch, you should think again about the basic ordering idea assigned to this branch. Now, you need to identify several concepts associated with the basic ordering idea. These concepts should be on the same level of generality but should be more specific than the basic ordering idea. It is like having the name of a set and now identifying members of this set. In this way, you create the several second-level branches with respect to the first primary branch and the basic ordering ideas describing these second-level branches. Next, images should be added to all the branches just created.

The entire process should be repeated for each second-level branch to create third-level branches with their own basic ordering ideas and images. Obviously, you can create fourth-level branches and even go to fifth-level branches, but that should be done later to avoid an explosion of details, which may overload your mind map at the expense of the "big picture" and of clarity.

When the process described above is completed for the first primary branch, it should be repeated for the remaining primary branches. In this way, the main structure (configuration) of your mind map is ready and you can use it to learn even more about your problem domain; that is, you can now focus on finding relationships existing within the created mind map.

5.3.2.6 Step 11: Finding relationships

This is the most interesting part of mind mapping. You have already acquired a lot of knowledge about your problem domain, identifying many basic ordering ideas and their hierarchical relationships within the individual primary branches. Now, it is time to use the entire mind map to improve your understanding of the problem domain and to find existing—or potentially existing but hidden—relationships among concepts or basic ordering ideas that belong to different primary branches. It is like using a big picture, or a bird's view, to find relationships among concepts that cannot be seen from the ground level or from within the individual primary branches. In this way, the initial drawing of your first mind map has been completed, but this is not the end of mind mapping.

5.3.2.7 Step 12: Modifications and improvements

Nothing is perfect in the real world, including your initial mind map, which is like a draft of a paper. It is wise to set aside it for several days and to analyze it again later. Usually, many potential modifications and improvements become immediately obvious; better words are found for the basic ordering ideas, better images become available, more relationships are discovered, even better colors could be used, and so on.

It is recommended that you start working again on your original mind map using an eraser, different colors, different shapes of lines, and so on. You will be surprised how many improvements simply come to you. In fact, many of them are products of your subconscious, which was active during the whole period between when you finished your mind mapping and when you began making improvements.

Most likely, you will need a second large sheet of paper to redraw your improved mind map and maybe even a third or fourth one. Going through a number of drafts is natural; it is even desirable, since it allows you to engage in a dynamic process of learning about your problem domain and in this way to perfect your problem identification. Depending on your time frame, motivation, and desired level of novelty, you may need to repeat Step 12 at least once, twice, three, four, or even more times. Ultimately, you need to be convinced that you have reached the end of your modifications and improvements, that you are happy with your mind map, and that you simply cannot improve it anymore.

5.3.2.8 Step 13: Professional visualization

Many mind-mapping experts claim (Buzan 2010) that using software is the best way to produce high-quality results that are already in digital form, that is, ready for easy distribution and use by other people. The Author, however, is of a different opinion. Doing mind mapping on a computer brings a fundamental contradiction:

> Mind mapping is about human spontaneity and creativity and about using powerful whole-brain thinking, the key to both.
>
> Using a computer program imposes all kinds of constraints associated with available colors, shapes, images, and so on. All these constraints must be satisfied and that requires systematic and deductive thinking mostly associated with the left side of the brain, subsequently that side becomes dominant with a potentially devastating impact on the effectiveness of the whole brain thinking, the essence of mind mapping.

There is a simple way to eliminate this contradiction: Conduct steps 1 through 11 manually, using paper and your knowledge and imagination to create the mind map and all the images. Only after that you should use one of the many mind-mapping tools that are commercially available. Probably the best and most developed one is iMindMap, developed by ThinkBuzan Ltd in the United Kingdom, which should be recommended for professional use. However, my students had also good experience with various free tools available on the Internet. These tools are much less sophisticated than iMindMap and offer only limited mind-mapping capabilities, but mastering them requires less time and they are free.

5.3.2.9 Example

When mind mapping is applied to problem identification, the central image is obviously a big question mark representing our subject, that is, problem identification. The first primary branch will be a branch called "Needs," leading to several specific needs to be addressed in our design.

The next primary branch will be called "Resources," with such branches on the second level as "Knowledge," "Humans," "Money," "Time," "Environment," and so on. In this case, we will also have branches on the third level, called "Knowledge," "Books," "Codes," "History," and so on. Also, the remaining second-level branches will lead to a number of third-level branches, as shown in Figure 5.9. The third primary branch will be called "Goals" and, in the case of a specific application, may have both second- and third-level branches, appropriately named. The fourth and last primary branch will be called "Filters," with two second-level branches called "Constraints" and "Requirements." These two branches will most likely have higher-level branches for a specific case.

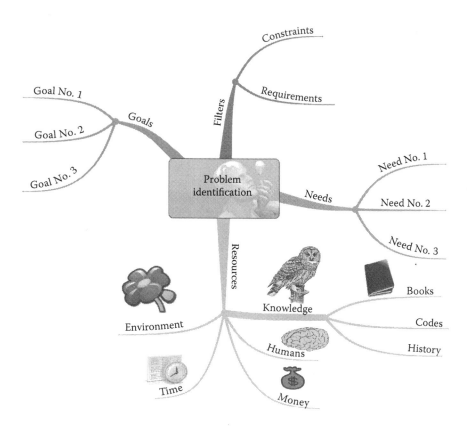

Figure 5.9 Problem identification: a mind map.

5.3.3 Innovation situation questionnaire

Knowledge is the key to inventive designing. It should be acquired in a systematic way and should be at least sufficient for the inventive problem considered. Working on the *innovation situation questionnaire* (ISQ) is a relatively simple way to acquire knowledge in a format that makes it easier to comprehend the acquired knowledge and to understand a given designing situation.

In the mid-1980s, Alla Zusman and Boris Zlotin established in Kishinev (Chisinau), Moldova, the Kishinev TRIZ School, their mission being to continue the development, promotion, and industrial applications of TRIZ. They both were prominent TRIZ scholars who previously worked with Altshuller, the creator of TRIZ. (For more details about Altshuller and TRIZ, see Chapter 9.) One of the most important methodological achievements of their school was the development of the ISQ, which provides a simple three-stage process for knowledge acquisition. The development of the ISQ was continued by its creators in the United States at Ideation International Inc. First, it was done in the 1990s in cooperation with John Terninko (Terninko et al. 1998) and next with Ron Fulbright in the 2010s (Fulbright 2011). The ISQ has been developed in the format appropriate for TRIZ users. However, the ISQ is a self-contained system and can be used by any inventive engineer to simply identify a problem that may be later formulated in the form required by the specific method to be used. The ISQ description presented in this chapter has been adapted and modified by the Author to make it useful for all inventive engineers and to be consistent with the book.

The entire process is time consuming and may require at least a day or more of work. In the case of difficult inventive problems, the process may take up to a week. However, it is not a waste of time; the inventive engineer learns about the designing problem and in the process structures the acquired knowledge. In this case, more means more; the more is learned about the designing problem, the better chance there is of producing novel results. The ISQ process has three main stages:

1. Briefly describe the designing problem
2. Prepare a detailed description of the designing problem
3. Identify available resources, constraints, and requirements

Individual stages are discussed with key heuristics intended to make the process more effective and easier to conduct for an inexperienced inventive engineer. There is one general heuristic to follow during the entire process:

> Use a casual language and avoid using any specific technical terms which might imply any specific solutions.

The process will be explained using as an example the design of a floor structural system in an industrial building.

5.3.3.1 Stage 1: Briefly describe the designing problem

Most likely, this is the most challenging and important part of the entire process. This brief description will not only initiate the process but will have a tremendous impact on the final results. It is also difficult from the psychological point of view for at least two reasons. First, engineers always proudly use their "secret" technical language. Therefore, it is truly difficult for them to describe problems without using their language, which usually not only gives them an advantage with respect to other people but also leads to specific potential solutions. For example, when they hear "transverse loading," their instant reaction is "beams." Such a reaction is obviously correct in routine designing, but in the case of inventive designing it immediately reduces the huge number of potential solutions to a small class of structural systems called beams, and that is simply harmful in this case. Second, engineers are trained in using their secret language and do not know how to describe their problems in ordinary, not technical words, particularly when a very general and abstract language is expected with metaphors and all kinds of poetic expressions. Learning how to use such a "poetic" language is not a trivial matter, but it is a necessity for inventive engineers as you will learn later in the chapter on *Synectics*, the most powerful inventive method. The poetic engineering language has the power to stimulate the brain, particularly its "creative" right hemisphere, and is often the key to inventions.

In the description, pairs of *active words* and *objects* should be used. Active words are verbs describing specific actions, for example "hold," "support," "keep," or "transfer." Nonactive words are such general verbs as "provide," "produce," or "secure."

In the case of our example, we could briefly describe it in at least two ways. In the first case, we will use relatively simple words, and in the second one we will use our new poetic language:

1. Our challenge is to develop a floor structural system in an industrial building. Machinery, people, and materials are held in place by our system, which is connected by columns.
2. Our challenge is to create a cloud holding machinery, people, and materials in space. Loads from this cloud will flow to columns.

In both descriptions, we have used words such as "challenge," which immediately suggests something different than traditional step-by-step routine designing. Also, using the verbs "develop" and "create" suggests the acquisition of knowledge and a complex process, definitely not routine designing. Our new poetic words "cloud" and "flow" are abstract and are intended to open and stimulate our creative minds while removing any rational restrictions we might have. Therefore, we have our next heuristic:

Use abstract words stimulating your imagination and implying various meanings.

It was Leonardo da Vinci (see Section 2.4.) who formulated the principle of *sfumato*, or ambiguity, to follow when seeking novel solutions.

5.3.3.2 Stage 2: Prepare a detailed description of the designing problem

5.3.3.2.1 Step 1: Name the designing situation

Use general terms; avoid at all costs using any specific technical terms, which might imply specific solutions. In our case, the name of our designing problem is

Designing the floor structural system in an industrial building.

5.3.3.2.2 Step 2: Determine the input and output of the system

The system will operate in a specific environment. This environment will interact with the system through input and output, which must be identified, understood, and described. We should realize that both input and output may be physical or nonphysical. Surprisingly, the nonphysical components of input and output may decide the final fate of the system, particularly if they are not identified and considered during the designing process. The best example illustrating the importance of the nonphysical components of the input and output is the history of the Embarcadero Freeway in San Francisco. It was designed using a purely technical approach without any consideration for its potential negative impact on the urban life of a significant part of San Francisco, which included blocking the priceless San Francisco Bay views. In the 1980s, an entire social movement emerged demanding the demolition of the freeway, which had cost billions of dollars to build and which had become a vital part of the entire bay transportation system. After a very long and complicated political battle, in 1991 the demolition of the Embarcadero Freeway began (Wikipedia, accessed on May 11, 2015).

In the case of our floor structural system, the input may have several components:

1. Weight of machinery
2. Weight of materials
3. Weight of dividing walls
4. Weight of installations (heating and cooling, electrical, suspended cranes, etc.)
5. Available materials

6. Available human resources
7. Design code requirements
8. Expected project completion date
9. Expected adequate working conditions without excessive vibrations

The output of our system may be described by its various components, both technical and nontechnical:

1. Bending moments, shear forces, and axial forces applied to the columns
2. Damping excessive vibrations
3. Adequate working conditions without excessive vibrations and deflections of the floor
4. Low maintenance

It is usually difficult to see the "big picture" in terms of input and output, but developing this ability is an important part of becoming an inventive engineer.

5.3.3.2.3 Step 3: Describe the function of the system and functions of its individual elements and subelements

This is an important step that usually requires a lot of effort and is rarely completed in a single pass. In the case of our example, the results of the functional analysis for our example may be as follows:

Entire system

> The function of the entire system is to support gravity and the dynamic loads of machinery, people, and materials and transfer them to the columns while limiting deformations of the system to the allowable values and keeping the system stable, that is, making sure that it will not collapse as a mechanism as a whole when loading is applied and none of its elements loses its stability.

This description is generic and it is not associated with any specific type of structural system. Even more importantly, it does not suggest any details for the solution, inspiring us to be open to all possibilities. The presented function is the desired function of the system and it is called the *primary useful function of the system* in TRIZ terminology (see Chapter 9), but we will call it simply the *positive function*. Unfortunately, most likely our system will also produce *negative functions*, which are not desired and may be even harmful but are simply unavoidable. In the case of our example, several obvious potentially negative functions can be specified:

- Using space within the building
- Increasing the total weight of the building

- Increasing the height of the building
- Increasing the cost of the building

These potentially negative functions of the system should also be incorporated into our problem identification. Knowing them makes us aware of their existences in our pursuit of novelty and ideality. Only so-called ideal systems (discussed in Chapter 9) do not have any negative functions while providing all desired or positive functions.

Positive functions of the individual elements and subelements of our system are as follows:

Element No. 1: Deck

The function of the deck is to be in direct contact with loads, to keep them in place, and to transfer them to Element No. 2, that is, to the part of the system that transfers loads from where they are applied to supports. The deck may also participate in the transfer of loads to supports and in providing stability, but these are its secondary functions.

Element No. 2: Load-bearing element

The function of this element is to transfer applied loads from where they are applied to supports while maintaining the required rigidity and stability of the entire system.

Subelement No. 2.1: Bending moment transfer

The function of this subelement is to carry out to the supports bending moments induced by the loads.

Subelement No. 2.2: Shear force transfer

The function of this subelement is to carry out to the supports shear forces induced by the loads.

Subelement No. 2.3: Axial force transfer

The function of this subelement is to carry out to the supports axial forces induced by the loads.

Subelement No. 2.4: Internal connections

The function of this subelement is to connect and integrate Elements 2.1 through 2.3 into a structural system.

Element No. 3: External connections

The function of this element is to connect Element No. 2 with the columns supporting the system and to transfer loads from the system to the columns.

Element No. 4: Bracings

The function of this element is to provide stability to the system, that is, to prevent it from becoming a mechanism when loading is applied.

5.3.3.2.4 Step 4: Describe the system's structure and identify all important elements and how they are connected

The description should be independent of any specific solutions and be prepared in a descriptive casual language (see our heuristic No. 1). In the case of our example, the system's structure can be described as follows:

The system must have an element called a "deck" (Element No. 1), which will be directly in contact with all loads applied to the floor system. The deck must be connected with the system's element (Element No. 2) that carries out internal forces induced by the loads. These internal forces are moved in the direction of supports where they are transferred by "external connections" (another element of the system, Element No. 3) to the columns supporting the floor system. Also, the element carrying out the internal forces must have a subelement carrying out bending moments (Subelement 2.1), a subelement carrying out shear forces (Subelement 2.2), and a subelement carrying out axial forces (Subelement 2.3). Within this element, we can also distinguish subelements (Subelement 2.4), which could be called "internal connections" since they connect subelements carrying our bending moments and shear forces. The final element (Element 4) of the system could be called "bracing."

All elements of the system should be shown in a simple diagram (Figure 5.10).

5.3.3.2.5 Step 5: Describe functioning of the system

The description should be also prepared in a casual, nontechnical language since it will not be used exclusively by engineers. The inventive design process may involve nonengineers who will share with engineers their nonengineering knowledge, and this knowledge may be the key to the solution. For example, biologists may be involved and share with engineers their understanding of how bamboo grows, and that may inspire the designers of tall buildings or steel structural designers working on new types of steel columns in industrial buildings. In the case of our example, this description may look like this:

> The considered industrial building is seven stories tall. It is a three-bay building. On each floor machinery will be installed and used by 20–30 people. Also, materials will be stored on each floor, which weigh approximately 20,000 pounds. Machines must be placed on a smooth and flat surface, which will sustain large concentrated forces applied where machines are located. Their operation should not cause any excessive deformations of the entire floor structural system and no excessive vibration should take place. To minimize the length and cost of columns, the depth of the floor structural system should be kept at a reasonable level but needs to be sufficient to run heating and cooling pipes through the system.
>
> The floor structural system is loaded by vertical forces, both gravity and dynamic forces. The gravity forces are caused by the weight of machinery, people, and materials. The dynamic forces are caused by the operations of the machinery.

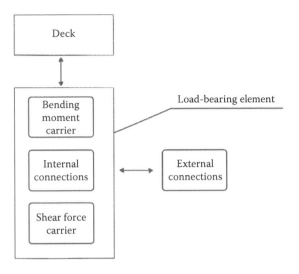

Figure 5.10 Structure.

All vertical forces are applied to the top of the floor structural system and they are transformed into internal forces: bending moments, shear forces, and axial forces. These forces are transferred by the system to the supports in the form of connections between our floor structural system and columns. Columns transfer these forces to the foundation.

5.3.3.2.6 Step 6: Describe the system's environment

The description should contain four parts, including descriptions of

1. Interacting systems
2. Nearby systems
3. Supersystems
4. The natural environment

In all cases, our descriptions should identify the nature of interactions, both positive and negative. This information will later help us to find unusual solutions in which we could use available resources to solve our problem.

In the first case of interacting systems, we should determine all other systems that may have even occasional impact on our system. For example, if the transportation system in Fairfax, Virginia, the United States, is considered, it can be determined that the power system and the water distribution systems will interact with our transportation system. (Obviously, we could find more systems interacting with our transportation system in Fairfax.) The power system provides power to all signals and monitoring devices, and this is a positive impact. The power system may fail, and this will be a negative impact. Similarly, when considering the water distribution system, its impact may be positive, like providing water to fight fires often associated with accidents; but it may be also negative, as when a water main breaks causing street flooding and forcing the rerouting of traffic.

In the case of our example dealing with a floor structural system, the most important interacting system is the heating and cooling system. The positive impact of this interacting system is very limited; eventually its large-diameter pipes might marginally improve the rigidity of our system. However, its negative impact is much more important in this case. The pipes impose strict spatial requirements on the geometrical configuration of our system and may increase its depth beyond structural requirements.

In the second case of nearby systems, our focus should be on all systems that satisfy two conditions: (1) they are in close proximity to our system, and (2) there is a possibility that they could interact with our system, although such interaction is not taking place right now.

In the case of our floor structural system, we could identify at least one nearby system, that is, the wind-bracing system in the entire building. Usually, wind-bracing systems are designed to be independent from floor structural systems, but they could be integrated, and such integration might be potentially beneficial for our floor system.

In the category of supersystems, we should identify at least one or two more general systems of which our system is a subsystem. For example, when a transportation system in an urban area is considered, its direct supersystem will be the infrastructure system. This infrastructure system will be a subsystem of the urban system for the urban area considered, and so on. In the case of our floor structural system, its immediate supersystem will be the structural system of the entire building. Its other more general supersystem will be this building considered as a system with such subsystems as the structural system, the heating and cooling system, the power system, the water distribution system, and so on.

Finally, by natural environment we mean the physical or actual environment in which our system will operate. For a submarine, it will be water; for a desert vehicle, it will be desert with its hot and dry air and high levels of fine dust. In the case of our floor structural system, its environment is the inside of an industrial building; that is, the air has high levels of industrial dust that is very corrosive and might be hot. We should notice that in the process of describing the natural environment of our system, we have discovered unique features of this environment, which may prove very important in our inventive designing. Operating in a hot and corrosive environment may mean that if we use steel in our system, we will need to minimize contact surfaces, and that may translate into a set of additional design requirements like using smooth surfaces or using closed sections.

5.3.3.3 Stage 3: Identify available resources, constraints, and requirements

5.3.3.3.1 Step 1: Identify available resources

There are six categories of available resources:

- Substance resources
- Field resources
- Functional resources
- Time resources
- Space resources
- Knowledge resources

The category of *substance resources* includes all kinds of available materials. In our example of a floor structural system, we have such materials available as steel, concrete, wood, paint, glass, and so on.

The concept of *field resources* is less familiar, but it may be the key to our inventive problems. By field resources we mean all kinds of fields that can be utilized in our project. For example, in our floor structural system, we could use a *stress field*, that is, we could use stresses to create a prestressed concrete or even rarely used but feasible prestressed steel structures. Also, we could use an electromagnetic field to transport the structural system to its location in the building or even to keep it in its desired position. We will have also such fields available as the *temperature field* or the *gravity field*, which both are easily available and eventually could be used.

Functional resources are all those functions that will be provided by our system or its supersystem. In our example, we have such functional resources as supports provided by the columns (which are a part of the supersystem) and machinery support provided by the deck. Knowledge resources simply means the state of the art in our problem domain, or what is known about the problem, including design codes, design manuals, related textbooks, the experience of our design team members, blueprints of various types of floor structural systems in industrial buildings, drawings of various types of structural systems in industrial buildings, and so on. Also, the sounds produced by a vibrating metal structure should be considered as a knowledge resource, since they provide important information about the state of our structure.

Time resources represent our understanding of the time factor in our designing situation. This kind of resource not only includes times when our team members are available and when experts are available but also information about the timing of the construction and manufacturing of the individual structural members. It may also contain information on how much time is available for our project. *Space resources* provide information about available space for our system, access to it from various directions, penalties for using more space than originally assigned, and so on.

5.3.3.3.2 Step 2: Identify allowable changes

In this case, we begin the process with the identification of one of the known structural systems, which are commercially available and could be used in our design case. Let us assume that in our case the reference system is in the form of trusses supporting a steel deck with concrete on the top of it. Such a system was used in the World Trade Center, but it is also widely used in industrial buildings because of its relatively low cost, simplicity, constructability, and rigidity.

We are working on the development of inventive design concepts, but we still need to determine the level of novelty of our final products. Therefore, we need to determine if we are looking for

- Minimal changes and improvements
- Modifications
- Major changes
- Revolutionary changes

The differences between individual levels of novelty are qualitative and difficult to formally define, but they can be explained with examples. In the first case of *minimal changes and improvements*, we would consider only changes like using a different welding technology for connections or a different grade of concrete. *Modifications* may mean changing the type of cross section of the steel deck, for example replacing the trapezoidal cross sections by smooth wave-type sections. *Major changes* may describe the situation in which we replace main steel trusses by prestressed concrete beams. Finally, when *revolutionary changes* are allowed or even encouraged, we are looking for an entirely different design, which would involve, for example, the use of new materials, a new configuration, new types of connections, and so on. Such a design should be definitely patentable and would mean a fundamental departure from the reference design representing the state of the art (SOTA). Obviously, in this case our new design would represent an advancement of SOTA.

5.3.3.3.3 Step 3: Identify constraints

The identification of constraints leads to the determination how our designing situation is constrained; that is, what features of the future engineering system are constrained and therefore cannot take place. For example, we cannot use steel as a material in our floor structural system. This absurd constraint was actually one of the most important design constraints in former socialist countries like Poland or Bulgaria. In such countries, there were always shortages of building materials and particularly of structure steel, which was mostly used for military purposes and was practically unavailable for civilian use.

We may have all kinds of constraints that are problem related. They may include, for instance, the constructability constraints preventing us from using connections requiring welding to be done on the construction site. In fact, we should construct a constraints tree with all kinds of constraints governing our designing process. Several major classes of constraints might include technological constraints, social constraints, political constraints, and so on.

5.3.3.3.4 Step 4: Identify requirements

In this step, we need to identify what features of our future system are required. In the case of inventive designing, requirements related to novelty may be particularly important and they may require that our designing must lead to a patent or a class of patents, which will become the intellectual property of our company. Similarly to the previous step dealing with constraints, our analysis of requirements should result in a requirements tree showing how these requirements are interrelated.

5.3.3.3.5 Step 5: Identify evaluation criteria

In the case of regular engineering designs, they are usually evaluated using such single criteria like cost or weight, and an evaluation based on these criteria is considered adequate. When inventive designs are evaluated, the cost-based evaluation is usually grossly inadequate, and we need to use a multicriteria evaluation with several classes of criteria, including such classes as

- Engineering criteria
- Economic criteria
- Time-related criteria
- Novelty criteria
- Social criteria

For each of these classes, specific relative criteria should be identified and defined in quantitative form. How this should be done is discussed in Section 5.5, "Design Concept Evaluation." The determination of evaluation criteria is not a trivial matter, because they reflect our priorities and their selection will have impact on the direction of our search for novel solutions. For example, if novelty is important, it should be reflected in a large value of the weights associated with all novelty criteria.

5.3.3.3.6 Step 6: Identify the evolution line leading to the present solution

This step may require a lot of effort and time, but it will give us not only specific knowledge about previous designs but also their sequence (evolution line) and, most importantly, the specific patterns of evolution that govern the evolution of our engineering system in addition to the known general patterns.

In the case of our example of a floor structural system in an industrial building, we should begin building our evolution line with floor structures in ancient Chinese and Egyptian buildings, followed by structures used by Greeks and Romans, during the Renaissance, in the nineteenth century, and in modern industrial buildings. To produce such an evolution line, we need to study books but also patents, which are usually the best source of information about previous designs.

Working on this step will result in the acquisition of knowledge about various previous designs, an important result in itself, but it will also help us to build an understanding of the entire evolution process, and that may be priceless. We will know patterns of evolution behind the emergence of the individual types, and we might even use these patterns in solving our inventive problem. Also, these patterns may help us to predict various possible future designs, the future evolution line, or even several evolution lines, which when taken together will create an envelope for future solutions. Such an envelope will give us an excellent understanding of what will most likely happen in the future.

5.4 PROBLEM FORMULATION

Many design scholars do not distinguish between problem identification and problem formulation. The Author believes, however, that there is a fundamental difference between these two activities that should be understood well by inventive engineers.

Problem identification may be understood as the process of the acquisition of knowledge that is relevant to a given inventive problem. This knowledge may be in multiple forms since it is intended for the humans who will use it to understand the problem before they attempt to solve it. This body of knowledge is universal, and it is not associated with any specific inventive design method.

On the other hand, problem formulation is a process of using this universal knowledge and transforming it into the form required by a specific inventive design method. Therefore, there is not one single procedure that could be presented here. For example, in the case of morphological analysis (Chapter 6), we will need to transform all the available knowledge into a design representation space, that is, into an organized collection of symbolic attributes and their known values. When the TRIZ method is used (Chapter 9), we will need our problem to be formulated as a single or several technical contradictions, which must be eliminated; this process constitutes the essence of TRIZ inventive problem solving. Only in the case of brainstorming and Synectics is a specific form of problem formulation not required because of the nature of these two methods.

For all these reasons, in the following chapter we will discuss problem formulation in the context of various inventive design methods.

5.5 DESIGN CONCEPT EVALUATION

As we will discover later (in the section on morphological analysis), it is possible to use a computer program to randomly generate in no time hundreds or even thousands of potential design concepts. The elimination of infeasible concepts is difficult and time consuming but possible for domain experts. However, any formal and quantitative evaluation of the remaining feasible design concepts is a true challenge. It is not only an intellectual challenge, as the Author discovered in the mid-1980s when he began working on automated computer systems for designing, including both conceptual and detailed designing. The challenge can be formulated as a contradiction between the nature of design concepts and the applicability of available evaluation methods:

> Design concepts are abstract and are presented in the form of sequences of symbolic attributes and combinations of their values. Obviously, these values are *qualitative* (like steel, timber, or paper).
>
> Engineering evaluation methods are *quantitative* and require descriptions of engineering systems by numerical attributes with quantitative values (like dimensions, weight, or speed).

When building an automated computer system for designing, a formal evaluation model must be developed and incorporated in the system. Its function will be to assign to all feasible design concepts various numerical/quantitative values, a measure of their "goodness." How to do it became a difficult research question, which has been addressed by two of the Author's Ph.D. students (Arafat et al. 1992; Shelton 2007; Shelton and Arciszewski 2008). Each used a different approach and produced impressive theoretical results and a sophisticated experimental computer system, but neither system was universal, and they were developed for applications in very specific areas of engineering. Dr. Arafat has developed a system for the evaluation of design concepts in the area of steel roof trusses in industrial buildings, while Dr. Shelton has developed a system for applications in the area of complex satellite communication systems. Unfortunately, their results are too complicated for easy everyday use, as often happens with doctoral studies and research.

We are talking about research initiated about 30 years ago, when computers were very slow; the automated generation of a structural system in a tall building required about 2–3 hours of computer time, while today the same process takes only a few seconds. In this situation, it is possible to avoid any formal evaluation of feasible design concepts and simply to transform them into complete detailed designs, which can be easily quantitatively evaluated. However, inventive engineers should know how to evaluate design concepts directly, even if progress in computing has made this skill less critical. The proposed method below is simple and can be used whenever evaluation of design concepts is necessary.

Let us assume that we have n design concepts (DC_i) to evaluate:

$$DC_1, DC_2, DC_3, \ldots DC_i, \ldots DC_n \quad \text{where } i = 1,2,3, \ldots n$$

We have also m evaluation criteria:

$$EC_1, EC_2, EC_3, \ldots EC_j, \ldots EC_m \rightarrow 0 < = EC_j = < 1.0 \quad \text{where } j = 1,2,3, \ldots m$$

and m weights:

$$w_1, w_2, w_3, \ldots w_j, \ldots w_m \rightarrow \text{where } 0 < = w_j = < 1.0 \quad \text{and}$$

$$\text{where } j = 1,2,3, \ldots m$$

The "goodness" E_j of a given design concept DC_i will be determined using the formula

$$E_i = \sum_{j=1}^{m} w_j EC_j \text{ for j from } j = 1 \text{ to } j = m$$

We need to know all evaluation criteria for a given evaluation process. Unfortunately, selection of these criteria is entirely problem related. However, such criteria as "expected cost," "manufacturability," "constructability," "patentability," and "novelty" may be used.

Also, before the evaluation process begins, the values of all weights need to be arbitrarily assumed. Assuming all weights are equal to 1 leads to an unbiased evaluation in which all criteria are equally important. Assuming that the weight w_j of a given evaluation criterion EC_j is equal to 0 simply eliminates this criterion from the evaluation. The inventive engineer may arbitrarily assume the values of weights, but it is wise to involve all stakeholders. Determination of weights may become highly political, particularly when the design concepts are related to environmental systems. Discussions of the weight of cost versus the weight of expected environmental pollution may never end.

Values of the individual criteria for a given design concept DC_i may be assumed by the inventive engineer using his or her limited understanding of the problem domain, but it is much better to use several domain experts. They will gladly review all the design concepts and assign as a team or individually the values of all criteria for the individual design concepts.

The described evaluation method is relative; that is, its results are valid only within the set of n design concepts. It will allow us to determine the "goodness" of the individual design concepts with respect to the other concepts from the same set. This result will also permit us to create a list of all design concepts ranging from the best one to the worst one. This list will be used for the selection of design concepts that will be converted into full designs (see Section 5.5.)

There is still a lot of research on concept evaluation, for example (Koziolek et al. 2010), but the proposed simple approach is usually sufficient, and improving its performance leads to a significant increase in the complexity and difficulty of the evaluation process.

5.6 DESIGN CONCEPT SELECTION

The last stage of conceptual designing is the selection of a concept, or several concepts, that will be used next to prepared final designs, including detailed designs. The previous stage, design concept evaluation, ended with the creation of the evaluated design concepts, which were listed in the order of their declining "goodness."

The simplest way to select these concepts is obviously to select the first several design concepts from the list. Unfortunately, nothing is truly simple or obvious in inventive engineering. Evaluation is usually done by engineers, who use relatively clear and technical relative evaluation criteria (expected weight, expected size, etc.). Selection may be done by administrators, whose priorities may be entirely different, and they use their own set of selection criteria, for example profit, PR impact, and so on. They may simply take the first several design concepts from the list (which are considered "best" from the engineering perspective) and select from them design concepts they will consider as "best." Their hierarchy of the "best" design concepts may be different that the one prepared by engineers, but that should be expected.

Chapter 6

Morphological analysis

6.1 CREATOR

Fritz Zwicky (Figure 6.1) is the best proof of the truth that great discoveries that change science and our lives require true Renaissance scholars and global men, or simply successful engineers, a concept we introduced in Section 3.2. Zwicky was born (1898) in Bulgaria to a Swiss father and a Czech mother. He spent his first 6 formative years in Bulgaria and the next 21 years in Switzerland. In 1925, he moved to the United States to work at Caltech, California. As a result, he lived in three different cultures and had to understand and accept them, and he spoke at least three languages fluently. Also, his academic education was exceptionally broad: He studied engineering, mathematics, and experimental physics at the ETH in Zurich, Switzerland. Therefore, his understanding of science was truly interdisciplinary, if not transdisciplinary.

From 1925 until his death in 1974, Zwicky lived in California, in the Los Angeles area. He was a professor of astronomy and astrophysics at Caltech and the research director at the Aerojet Engineering Corporation for nearly 20 years. He was a brilliant scientist, maybe even a genius, who made fundamental contributions in astrophysics, astronomy, and cosmology. For us, however, Zwicky is one of the "founding fathers" of inventive engineering, and he felt the same way. In 1948, in the year when your humble author was born, Zwicky was invited to deliver the prestigious Oxford University Halley Lecture. He decided not to talk about his discoveries in astronomy or astrophysics. Instead, he proposed the method of morphological analysis. He believed that it would become the key to many discoveries and inventions. Therefore, the method was much more important than one or two discoveries he could present in his short lecture. Later, he wrote, "I feel that I have finally found the philosopher's stone" (Panek 2009). In fact, Zwicky was not only the scholar and creator of Inventive Engineering. He himself was also an important inventor with about 50 significant patents, mostly related to the development of jet and rocket propulsion systems, and was even considered the "father" of the modern jet engine.

Figure 6.1 Fritz Zwicky.

As a true Renaissance man, Zwicky believed in the human dimension of our lives and he was involved in various charitable activities related to education. These emotionally charged activities complemented his mathematics- and physics-based research and made him a complete human, balancing his emotional and rational thinking. Most likely, it was one of the keys to his extraordinary achievements in science.

6.2 HISTORY

We may trace the philosophical roots of the method to two ancient civilizations. In the sixth century BC, three schools of philosophers in India (Jain, Ajivika, and Carvaka) promoted the concept of Atomism. About a century later, the same concept emerged in Greece. Atomists argued that the universe is composed of atoms and void (vacuum). "If the world is divided into its smallest indivisible parts (*Atoms* in Greek) that cannot be broken up, the reality (understanding) will be reached," claimed Leucippus, the founder of the Greek School of Atomists (Arciszewski 1988b). The principle of division and unity, which is at the heart of the method, was born.

In the seventeenth century, more than two millennia after the School of Atomists was established, René Descartes published *Discourse on the Method*. He was a true French Renaissance person of his time. He was a philosopher, a mathematician, and an engineer.

His book reflected his various interests, and it was truly transdisciplinary, providing a methodological foundation for science and engineering for centuries to come. In part two of the book, he described the *scientific method* and its four governing rules, which can be considered as a philosophical foundation of morphological analysis. These rules are as follows (Descartes 1960):

1. Accept nothing as true which you did not clearly recognize to be so: that is to say, advance carefully to avoid precipitation and prejudice in judgment, and to accept in them nothing more that what was presented to my mind so clearly and distinctly that you could have no occasion to doubt it.
2. Divide up each of the difficulties, which you examined into as many parts as possible, and as seemed requisite in order that it might be resolved in the best manner possible.
3. Carry on your reflections in due order, commencing with objects that were the most simple and easy to understand, in order to rise little by little, or by degrees, to knowledge of the most complex, assuming an order, even a fictitious one, among those which do not follow a natural sequence relatively to one another.
4. In all cases make enumerations so complete and reviews so general that you should be certain having omitted nothing.

Zwicky, the creator of morphological analysis (see Section 6.3), claimed that he was inspired by Johann Wolfgang Goethe (Wikipedia, used October 2, 2014). Goethe is best known as the most influential German romantic poet, inspiring generations of poets in several countries. He was also an unusual scientist interested in the philosophical and methodological aspects of science. Furthermore, he studied biology, in particular the morphology of plants. Most likely, that inspired Goethe to create his own holistic philosophy of science with elements of order resembling the morphology of plants. Zwicky was a sophisticated scholar and familiar with the scientific contributions of the German-speaking scientists. He found Goethe's ideas attractive and still relevant, although they had been developed several hundreds of years previously. In this way, Goethe's poetic and abstract ideas became integrated with Zwicky's mathematician's rationality and led to morphological analysis.

Morphological analysis was formally introduced by Zwicky in 1948 during his Oxford University Halley Lecture. Until his death, Zwicky worked on the development of his method. He also established in Pasadena, California, the Society for Morphological Research promoting the method. Today, however, the Swedish Morphological Society is the leading scholarly organization focused on the method, providing access to various publications on the subject (e.g., Ritchey 2010) and to the software supporting the method.

In the 1960s, when the design research revolution in Europe began, morphological analysis became the subject of many studies, which led to its practical application. The research began in the United Kingdom and gradually continued in Switzerland, the Netherlands, Germany, and Poland. In the early 1970s, the method became for the Author the key to his future, first to the inventive engineering and later to his global engineering research. He successfully used it to produce his first invention, patented

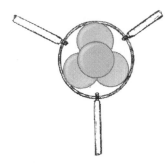

Figure 6.2 Patented connection in steel space structures.

in Poland (beam-column connections in skeleton structures). Later, while teaching at the University of Nigeria, he developed a mathematical model of the morphological analysis process, called *stochastic form optimization* (Arciszewski 1987). In the United States, he developed with the support of the National Science Foundation (NSF) a computer program based on this model and used it to develop a connection in steel space structures (Figure 6.2), which was patented in Canada and in the United States (Arciszewski 1989, 1991).

6.3 ASSUMPTIONS

The method (Zwicky 1969) was based on the fundamental principle of division and integration, which may be dated back to the Atomists. This principle and several more specific assumptions create the methodological foundation of the method:

1. It is a *closed-world* approach; that is, the problem domain knowledge is acquired and *stored* in the problem's representation space (called by Zwicky the *morphological table*). This closed space becomes the equivalent of the acquired knowledge and is exclusively used to produce solutions.
2. A design concept of an engineering system (which may be actual or abstract) is described by a finite number of symbolic attributes and their values. Such a description should be at least necessary and sufficient to identify all known concepts and to distinguish between them.
3. Each symbolic attribute identifies a different abstract feature of a design concept of the future engineering system.
4. Any complex conceptual design problem can be divided into a finite number of *elementary subproblems*, which cannot be any further divided.

5. Each subproblem must be considered as independent from the remaining subproblems, and its relationships with the other subproblems must be temporarily suspended.
6. All subproblems and their solutions can be represented in a systematic way in a single table with a number of rows and columns. Such a table becomes the problem's design knowledge representation space.
7. In the table, each row is temporarily independent from all other rows.
8. Each row in the table represents an elementary subproblem and contains a different symbolic attribute and a number of its feasible values within the context of the state of the art and with full disregard for values of other symbolic attributes (see assumption 7).
9. Any potential solution to the entire problem is represented by a sequence of symbolic attributes from all rows in the table and a combination of their values, taking one value from each row.
10. A potential solution to the entire problem is generated in an unbiased way through the random generation of combinations of symbolic values from all rows in the table, taking one value from each row.

Assumptions 1 through 8 are related to the process of division, which in this case leads from the entire problem to its elementary subproblems and to the associated symbolic attributes and their values. The last two assumptions (9, 10) concern the reverse process, that is, the integration of solutions to the elementary subproblems into a potential solution to the entire problem; the values of symbolic attributes from the entire table are brought together into a combination that may describe a potential new solution.

These assumptions have been modified and expanded from Zwicky's original simple assumptions, based on more than 40 years of the author's design research and various practical applications of the method (Arciszewski and Pancewicz 1976; Arciszewski and Kisielnicka 1977; Arciszewski 1984, 1985).

6.4 PROCEDURE

The procedure has four stages, each with several steps. This procedure is provided below, and its individual steps are discussed later.

Stage 1: Problem identification and formulation
  1. Formulation of the inventive challenge
  2. Identification of the problem domain
  3. Determination of the problem's boundaries
  4. Problem formulation

Stage 2: Analysis
  1. Identification of symbolic attributes
  2. Determination of values of symbolic attributes

   3. Building the morphological table
   4. Verification of results

Stage 3: Synthesis
   1. Random generation of potential design concepts
   2. Feasibility analysis

Stage 4: Presentation of results
   1. Presentation of problem formulation
   2. Presentation of selected design concepts

6.4.1 Stage 1: Problem identification and formulation

The objective of this stage is to prepare a definition of a given inventive problem, that is, to transform our *inventive challenge* into a properly prepared *problem formulation*. In the case of morphological analysis, this is a four-step process, discussed here.

6.4.1.1 Formulation of the inventive challenge

An inventive challenge is a short statement describing what is to be accomplished as a result of our inventive process. For example, we may have challenges like "Develop a flying machine using solar power" or "Develop a car braking system that will not overheat under extreme racing conditions."

Usually, an inventive challenge is given, but sometimes it must be formulated as a result of a conversation with an investor or with a group of domain experts. Even if an inventive challenge is provided, it may need to be reformulated considering a simple four-part heuristic.

An inventive challenge should be

1. Short, not more than 20 words
2. A single sentence
3. Written using a casual language, no technical terms
4. Use no words that might suggest any specific solutions

6.4.1.2 Identification of the problem domain

The objective is to determine what body of engineering knowledge is the most useful to address our inventive challenge. It should be done in a specific way. Instead of identifying only a direct problem domain, a wider domain should be considered. For example, if our challenge is to develop a novel connection in a structural system, it is wise to assume that the problem domain covers both structural and mechanical engineering. These two areas are closely related, and such expansion will allow us to access a body of potentially useful knowledge. Very often, when looking for inventive design concepts, we need knowledge from not only the problem domain but also from other domains, which could be called *secondary domains*. In our example of connections in a structural system, such secondary domains could be, for example, material engineering (dealing with new materials for bolts and stiffeners) or even chemical engineering, which deals with new kinds of adhesives.

6.4.1.3 Determination of the problem boundaries

This step is unique for the method and must be appropriately explained. Let us visualize a morphological table, or a field, as an allegoric field in which our body of knowledge is blooming and waiting for us. We would like our field to be as large as possible in order to maximize the probability that it will contain knowledge leading to inventions. At the same time, we know that our allegoric field and our body of considered knowledge must have boundaries. They represent a compromise between the need to maximize innovation and the need to minimize efforts. Therefore, we will use a procedure that is usually called *pegging*, which will allow us to find the boundaries of our knowledge.

When a surveyor determines the boundaries of a field, he walks in various directions through the field and puts pegs at the ends of our property (Figure 6.3). These pegs clearly determine where our property ends and identify a piece of land belonging exclusively to us and where we are allowed to cultivate the land. In the case of morphological analysis, pegging is an abstract process of knowledge acquisition, but it is conceptually similar to the activities of a land surveyor working for a landowner. In our case, our mission is to determine the body of knowledge available to us and which is to be used for the generation of potential design concepts. In a certain way, it is a closed-world approach. We acquire knowledge over a period of time, but finally we put this knowledge in a *box* or in a *field*, and after that we *forget* about the remaining part of the universe of knowledge. It has been reduced to our box, and only this knowledge will be used for the generation of potential design concepts.

Figure 6.3 Pegging.

Pegging is critical for the novelty of the final results, but it is also time consuming and difficult. The best results will be produced using both traditional and modern methods of knowledge acquisition, that is, conducting several parallel lines of activities, the results of which will be integrated.

Traditional knowledge acquisition may include writing down first what the inventive engineer knows about the problem domain and then talking to the domain experts and recording the results of such interviews. It also includes reading the related books and textbooks, professional journals and magazines, and so on. The Author once spent several weeks working on problem identification and formulation in the inventive designing of connections in large-span roof-space steel structures. Although he was a structural engineer, it took him a lot of effort to learn the basics of such connections and to be ready for interactions with several very experienced structural engineers, who spent long years specializing in designing such connections. Next, there was a period of at least two or three weeks during which the Author worked on the organization of the acquired knowledge and discussed his results with the experts. Their understanding was mostly intuitive and difficult to translate or transform into a clearly organized body of knowledge. Such cooperation guaranteed, however, that the acquired knowledge was complete in a given situation and well structured, at least in the opinion of the experts who helped to acquire it.

All these traditional processes of learning are usually recursive; that is, we may need to repeat them several times until we are able to proceed to the next activity, which could be called *patent search*. A patent search usually results in a number of patents related to our problem domain. We may find only several such patent or tens or even hundreds of them. In practical terms, we will be able to focus on a maximum of between five and ten patents, and we need to select them from all found patents. This can be done using various selection criteria, for example only the most recent patents or only the patents from a specific country. In the United States, the most important are American patents because our own inventions and the related patent claims will be examined and compared with the inventions behind these patents.

Unfortunately, patents are described in a very difficult legal language and using abstract technical terms, which are clear almost exclusively to their inventors and patent examiners. Therefore, we need first to develop at least a good domain understanding to benefit from the patent analysis and to be able to acquire from that analysis knowledge relevant to our problem. Obviously, we need to know the goals of our analysis. In fact, we have four such goals:

1. Identify all major concepts, including their names
2. Identify names of symbolic attributes associated with these concepts
3. Identify all known values of these attributes

4. Compare available patents, preferably in terms of the used patent claims (symbolic attributes) and their values

Finally, in the case of an important and complex domain, particularly when the potential value of inventions leading to patents is very high and time is a factor, an inventive engineer may hire a consultant, or a consulting company, with a lot of experience in the problem domain. Their goal will be to prepare the state of the art in a short time period and, in the process, save us time. Hopefully, such a practice should also produce results that are more complete and consistent than those that could be eventually produced on his or her own by the inventive engineer. He or she has a strong disadvantage with respect to the domain experts because he or she does not have many years of experience in the problem domain. Unfortunately, the quality and extent of learning are always somewhat time dependent, although this relationship is not entirely understood.

When an inventive engineer operates within a large corporation, a unit specializing in knowledge acquisition may produce the state of the art, although this must be done legally, without the use of industrial intelligence. In fact, inventive engineers should be aware of the ethical implications of inappropriate practices in knowledge acquisition. We are living at a time when knowledge is simply a commodity and is usually easily available on the Internet. It can be copied and, without any technical difficulty, used for various purposes. However, we need to remember that it is the intellectual property of other people, and that fact should be always recognized even if the copyrights have not been violated.

Modern methods of knowledge acquisition should be used only after the rate of progress with the traditional methods is significantly going down (i.e., we have reached the final stage of decline on our S-curve describing the process of learning). We may assume that at this stage we know enough about our problem and its domain. Therefore, we are able to prepare a list of key words associated with our inventive designing situation. These key words may be, obviously, the names of the symbolic attributes that have already been identified as a result of our earlier work.

When our key words are known, they become the key to the Internet and its resources. The process may begin with the use of any search tools, for example Google or Google Scholar, or any other specialized search machine or agent. In fact, although the selection of a search machine is important, how we use it is even more important. Here we have several helpful heuristics in our quest to learn about the problem domain:

1. Remember, our Internet search has the same four specific goals as our knowledge acquisition conducted earlier, in which the traditional methods were used.

2. Determine the importance of your key words and order them appropriately.
3. Begin your search with the most important key word and gradually use all the remaining key words.
4. More is better. Work hard to acquire the "maximum" amount of knowledge within the imposed time limit—or, for example, an equivalent of about 15,000 words—and store it all electronically. Print about 30% of the acquired pieces of text, drawings, tables, references, and so on. Make sure that you select carefully balanced pieces, including
   a.   The most important knowledge
   b.   The most unknown and surprising knowledge
   c.   The most promising knowledge
   d.   The most useless and absurd knowledge

In general, you should use your understanding of the problem domain, developed earlier when working with the traditional methods, to select pieces of knowledge to become your knowledge foundation. When both the traditional and modern knowledge acquisition processes are completed, it is time to integrate their results. The best way is to conduct a mind-mapping process to show major concepts in the problem domain, their relationships and hierarchies, and the body of the acquired knowledge. Mind mapping may require a lot of time and effort, but when after several iterations we feel comfortable that our results reflect the state of the art and are complete, the final pegging may begin.

As we know, the main mission of pegging is to determine the boundaries of our problem domain knowledge, but its secondary mission is to make sure that these boundaries contain a sufficient body of knowledge to develop inventive design concepts. The completeness of our acquired knowledge may be verified using at least two approaches. First, it can be done trying to find within it all the designs and patents known to us. When we discover that a well-known design concept or a patent cannot be identified, we need to repeat the knowledge acquisition process to expand our knowledge. In extreme case, we may need to repeat such a loop several times until we are convinced that our knowledge is "complete." Second, we may show our acquired knowledge to the problem domain experts. Such experts are usually very reluctant to articulate their personal knowledge but are eager to critique us and usually love to criticize the results of our work. They will provide many comments about missing knowledge, about wrong or missing relationships in our mind map, about missed sources of knowledge, and so on. Their comments may hurt our ego (the Author has had this experience) but may be potentially very useful.

Unfortunately, there are no formal completeness criteria, and we must depend on our judgment. On the other hand, inventive engineers using morphological analysis are aware that the completeness of their morphological tables is the necessary condition for ultimate success, and usually they pay a lot of attention to acquiring knowledge.

6.4.1.4 Problem formulation

As a result of our work on the two previous steps, we have accumulated a large body of information. It is potentially useful, but it needs to be properly formulated to make it suitable for the identification of symbolic attributes and determination of their values. We need to read and analyze all our acquire sources of information and to prepare three parts: the *inventive challenge*, the *problem domain*, and *specific information*.

As discussed earlier, an inventive challenge is a short statement, at most 20 words, capturing the nature of our challenge. The final products of our inventive problem solving will be sophisticated engineering design concepts, but an inventive challenge should be presented in a casual language, easily understood by everybody.

The *problem domain* description should be relatively short, on the level of a maximum of 200–300 words. It should identify the main problem domain and several domains that are expected to contain knowledge potentially relevant to our problem. Also, this domain should be described in simple engineering terms. For example, if we are looking for novel joints in timber floor trusses, the problem domain is "timber structures" and the related domains may be "structural engineering," "steel structures," or even "bone engineering," because we may find bioknowledge about joints in living organisms that could be used in our inventive design. In this case, the problem domain could be described as the area of structural engineering dealing with structural system in which timber is used as a structural material. The area of bone engineering is more unusual for engineers. However, it can be also described in simple terms as the area of engineering dealing with bones and joints in living organisms. These are considered from a structural engineering perspective as structural systems; that is, the focus is on their behavior under loading, including the analysis of stresses, strains, deformations, collapse mechanisms, fatigue, and so on.

The third part, titled *specific information*, could be quite lengthy, and no specific length limitations should be imposed. In this part, short summaries of information acquired from individual sources should be provided. It is important to write these summaries in such a form that it will be easy later to find major concepts and known realization of these concepts (which will be related to symbolic attributes). The identified concepts will eventually lead us to symbolic attributes, while examples of realizations will provide information about the feasible values of these symbolic attributes. For example, in the case of joints in timber floor trusses, we may identify such concepts as "truss joint" or "joint plate" and such examples of realizations as "steel joint plate" or "nails." These concepts may lead to such symbolic attributes as "joint type," "joint plate material," and "connectors type."

The first symbolic attribute, joint type, may have such values as "hinged" or "rigid," and the second attribute may have such values as "steel," "aluminum," or "copper." The third attribute, connectors type, may have such values as "nails," "adhesive," or "no individual connectors."

6.4.2 Stage 2: Analysis

6.4.2.1 Identification of symbolic attributes

Identification of symbolic attributes is probably the most challenging activity within morphological analysis. It requires a rare ability to think in abstract terms of symbolic attributes. Even more rare and difficult is the ability to transform traditional descriptions of engineering systems (words, drawings, dimensions, etc.) into their abstract description using the language of symbolic attributes. Fortunately, these two abilities can be developed as a result of practice.

The Author's simple procedure should definitely help. It is intended to make it easier for an inventive engineer to transform a traditional description of an engineering system into an equivalent description in the form of a number of symbolic attributes. The Author developed it in the 1970s (Arciszewski 1976, 1977a,b) as a result of his frustration with the intuitive or trial-and-error approaches to building a morphological field. The procedure can be described as a functional analysis as it is presented here.

Each engineering system exists to provide a specific function, which we can call the system's *main function*. Such a function is usually well understood and is identified by a single term or described by a short sentence. It can be denoted for a given system as MF. The main function MF is analyzed next, and several of its subfunctions are distinguished and denoted by $SF_1, SF_2, \ldots SF_i \ldots SF_n$, where i is an index changing from 1 to n, and n is the number of the distinguished subfunction. The next step involves the analysis of the individual subfunctions SF_i and finding their subfunctions, if they exist. For example, for SF_i we may have $SF_{i1}, SF_{i2}, \ldots SF_{ij}, \ldots SF_{ik}$, where j is an index changing from 1 to k, and k is the number of the distinguished subfunction for SF_i. The process of the functional analysis should be continued until the moment when no further functional division is possible; that is, we have reached the level of elementary functions $EF_1, EF_2, \ldots EF_e$, where e is the total number of elementary functions. In the process, a functional structure of our engineering system has been identified, with the main function MF on the top and the elementary functions on the bottom. An example of such a structure is shown in Figure 6.4.

When all elementary functions are known, from EF_1 to EF_e, they should be analyzed separately. For each elementary function, a characteristic, or several characteristics, of our engineering system should be determined, all associated with this elementary function. For example, if an elementary function is "transfer axial force from point A to B," associated characteristics

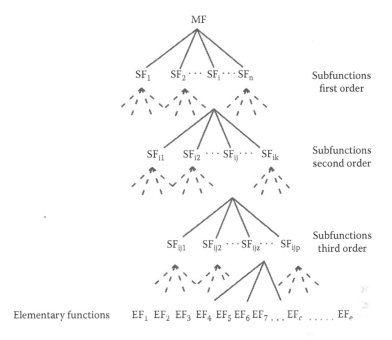

Figure 6.4 Functional structure of an engineering system.

could be "material," "shape of a structural member," "cross section type," "solid or hollow member," and so on. Usually, on this level of functional specificity we do not have trouble finding many associated characteristics.

Finally, we have a long list of characteristics associated with the individual elementary functions. Together, they constitute the collection of all characteristics describing our system. These characteristics could ultimately be used to identify symbolic attributes. A single characteristic may be transformed into a single symbolic attribute or into several attributes. The rule is "more is better." It is easy to eliminate redundant symbolic attributes, but it is very difficult, and sometimes even impossible, to add missing attributes, particularly when our understanding of the domain knowledge is incomplete.

The procedure does not guarantee that the complete set of symbolic attributes will be identified. However, using it at least increases the probability of the transformation from a traditional verbal description to an equivalent set of symbolic attributes. The procedure is difficult and time consuming, but as a byproduct of its usage, an inventive engineer will also improve his or her understanding of the engineering system considered. In addition, the system's entire functional structure will be identified. It can be always revisited and eventually modified, particularly in the case when the results coming from the morphological analysis are insufficient or even disappointing in terms of their novelty. Such an outcome is usually a signal that the morphological table is incomplete, and much more knowledge should be added through the process of expanding the table, adding missing attributes and provided additional values to the existing attributes. Expanding a table increases the probability of finding inventive design concepts and is sometimes required.

6.4.2.2 Determination of values of symbolic attributes

When symbolic attributes are known, finding their values is a relatively simple activity but may be quite time consuming. We should remember that each symbolic attribute must be considered entirely separately from the other attributes since this is one of the most fundamental assumptions of morphological analysis (see Section 6.3, "Assumptions"). Suspending relationships among symbolic attributes and their values is much more difficult than it might seem. It goes against the engineering practice of always looking for relationships and trying to develop a better understanding of the system we are analyzing or designing.

A good solution is to write down the names of the individual symbolic attributes on separate sheets of paper. We will work on them in a random order to minimize the impact of our engineering tendency to always see things in the context of our engineering knowledge, that is, to prematurely judge the feasibility of the individual values and relationships among symbolic attributes.

Before we begin our knowledge acquisition process, we need to address an important dilemma:

> Should we use in the table only very few truly meaningful to us values of symbolic attributes or as many as possible?

As engineers, we would like to eliminate in the process all *crazy* values and to focus only on the *meaningful* values in the context of our background engineering knowledge (we are respectable and serious engineers and we do not do any crazy stuff). Such a strategy would lead to a relatively small number of values and would make the verification of the table and, more importantly, the generation of combinations of attributes and their values easier. On the other hand, we are inventive engineers and we know the powerful heuristic "more is better." From this perspective, preselection of values is simply wrong. It brings our personal bias into play and most likely will hurt the novelty of our final results.

For example, let us consider the attribute "moving vessel" describing a vessel used in your office to boil water. When we look for values of this symbolic attribute without thinking that it will be an office device, we immediately identify the value "steam engine." Our first thought is that this is a crazy value since nobody would use a steam engine to move an office device. However, our second thought is different. We realize that our device will boil water and produce steam, which could be used to move our device when boiling is completed and a sufficient quantity of steam has been accumulated. Such reasoning is particularly attractive when we are looking for a novel device with a strong conversional value, and suddenly the value "steam engine" makes a lot of sense, as we surprisingly conclude.

Determination of values of symbolic attributes is obviously a process of knowledge acquisition, and all available sources of knowledge should be used. The sequence of using individual sources is basically immaterial, but it is natural to begin with values well known to the inventive engineer. Next, values from the available patents should be used. After that, a talk with the problem experts is recommended and their values should be elicited. Finally, after all *easy* values are found and put in the table, the more difficult part of finding values begins.

The inventive engineer should analyze again the acquired earlier background knowledge in the form of known design cases, books and textbooks, design manuals, and so on. It may be even advisable to run an Internet search again but this time looking specifically for known designs in order to find values of attributes necessary to discern between them.

When the inventive engineer finally believes that all values have been found and used in the table, the magic value "other" should be added to each row. This should be done for two reasons. First, this word will remind us that more values could be eventually found. Second, and probably more importantly, this word activates complex psychological mechanisms in our brain. These will cause the inventive engineer to continue subconsciously thinking about the representation and trying to finish the table and to eliminate the value "other." Such thinking may produce surprising results—not only missed values but also values with great inventive potential in the context of the entire table.

We have already conducted pegging, which can be considered as an initial verification of the completeness of our body of knowledge. Now it is time to verify our symbolic attributes and their values. That can be done using a procedure similar to pegging, but in this case we want to find all known solutions (design concepts) in our new morphological table. If a known solution—particularly a solution that has been patented—cannot be found in the table, it can be determined that an attribute is missing or a value is missing, and we need to repeat the process of building the table.

6.4.2.3 Building a morphological table

When all symbolic attributes and their values are identified, we are ready to construct a morphological table. This table will have a number of rows equal to the number of attributes and a number of columns equal to the maximum number of values any attribute may have. For example, if we have five symbolic attributes having two, three, six, four, and seven values respectively, our table will have five rows and seven columns. Such a table is shown below (Table 6.1). The symbols A_1 to A_5 represent our individual

Table 6.1 A morphological table for five symbolic attributes with the maximum number of values equal to seven

Attributes	Attribute values						
A_1	A_{11}	A_{12}					
A_2	A_{21}	A_{22}	A_{23}				
A_3	A_{31}	A_{33}	A_{34}	A_{35}	A_{36}		
A_4	A_{41}	A_{42}	A_{43}	A_{44}			
A_5	A_{51}	A_{52}	A_{53}	A_{54}	A_{55}	A_{56}	A_{57}

symbolic attributes, while A_{ij} symbols (where i is the number of attributes and j the number of value for a given attribute) are used here to represent values of these attribute.

Usually, a morphological table is gradually built and expanded. In the process, we add more attributes (and sometimes eliminate them) and add values to the individual attributes. This is a dynamic process, and the best practice is to build a morphological table from scratch using a word processor or a spreadsheet.

6.4.2.4 *Verification of results*

We should distinguish between the verification and validation of results. In the first case, we consider the consistency of results with a given body of knowledge, while in the second, our reference is the state of the art. Obviously, validation is much more demanding and difficult, not to mention much more time consuming and ultimately expensive. In the case of morphological analysis, we usually only verify our results to make sure that we have correctly and completely transformed our body of acquired knowledge into a morphological table.

The verification of results is a three-step process. Only the single morphologist working on a problem can conduct it, or experts may be invited, as this is a good and recommended practice. The process involves three tests, and each must bring positive results. It is a *binary* process. If only a single test fails, the entire analysis stage must be repeated to expand the body of acquired knowledge and to introduce new attributes and their values or to add additional values to existing attributes.

All three tests can be described as finding various concepts in the table. In the first test, several well-known design concepts should be found. In the second one, we are looking for relevant patents. Finally, in the last one, we are searching the table for test design concepts just created by us or by the experts.

In the case of a complex problem and a large morphological table (featuring more than 20–30 attributes), the verification of results may even take several weeks, as the Author once experienced, and may lead to significant changes and improvements in the table. We need to understand, however, that the process could take months without moving to the next stage, that is, the generation of potential design concepts. Such a situation is simply unacceptable. Usually, we have our timeline and if the time assigned for the verification of results is over, we should, at least temporarily, freeze our table and move on to the next stage. Such a decision may seem to be contrary to engineering practice, but in the case of morphological analysis, it is at least defensible, if not justified. Even if our table is incomplete or imperfect, it still may contain valuable combinations or potential design concepts, and delaying finding them would not be a good practice.

6.4.3 Stage 3: Synthesis

6.4.3.1 Random generation of potential design concepts

Finally, we are ready for the most exciting part of morphological analysis: the generation of potential design concepts. These concepts may become the key to our fame and fortune.

As we remember, we are looking for sequences of symbolic attributes and combinations of their values. Each attribute from the table must be used but only once in a given run. Also, for each attribute, only its single value must be randomly selected. For example, if a table with four rows is considered, a sequence of four symbolic attributes and their values may look like this:

$$A_{14}, A_{22}, A_{35}, A_{43}$$

Combinations of symbolic attributes values must be randomly selected, and the importance of the randomness cannot be simply overstated. If we select combinations in a way that is not entirely random, for example, while looking at our table, we will instill our bias consciously or subconsciously. This will result in the reduced probability of finding novel design concepts, and in practical terms we will be denied the most important benefit of using morphological analysis, that is, finding truly surprising but patentable design concepts, our key to fame and fortune.

We should be particularly aware of a typical error when using the method. When an inexperienced morphologist wants to create a random combination, often he or she simply looks at the table and subsequently selects values from the individual rows. In the process, the morphologist will see adjacent rows and realize the names of attributes residing in these rows. As an engineer, he or she will immediately begin analyzing the relationship between these two attributes and the feasibility of various combinations of their values. The choice of values will almost certainly be affected by this analysis, and the results of such generation will be strongly biased by the morphologist's understanding of the problem domain. The hated *vector of psychological inertia* will become involved and will do its best to make sure that only trivial combinations are generated.

We can produce random combinations manually or using various computer tools. Both paths are good; using a computer tool saves time and nearly guarantees random generation, but it is also less fun.

In the first case, we need to assume that each row is considered independently from the remaining rows and that the generation must be repeated for all rows. Let us consider the row i, which has m cells, each with a different value of the symbolic attribute A_i, which resides in this row.

Our first step is to prepare small balls with numbers from 0 to k, where k is the maximum number of values in a row for the entire table. For our row number i under consideration, we will use only the first m balls, numbered from 1 to m, and we will put them in an open vase. Obviously, when we have an appropriately trained animal, a parrot or a monkey, we will ask our animal to draw a ball. Next, we will read the number on it and this number will become the number of a value to be used in our combination. For example, if the number is 4, that means that we will use the value A_{i4} or the fourth value in the row i. If it happens that we do not have any trained animals, we should be able to draw a ball ourselves (but with covered eyes) and record the drawn number as in the previous case.

If we want to use a computer, the simplest solution is to write a program for the random generation of integer numbers within the range 1–m, where m is a variable representing the number of cells (values) in a given row i. This program should be run n times, when n is the number of rows in a given table. After that, a sequence of numbers of cells A_{ij} is obtained, and these numbers identify the sought combination of attributes and their values. This combination obviously represents a potential design concept.

At the other end of spectrum of computer programs for the random generation of design concepts is a computer program based on a mathematical model of the entire process of morphological analysis, which was originally developed by the Author while in Nigeria (1982) but published much later (Arciszewski 1988). The program uses the stochastic simulation of morphological analysis, a Markov chain with a number of random transitions corresponding to the number of rows in the table. The program was successfully used to generate a novel type of wind bracing in a skeleton structure of a tall building (Arciszewski 1985) and an inventive design concept for a spherical connection in steel space structures (Arciszewski 1984), which was patented in 1989 in the United States (Arciszewski 1989) and 1991 in Canada (Arciszewski 1991). Students at George Mason University regularly used the latest version of the program from 2007 to 2014.

6.4.3.2 Feasibility analysis

The objective of the *feasibility analysis* is to eliminate from any further consideration all sequences of symbolic attributes and combinations of their values that we believe are entirely useless for us. It is a highly subjective process and must be performed carefully. It can be compared to a process of filtering potential design concepts and eliminating useless ones. If the openings in the filter mesh are too large, we will be flooded with too many combinations, including many that are actually infeasible. On the other hand, if we are too strict and eliminate too many combinations, we may be left only with trivial combinations and in the process we may miss our "golden egg"—a combination, or a potential design concept, that is inventive and patentable and may lead to engineering applications.

The feasibility analysis is a dynamic process; that is, working with the same collections of combinations today and again next year may produce entirely different results. We are not doing any objective assessment in absolute terms but an assessment driven by our current priorities and our understanding of the state of the art, which obviously changes with time.

Understanding our priorities is crucial for doing the feasibility analysis the right way, that is, in a way that leads to meaningful results. Let us assume that our priority is a resultant of two factors, including the novelty and time factors. In the first case, we need to determine the desirable level of novelty within the range from very low to very high. In the second case, we ought to know when our invention will be eventually considered for implementation within the range from "0.5 of a year" to "more than 5 years." Both factors may be presented in the graphic form as our priority square (Figure 6.5).

The dot on the left represents a case in which we need a very low level of novelty (an improvement of an existing patent) and we need it soon. The dot on the right represents a situation in which we are looking for a very high level of novelty but where this invention will be considered for implementation in the future, more than five years from now. In these two extreme cases, our feasibility assessment of a given potential design concept

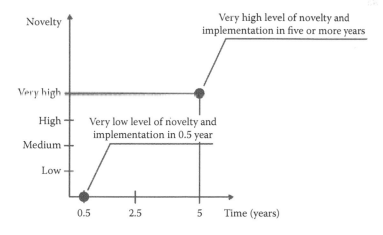

Figure 6.5 Priority square.

will be different. In the second case, we should be much more risk-taking or accepting of a much higher level of uncertainty than in the first case. Our priority gives us the context of the feasibility analysis, but unfortunately it is up to us to translate our priority into a specific decision to accept or reject a given combination.

There are two main reasons why a given combination may be useless for us. First, we eliminate combinations violating laws of nature, for example a design concept for a water distribution system in which water is flowing uphill in an open channel. Next, we eliminate combinations representing design concepts that we believe that cannot be implemented at the required time. In this case, we must be particularly careful, because our immediate responses very often are wrong. Particularly when we see an unusual combination, our first and natural reaction will be to reject it as infeasible. However, if we think about it for a moment, we may realize that this combination could actually be feasible and should be preserved.

It is a good practice while doing the feasibility analysis to use a simple procedure, which could be called "One NO and two YES." All the steps must be conducted, remembering our priority, which was determined earlier; in other words, the procedure needs to be performed as a situated action in which all specific decisions are done in the context of our priorities, and this has a tremendous impact on our decisions. This is a four-step procedure:

1. Sketch the design concept described by the combination. If you are able to sketch it and it looks reasonable to you, continue to the next step. If you are unable to sketch it or it does not make any sense to you, go on to the next combination.
2. Determine if any laws of nature are broken. If the answer is no, continue to the next step, No. 3. If the answer is yes, go to the next combination.
3. Use your engineering judgment to determine if the combination describes a design concept of the desired novelty. If the answer is yes, go to the next step. If the answer is no, go to the next combination.
4. Use your engineering judgment to determine if this design concept can be developed into a product within the imposed time limit. Implementation criterion: Is it possible to implement this design concept within the imposed time limit? If the answer is yes, the combination is feasible. If the answer is no, go to the next combination.

When the procedure is completed with one "no" and two "yes" answers, we have a winning combination—a potential design concept.

Steps 1 and 2 are relatively easy, although they require basic engineering knowledge, an ability to visualize abstract solutions, and some sketching skills. In contrast, steps 3 and 4 are difficult. They require not only an understanding of the problem domain but also a good grasp of the detailed designing and manufacturing of systems similar to the system just conceptualized. Very often, a morphologist is tempted to ask experts for advice regarding the novelty of the design concept described by the just generated combination. The Author has had mixed experience with using experts in this capacity. Usually, they are more interested in satisfying their ego than in helping us, and their typical response is "I know this concept; it is not new to me." However, a true expert may give us an honest and useful advice. In any case, we should not entirely depend on experts when assessing the novelty of a new combination.

It is easier to determine the manufacturability of a new design concept within a given time frame. If we are not able to do this, we can always contact engineers working in the area of manufacturing, and they should give us reasonable answers.

Unfortunately, the feasibility analysis is always subjective and conducted using our engineering judgment. The only heuristic available is "Always be optimistic and never underestimate the progress in engineering." Therefore, as inventors, we should do everything to avoid the error of omission, which is much more dangerous for us than the error of overcommission.

6.4.4 Stage 4: Presentation of results

Final results should be presented in a report called, appropriately, "Final Results Report." It should contain a body of knowledge about the inventive problem and results produced during the entire morphological analysis process. The size of this report will obviously depend on the complexity of the inventive problem, the nature of the domain, the specific expectations of the sponsors of the project, and so on.

At the beginning of the morphological analysis, a problem formulation was prepared. It has three main parts, including "Inventive Challenge," "Problem Domain," and "Specific Information." This report will become the second part of the final results report after the "Executive Summary." The last part will be called "New Design Concepts."

The executive summary should be relatively short, definitely not longer than one or two pages. It should provide a very comprehensive description of the problem formulation and of new design concepts. This part of the final report will be read by many people, including decision makers who do not have any engineering background but are very influential. These people are particularly important readers because they could help you to transform your design concepts into products. Therefore, the executive summary should be prepared in simple, preferably nontechnical language and without any drawings, which could easily confuse nonengineering readers. Also, it should be carefully edited for clarity.

At this stage of the morphological analysis, the problem formulation is available. We need to remember, however, that it was prepared at the beginning of the morphological analysis process, when our understanding of the inventive challenge and its domain was much worse than it is at the end of this process. For this reason, it is a good practice to read carefully the entire "Problem Formulation" part again and update it reflecting our improved understanding.

Probably the most important part of our final report is the last part, "New Design Concepts." It should be divided into six sections:

1. Morphological table
2. Concept generation process
3. Feasibility analysis
4. Promising design concepts
5. Useful design concepts
6. Interesting design concepts

The first section, the *morphological table*, should present the developed table and provide definitions of the used symbolic attributes and their values. In the second section, the *concept generation process*, the used process should be described with clear explanation how the morphologist ensured that the process was random and if any randomness tests were conducted. Section No. 3, the *feasibility analysis*, should present the feasibility criteria and should explain why these criteria were used.

Sections 4 through 6 should present concepts considered by the morphologist as promising, useful, and interesting, respectively. The first of these three sections should list at most three design concepts that were found most promising in terms of their novelty and manufacturability. Section No. 5 should provide concepts (between three and five) that are worth further investigation but are less novel than concepts from the previous group. Finally, the last group of concepts, *interesting design concepts*, should include concepts–no more than five–that may not be feasible at this time but have the potential to inspire inventors or need to wait for further progress in technology before they become entirely feasible.

Each design concept should be presented in the same format for clarity and easy comparison with other concepts. Basic components of such presentation should be

1. Description
2. Sketches
3. Combinations of attributes and their values
4. Novelty
5. Comparison with existing solutions
6. Manufacturability
7. Comments

It is important to begin with a short verbal description of a new design concept (not more than three to five sentences) using a simple language to engage the reader and to force him or her to use their imagination to visualize the developed design concept (Section No. 1). Next, several concept sketches should provide the reader with an improved understanding of a given design concept (Section No. 2). Section No. 3 should provide not only the sequences of attributes and combinations of their values but should also show the unusual aspects of this combination, which constitute the key to the novelty of this design concept. The section on novelty (Section No. 4) should explain novelty not in the context of a sequence of attributes and combinations of their values but in technical terms. It should be like a famous "elevator speech," that is, explaining novelty in 35 seconds. In Section No. 5, the generated design concept should be compared with existing solutions in terms of attributes and their values but should also use casual language, preferably nontechnical if possible. The section on manufacturability should present at least a preliminary discussion of how the developed design concept could be transformed into a product. If opinions from experts are available, these opinions, or at least the experts' most important findings, should be incorporated. The last section, "Comments," should provide the opinion of the morphologist about the given design concepts and all kinds of information that he or she considers appropriately interesting or potentially useful.

6.5 EXAMPLE

6.5.1 Introduction

In the early 1970s, the Author was a junior faculty in the Department of Metal Structures at the Warsaw University of Technology. During this time he participated with a senior faculty from the same department and an architect (now a famous designer of mansions in Toronto, Canada) in a national competition in Poland. This was a rare competition for structural engineers to develop a new universal steel structural system for applications in tall office and residential buildings. The system was required to allow the design of buildings in the height range of 12–16 stories but also of much taller buildings up to 36 stories.

There was a dual challenge: to develop a universal system of wind bracings and a universal beam–column connection. The connection was to be applicable in rigid steel frames (as a moment or rigid connection) and also, with some minor modifications, in buildings braced by various truss systems. From the perspective of this book, the second challenge was much more interesting; the morphological method was used, and an inventive design concept was developed and patented by the Author in 1976. Therefore, the subject of our example is the development of a new type of a beam-to-column connection in steel skeleton structures.

6.5.2 Stage I: Problem identification and formulation

6.5.2.1 Formulation of the inventive challenge

At that time (the 1970s), the Author was involved in research on steel skeleton structures and had some understanding of designing beam-to-column connections. But this was only the beginning of the process of knowledge acquisition. It started with careful analysis of the competition requirements, followed by discussing the challenge with much more experienced faculty in his department. Next, a comparison of connections known at that time was conducted, which revealed that there are basically two separate classes of types of connections: flexible or hinged connections, and moment or rigid connections. No universal connections were available at that time that could be used as a source of inspiration. That made the Author very happy, because while the challenge was much more difficult than expected, the potential pay off would also be much greater. In this situation, a bold challenge was formulated:

> Develop an entirely new design concept of a universal beam-to-column connection in steel skeleton structures, which could be used as a moment connection or, with minor changes, as a flexible connection.

The senior faculty and a team member enthusiastically approved such a challenge, but the other senior faculty in the department only smiled mysteriously and wished us good luck. They *knew* that it was an impossible challenge and that the junior faculty would fail, in the process learning an important lesson: never to think big again, as the vector of psychological inertia always tries to prevent us from doing. However, they did not know the Author's secret: morphological analysis.

6.5.2.2 Identification of the problem domain

Designing connections in steel skeleton structures is the domain of both structural and mechanical engineers. Structural engineers develop concepts of connections (or usually use well-known concepts) and analyze and optimize them, considering three areas. First, a connection must provide a smooth flow of internal forces between a beam, or beams, and a column and all design code stress-related requirements must be satisfied. Next, all components of a connection must be locally stable under all combinations of loading, and obviously all design code stability-related requirements must be satisfied. Finally, a good connection does not only satisfy all design code requirements. It is a part of a structural system and contributes to its overall stability. Preferably, such a connection will also contribute to the spatial distribution of loading, particularly of wind forces. In this way, if a hurricane hits the building from a specific direction, the entire three-dimensional structure will respond, reducing the probability that the members directly impacted by the hurricane will be overloaded.

Mechanical engineers are focused on the manufacturability and constructability of connections. In other words, they analyze the designed connection from the perspective of a manufacturing plant, which will produce all the components of the connection. A single skeleton structure may have easily hundreds of connections, and their manufacturability has a tremendous impact on the cost of the structure. Also, all connections must be also installed on the construction side, often at high elevation and with full exposure to the elements. Therefore, their constructability is quite important considering safety, costs, and construction times.

Our challenge required knowledge from two engineering domains. Obviously, such a large body of knowledge was not available to the Author and some cooperation with various domain experts from the areas of designing connections in steel structures and their manufacturing was required.

6.5.2.3 Determination of the problem boundaries

The first step was to read carefully the part of the competition requirements that dealt with the development of connections. Five important pieces of information were found and translated into the problem boundaries. First, the connections were to be restricted to steel structures only, and they did not need to be applicable to the concrete skeleton structures, although this possibility was not precluded. Second, the connections were to be developed assuming the use of a mild structural steel (low carbon steel), an equivalent of ST 36 steel in the United States. Third, both welding and bolting could be used separately or together, and no preferences were provided. Fourth, the components of the connections were to be easy to manufacture (high manufacturability), easy to transport, and easy to install on the construction side (high constructability). Finally, there was also a general requirement that the connections should be designed in accordance to the state of the art at that time (the mid-1970s).

The reading of the competition requirements has resulted in the first two problem boundaries. First, only a mild structural steel should be considered. Second, only bolting or welding should be used. In other words, no glued connections would be accepted. As a matter of fact, glued connections in structural systems in tall buildings were considered more a dream than a reality in the 1970s, as they are still today, but an inventive engineer should always consider the impossible and purposefully pretend that he or she does not even know the term "feasible" when building a morphological table.

The second step was to identify known concepts for connections, and about eight types were found. Figure 6.6 shows two examples of typical

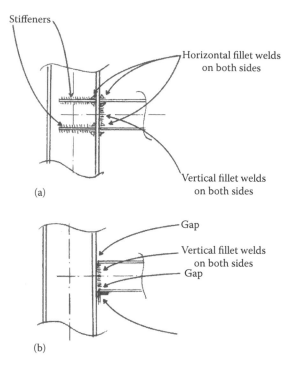

Figure 6.6 Typical beam-to-column connections in steel skeleton structures. (a) Moment connection, (b) flexible connection.

connections, both welded. A rigid one is on the top and a flexible one is on the bottom.

When these two concepts were examined, the Author realized that several attributes were necessary to discern between these two concepts. The attributes would become later a part of the morphological table and would be used in Step 2.1, "Identification of symbolic attributes." The following symbolic attributes and their values were identified:

A_0 Material	Mild structural steel
A_1 Direct connection upper flange-column	None, yes
A_2 Weld type for direct connection upper flange-column	None, fillet, butt, other
A_{16} Angle seat	Yes, none
A_{17} Connection angle seat-column	None, bolts, welds
A_{18} Gap between column and flanges	None, yes

The numbering of attributes is consistent with the morphological table (see Table 6.2).

The attribute A_0, "Material," has a single value (mild structural steel), the same for all combinations of values of the remaining attributes. Therefore, its discerning power is nonexistent, and there is no reason to use it in the table.

The remaining attributes must be explained from an engineering perspective. The angle seat is a short piece of a rolled steel member with an angle section. Such a seat is often welded horizontally to a column to provide a seat (support) for a beam during construction, but it is also used in hinged connections, not only to make construction easier but also to serve as a permanent support of a beam and to carry out part of the beam's weight to the column.

Vertical fillet welds are used to transfer the shear force from the beam to the column. This transfer takes place both in rigid and hinged connections, but the transfer may be accomplished in different ways, and fillet welds are only one of them (e.g., using vertically positioned angles, which are welded to the beam and bolted to the column). When fillet welds are used, they are situated symmetrically on the both sides of the beam's web. *Horizontal fillet welds* are located horizontally and perpendicularly to the beam's bending plane (its vertical plane of symmetry). They are used only in the connections carrying our bending moments.

Finally, gaps between the column and the beam's flanges are created in the hinged connections only to allow deformations of the beam and to prevent any moment transfer between the beam and the column.

The conducted analysis revealed the problem boundaries (only steel structures, only traditional welded or bolted connections) but also led to the identification of several attributes, leading to the construction of the morphological table.

6.5.2.4 Problem formulation

As we remember from Section 6.4, "Procedure," a problem formulation should have three separate parts, including the inventive challenge, the problem domain, and specific information.

In the case of our example, the inventive challenge has already been formulated:

6.5.2.4.1 Inventive challenge

Develop an entirely new design concept of a universal beam-to-column connection in steel skeleton structures, which could be used as a moment connection or, with minor changes, as a flexible connection.

As a result of our analysis conducted within the previous steps—2.1, 2.2, and 2.3—a good understanding of the problem domain has been developed. It can be summarized in the following short statement:

6.5.2.4.2 Problem domain

The inventive problem is in the domain of structural engineering, specifically in the area called "steel structures." The secondary domain is mechanical engineering.

The problem can be classified as a beam-to-column connection problem in steel skeleton structures of tall buildings.

Finally, specific information has been also prepared and formulated:

6.5.2.4.3 Specific information

The new connection needs to be adaptable; that is, it should be in two forms, including a connection that allows the transfer of bending moments between beams and column (moment or rigid connection) and a connection that does not allow such transfer (flexible or hinged connection). Also, only mild structural steel should be used. Finally, bolts or welding, or both, could be used to connect the individual elements of the connection as well as to connect the beam to the column.

6.5.3 Stage 2: Analysis

6.5.3.1 Identification of symbolic attributes

Our first step is to conduct a functional analysis of our system in order to find its all elementary functions (EF_j, j = 1,2,3, ... e). Knowing them will allow us to find symbolic attributes and their values associated with the individual elementary functions.

The functional analysis (see Step 2.1. in Stage 2 of the procedure) begins with the determination of the main function (MF) of our system. In our case, it is the transfer of internal forces from the beam to the column in a skeleton steel structural system:

MF = Internal Forces Transfer

Next, we need to find subfunctions (SF_i, i = 1,2,3 ... n). For our system, we identify four such subfunctions (Figure 6.7):

SF_1 = Bending Moment Transfer

SF_2 = Shear Force Transfer

SF_3 = Axial Force Transfer

SF_4 = Contributing to Building's Stability and Safety

Now, each subfunction must be analyzed separately to divide it into subfunctions of lower level (SF_{ij}) or simply into elementary subfunctions (EF_j).

In the case of the first subfunction, $SF_1 =$ Bending Moments Transfer, we have two subfunctions, which will become elementary functions (EF_j):

EF_1 = Compressive Force Transfer

EF_2 = Tensile Force Transfer

When the second subfunction, $SF_2 =$ Shear Force Transfer, is analyzed, it may be assumed to be an elementary function.

$SF_2 = EF_3$ = Shear Force Transfer

The third subfunction is $SF_3 = $ *Axial Force Transfer.* For the simplicity of our presentation, this subfunction will not be discussed as it is a secondary issue.

$SF_3 = EF_4$ = Axial Force Transfer

The fourth subfunction is $SF_4 = $ *Contributing to Building's Stability and Safety.* In this case, the following elementary functions have been found:

EF_5 = Contributing to Building's Stability

EF_6 = Contributing to Building's Safety

Neither of these elementary functions will be discussed here because of the complicated technical nature of these two functions and because it is not necessary to explain morphological analysis in action.

As a result of our functional analysis, five elementary functions have been identified (Figure 6.7). Now it is time to identify the symbolic attributes associated with the individual elementary functions, with the exception of EF_4, EF_5, and EF_6. There is no mechanistic way to do this, but the provided analysis should show how this could be done.

Our first elementary function is $EF_1 =$ Compressive Force Transfer. As we know, there must be a subsystem providing this elementary function, and our task is to determine its main features, which are identified by symbolic attributes. This must be done using our knowledge and by identifying

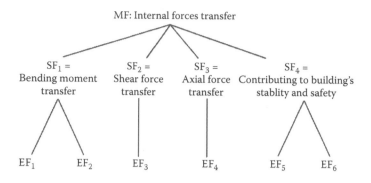

Figure 6.7 Example: Functional structure.

individual symbolic attributes. It is a difficult process with unpredictable results, but eventually brainstorming could be used if time and resources allow.

Let us explain the emergence of this compressive force and how to deal with it. When a bending moment is considered, it can be understood as the result of a stream of normal stresses that are distributed over the entire cross section of the beam. For example, in a fixed beam in the neighborhood of supports, which is under downward-directed transverse loading, the top part of the cross section (and the beam) is under tension and the bottom part under compression. We may deduce that all these tensile stresses in the top part produce the tensile force, while the compressive stresses in the bottom part produce the compressive force. In this way, we have a couple of forces whose moment is obviously our bending moment to be transferred. We may simplify the situation slightly and assume that both forces are located in the flanges of our beam (in fact, the web participates in the carrying out bending moment, but its contribution is insignificant when compared with flanges).

Our reasoning has led us to the conclusion that the elementary function EF_1 means the transfer of the compressive force (hidden in the beam) to the column. Determining how this will be done and what attributes describe this subsystem is our next challenge.

First, this compressive force could be transferred through direct contact with any connections, but this is not done because of concern about the structural integrity of the entire building (see the elementary function EF_5 and EF_6), not to mention that the bending moment's sign could be reversed. For all these reasons, top and bottom flanges in a fixed beam are usually connected in the same way with the column.

A flange can be connected with the column using welding (butt or/and fillet welds), or a transitional member can be used between the column and the web. This transitional member must be connected with the column (by

Table 6.2 Example: Morphological table

Symbolic attributes		Symbolic attributes values		
A_1 Direct connection upper flange-column	None	Weld		
A_2 Weld type for direct connection upper flange-column	None	Fillet	Butt	Other
A_3 Transitional upper member	Yes	No		
A_4 Kind of transitional upper member	None	Horizontal steel plate	Other	
A_5 Weld type for connection transitional upper member-flange	None	Fillet	Butt	Other
A_6 Weld type for connection transitional upper member-column	None	Fillet	Butt	Other
A_7 Bolt type for connection transitional upper member-flange	None	Regular	Prestressed	Other
A_8 Bolt type for connection transitional upper member-column	None	Regular	Prestressed	Other
A_9 direct connection lower flange-column	None	Weld		
A_{10} Weld type for direct connection lower flange-column	None	Fillet	Butt	Other
A_{11} Transitional lower member	Yes	No		
A_{12} Kind of transitional lower member	None	Horizontal steel plate	Other	
A_{13} Weld type for connection transitional lower member-flange	None	Fillet	Butt	Other
A_{14} Weld type for connection transitional lower member-column	None	Fillet	Butt	Other
A_{15} Bolt type for connection transitional lower member-flange	None	Regular	Prestressed	Other
A_{16} Angle seat	None	Yes		
A_{17} Angle seat-column connection	None	Weld	Bolts	
A_{18} Gap between column and flanges	None	Yes		
A_{19} Direct connection web-column	None	Yes		
A_{20} Weld type for connection web-column	None	Fillet	Butt	Other
A_{21} Transitional vertical member	None	Yes		
A_{22} Kind of transitional vertical member	None	Vertical steel plate	Angles	Other
A_{23} Weld type for connection transitional vertical member-web	None	Fillet	Butt	Other
A_{24} Weld type for connection transitional vertical member-column	None	Fillet	Butt	Other
A_{25} Bolt type for connection transitional vertical member-web	None	Regular	Prestressed	Other

welding or other means), with the upper flange using welding or bolts. Since we know this, we also know the next several attributes describing the discussed part of the structural design (attributes A_1, A_2, and A_{17} are repeated):

A_1 Direct connection upper flange-column	None, yes
A_2 Weld type for direct connection upper flange-column	None, fillet, butt, other
A_3 Transitional upper member	None, yes
A_4 Kind of transitional upper member	None, horizontal steel plate, other
A_5 Weld type for connection transitional member-flange	None, fillet, butt, other
A_6 Weld type for connection transitional member-column	None, fillet, butt, other
A_7 Bolt type for connection transitional member-flange	None, regular, prestressed, other
A_8 Bolt type for connection transitional member-column	None, regular, prestressed, other

The analysis of the elementary function EF_2 is practically identical to the analysis described above for EF_1, and it leads to the attributes A_9–A_{18} (see Table 6.2).

Identification of attributes associated with the elementary function EF_3, Shear Force Transfer, requires explanation. We have already identified attributes A_{16}, A_{17}, and A_{18} related to the transfer of shear force. The attributes A_{19} and A_{20} deal with the so-called direct transfer, but the shear force may be also transferred using transitional elements situated on both sides of the web, and these may be in various forms. Therefore, we need attributes describing all these possibilities:

A_{19} Direct connection web-column	None, yes
A_{20} Weld type for connection web-column	None, fillet, butt, other
A_{21} Transitional vertical member	None, yes
A_{22} Kind of transitional vertical member	None, vertical steel plate, angles, other
A_{23} Weld type for connection transitional vertical member-web	None, fillet, butt, other
A_{24} Weld type for connection transitional vertical member-column	None, fillet, butt, other
A_{25} Bolt type for connection transitional vertical member-web	None, regular, prestressed, none, other

All identified attributes and their values are presented in Table 6.2.

6.5.3.2 Determination of values of symbolic attributes

In our particular case, the determination of the values of symbolic attributes was conducted as a part of the process of identifying symbolic attributes.

In fact, in the case of complex problems, such integrated acquisition of symbolic attributes is not only allowable but even recommended. It requires working with experts, and it is easy to ask them at the same time about symbolic attributes and their possible values.

6.5.3.3 Building the morphological table

In our example, building the table was simple and basically came down to putting together all the identified attributes and their values, as shown in Table 6.2.

6.5.3.4 Verification of results

When the table was available, the fun part began. All the members of the team tried to find in the table all types of connections that were known to them. Since they could not find a connection that would not be possible to find in the table, it was concluded that the developed morphological table was sufficient for our purpose.

6.5.4 Stage 3: Synthesis

6.5.4.1 Random generation of potential design concepts

The reported application of morphological analysis took place about 40 years ago. Obviously, at that time, writing a computer program for the generation of random combinations of symbolic attribute values was not a trivial matter for structural engineers. For this reason, a manual generation of these combinations was conducted using the "ball method" described earlier in Section 3.1.

One combination was found to be particularly attractive: A_1 = None, A_2 = None, A_3 = Yes, A_4 = Horizontal Steel Plate, A_5 = None, A_6 = None, A_7 = Regular, A_8 = None, A_9 = None, A_{10} = None, A_{11} = Yes, A_{12} = Horizontal Steel Plate, A_{13} = None, A_{14} = None, A_{15} = Regular, A_{16} = None, A_{17} = None, A_{18} = Yes, A_{19} = None, A_{20} = None, A_{21} = Yes, A_{22} = Vertical Steel Plate, A_{23} = None, A_{24} = None, A_{25} = Regular.

This combination describes a simple connection in which both horizontal and vertical steel plates are used. Horizontal plates act as transitional members transferring axial forces in the flanges of beams to the column. Flanges are connected to the horizontal plates through bolts. Vertical plates transfer the shear force to the column without the need to use any connectors between beams and the vertical plates.

When a specific floor and an internal column are considered in a skeleton structure, four beams are connected to this column. In the case of the connection produced by the morphological analysis, that would mean four separate upper horizontal plates and four lower horizontal plates, not to mention eight vertical plates. Such four separate

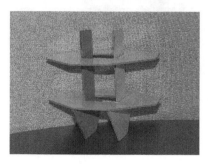

Figure 6.8 Example: Patented beam-to-column connections in steel skeleton structures.

beam-to-column connections are possible (and in fact are often used), but their manufacturability and constructability would be bad and their structural behavior would be clearly planar; that is, loading brought by the individual beams would be distributed mostly in the planes determined by columns and these beams. In other words, the skeleton structure would work as a collection of many independent planar structures, and very few, if any, highly desirable spatial effects would take place. As a structural engineer, the Author was aware of all these deficiencies of the mechanistically produced design concept, but it inspired him to introduce several modifications and to develop an inventive design concept (Arciszewski and Pancewicz 1976), which was patented in 1976 and is shown in Figure 6.8.

The modifications were related to the plates used in the produced design concept. All four upper plates were combined into a single upper steel plate, and similarly all four bottom plates were combined into a single bottom steel plate. Next, all eight vertical plates were substituted by only two vertical steel plates but were shaped in such a way that part of them would support the bottom horizontal plate and would be used for the transfer of shear forces. As a result of all these modifications, an elegant and simple spatial system of four steel plates was created with two identical horizontal plates and two identical vertical plates.

6.5.4.2 Feasibility analysis

A feasibility analysis was conducted immediately after the potential design concept was generated and for a second time about 35 years later as a part of a different project conducted in the United States. In the first case, the connection was shown to a number of experienced structural designers specializing in steel structures, and they all agreed that the connection, although unusual, was actually feasible and that they did not expect any major problems with manufacturing individual elements and connecting them together. They were only concerned about constructability, that is, how easy it would be to put a connection on a column. In this case, however, only actual tests would provide the answer.

About 35 years later, the Author participated in a project dealing with the design of steel skeleton structures for blast, that is, in the context of infrastructure security. Since the invented connection integrates all beams with the column and, at least partially, contributes to the spatial behavior of the entire steel skeleton structure, it should potentially provide a desired behavior during the blast contributing to the increased safety of a skeleton structure. For this reason, Urgessa and Arciszewski (2011) conducted a comparison study of the behavior of various types of connections in a steel skeleton structure under blast conditions. The Author's connection performed surprisingly well, and its behavior was comparable with a very complicated connection specifically developed for blast-resistant steel skeleton structures. It was the final feasibility proof of the developed connection.

6.5.5 Stage 4: Presentation of results

6.5.5.1 Presentation of problem formulation

The problem formulation was presented in Section 1.4, "Problem Formulation," and there is no need to present these results here again.

6.5.5.2 Presentation of selected design concepts

The selected design concept was presented earlier in Section 6.5.4.1, "Random Generation of Potential Design Concepts."

6.6 POTENTIAL APPLICATIONS

Morphological analysis has been presented as a method for the generation of inventive design concepts. This is obviously the method's most important application, but it can be also used for five other applications, which are briefly discussed here.

6.6.1 Classification

When we work with complex engineering systems, their formal classification is often difficult, but morphological analysis may help. We can develop a morphological table to generate inventive design concepts, but we can also develop a morphological table only for classification purposes. In fact, when the Author worked in the construction industry in Poland in the 1970s, one of his first tasks was to formally classify several existing systems of joints in steel space structures. In this case, descriptions using casual words would have been too long and difficult to comprehend, and even drawings would have been confusing considering the significant complexity of these joints. A better way was necessary, and the morphological table was the answer.

The table was developed for joints in steel space structures. Then individual types of connections could be easily identified by the appropriate combinations of attributes and their values (Arciszewski 1984).

The area of joints in steel space structures is too complicated for nonstructural engineers to be used as an example; therefore, the process will be explained using a simple example of a wind bracing in the wall of a steel industrial building. Such bracing provides three main functions: (1) It makes a given wall stable (geometrically invariable), (2) it contributes to the stability of the entire building, and (3) it transfers wind forces from the wall to the foundation.

A bracing may be adequately described by four attributes and their values, as shown in Table 6.3. This table provides 64 combinations of values of symbolic attributes, but obviously not all such combinations are feasible. For example, if a given bracing is in the form of a shear wall, this wall must be of concrete and the combination "A_1 = Steel, A_2 = Shear wall, A_3 = No joints, A_4 = No diagonals" is infeasible.

The table can be used to formally define or classify four main types of bracing. First, the combination

$$A_1 = \text{Steel}, A_2 = \text{Linear Members}, A_3 = \text{Rigid}, A_4 = \text{No Diagonals}$$

defines a wind bracing in the form of a rigid frame, as shown in Figure 6.9a. The combination

$$A_1 = \text{Steel}, A_2 = \text{Linear Members}, A_3 = \text{Hinged}, A_4 = \text{Diagonal}$$

defines a wind bracing in the form of a truss (Figure 6.9b). The combination:

Table 6.3 Morphological table for wall wind bracings in a steel industrial building

Symbolic attributes	Symbolic attributes values			
A_1 Material	Steel	Concrete		
A_2 Structure type	Linear members	Shear wall		
A_3 Joint type	Rigid	Hinged	Rigid and hinged	No joints
A_4 Member configuration	Diagonal	X-type	K-type	No diagonals	No steel bracing

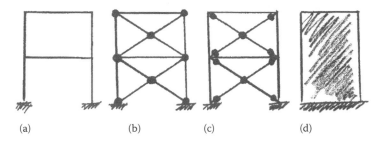

Figure 6.9 Basic types of a wall bracing in a steel industrial building: (a) rigid frame, (b) truss, (c) braced frame, and (d) shear wall.

A_1 = Steel, A_2 = Linear Members, A_3 = Rigid and Hinged, A_4 = Diagonal

describes a bracing known as a braced frame (Figure 6.9c).

Finally, the combination

A_1 = Concrete, A_2 = Shear Wall, A_3 = No Joints, A_4 = No Diagonals

describes a bracing called a shear wall, that is, a bracing in the form of a concrete wall (Figure 6.9d).

6.6.2 Comparison of design concepts

When a morphological table is known for a class of engineering systems (like it is known for a class of wall wind bracings in an industrial steel building, which is discussed in the previous section), it may be used to formally compare these systems. The Author often uses such comparisons for two purposes.

First, it is the best way to explain the differences between two design concepts, which may be subtle and difficult to grasp when reading a text or even looking at drawings. For example, we may need to explain the difference between a wall wind bracing in the form of a truss and in the form of a braced frame, discussed in the previous section.

A truss bracing is described by

A_1 = Steel, A_2 = Linear Members, A_3 = Hinged, A_4 = Diagonal

and a braced frame is described by

A_1 = Steel, A_2 = Linear Members, A_3 = Rigid and Hinged, A_4 = Diagonal

These descriptions differ only in one value of the single symbolic attribute A_2 Joint Type, that is, "Hinged" versus "Rigid & Hinged." On the surface it looks like a minor difference, but in fact, from the structural point of view, the difference is huge. In the first case, the entire bracing is in the form of a truss; that is, all its joints are hinged (or flexible). In the second case of a braced frame, we can describe the bracing as a rigid frame, which has all rigid joints (or inflexible). In the case of this design concept, additional bracings are connected with the rigid frame through added hinged (or flexible) joints located close to rigid joints in the rigid frame. The Author used to teach the design of steel structures, and many times he needed to explain the difference between trusses and braced frames since such structures are used for various purposes, not only as wall wind bracings in steel industrial buildings. Only when he began using symbolic attributes and a morphological table did the students' confusion disappear and smiles would emerge on their happy faces.

The second important application of a morphological table in the context of comparison is the determination of novelty of a new design concept with respect to a known one. In fact, the Author has conducted research on the subject, and his PhD student Kenneth Shelton (2007) has developed a sophisticated novelty assessment method (Shelton 2008) with roots in morphological analysis. This method has been applied in the area of the conceptual design of global communication satellite systems. Unfortunately, because of the complexity of such systems and their sophistication, this application cannot be presented here, but it can be found in (Shelton 2008).

As an example, we will use here a simplified comparison of two concepts for a car vehicle, reduced to their two most important features (Table 6.4).

Our reference concept is a well-known type of car vehicle in the form of a car with a gasoline engine and drum brakes. It is described by the combination

A_1 = Gasoline, A_2 = Drums

Let us consider a new concept of a vehicle with a gasoline engine but with disc brakes. It is described by the combination

A_1 = Gasoline, A_2 = Disc Brakes

Table 6.4 Morphological table for car vehicles

Symbolic attributes	Values of symbolic attributes		
A_1 Source of energy	Diesel fuel	Gasoline	Compressed oil
A_2 Brakes type	Drums	Disc brakes	

When it is compared with a reference concept, the difference is in only one symbolic attribute, A_2, whose value is changed from "Drums" to "Disc Brakes." In this case, we could say that the morphological distance is 1.

When another new concept is compared with the reference concept, it is for a vehicle with disc brakes and with a novel engine using the energy of a compressed oil. It is described by the combination

$$A_1 = \text{Compressed Oil},\ A_2 = \text{Disc Brakes}$$

This concept differs from the reference concept in the values of two symbolic attributes. The attribute A_1 changes from "Gasoline" to "Compressed Oil" and the attribute A_2 changes from "Drums" to "Discs." In this case, the morphological distance is equal to 2, and obviously this concept is more novel than the other new concept.

Our examples are intended only to explain the principles and are very simple, but they illustrate two large classes of practical applications. Other details of the second class can be found in Shelton (2008).

6.6.3 Finding patent holes and building patent fences

The language of morphological analysis is also the language of patent applications and patent claims. Large industrial corporations are aware of the importance of intellectual property in general and especially patents and the inventions behind these patents. They want to protect their own inventions but are also always looking for legal opportunities to acquire and use knowledge associated with the patents developed and owned by the competition, preferably free. Sometimes such activity is called *fishing* or *finding patent holes*.

Finding patent holes can be described as a careful analysis of existing patents. It is focused on finding missing patent claims, that is, values of symbolic attributes that are missing from the description of a competition's patent. Such values could be used to allow the legal free utilization of major ideas pertaining to a given invention. For example, let us assume that a hypothetical patent has a claim that the contact surface must be smooth. If we find out that the invention would also work with a rough contact surface, then we could design a device based on all competition's patent claims but the one related to the smoothness of the contact surface. In this way, we could legally avoid buying a license or paying royalties for the use of the competition's patent, although there are ethical issues here that need to be addressed.

Building a patent fence is a process in which we forecast future inventions and patents. Next, we use this knowledge to protect our future interests through patenting a class of inventions. Such inventions are hypothetically feasible, although at the time of patenting them we do not have the technology to implement them. An even more important situation is when we do not need to use these patents now, but we want to prevent our competition from using them because we may need them in the future. In Chapter 9, we discuss the concept of a line of evolution of an engineering system. When we have such a line of evolution, or an envelope of several such lines, we may need to decide first which line should be selected and next to construct a patent fence around this line.

6.6.4 Design knowledge acquisition

Learning and acquiring knowledge is the most important activity for all engineers who are focused on advancing their careers and on changing the world in the process. One of the most interesting and effective but also least known forms of learning is through the process called by the Author *morphological learning*. Such learning is particularly recommended when the objective is to develop an understanding of the rules behind conceptual designing, for example rules behind structural shaping; that is, the rules guiding a structural engineer through the process of establishing a configuration of a structural system (establishing locations of joints, assuming the nature of joints, selecting types of materials, making decisions about the locations of the individual joints, etc.). Such rules are rarely, if ever, revealed in an explicit form. Usually, they are carefully guarded by experienced engineers as their top secrets and as the core of their experience and the key to their competitive advantage. Therefore, young engineers very rarely have a chance to learn such rules from the other engineers. Fortunately, *morphological learning* gives all engineers, including young engineers, a unique opportunity to learn them directly from a cluster of design examples. This is a five-stage process, which is presented below.

6.6.4.1 Construction

The process of constructing a morphological table has been described earlier. In this case, the process should be conducted particularly carefully to make sure that all necessary symbolic attributes and their values have been included in the table. Elimination of redundant attributes, and those unnecessary for discerning between any two design concepts, is relatively easy. However, adding missing attributes later, when the process of knowledge acquisition has already begun, is very difficult and should be avoided at any cost.

6.6.4.2 Verification

Before a given table is used for acquiring knowledge, it must be verified considering its completeness and simplicity. Therefore, first we need to know that the table contains all known design concepts (the required completeness). Such concepts can simply be provided by experts or found in books, in design guidelines, and, most importantly, in patents. Second, the table should be prepared using only the minimum number of symbolic attributes and their values but a sufficient number to allow the user to discern between any two design concepts (required simplicity). Both objectives of the verification stage can be summarized as a heuristic: "Make sure that the table contains only the necessary and sufficient number of attributes and their values."

6.6.4.3 Identification of design concepts

This is the most interesting and challenging but also the most difficult stage. It has two objectives, both related to searching for design concepts *hidden in the table*. The first one is to find all known design concepts while the second one is to find all unknown ones.

In the first case, the search is relatively simple and begins with sketching the first best known design concept in a given domain and describing it by a specific and unique sequence of symbolic attributes combinations of and their values (formal identification), which is associated with this concept and identifies it. When this first concept is known, the second one is sketched and formally identified, and so on. Very soon, we have several sketches associated with the known design concepts and their formal identification. This simple process produces sketches and combinations of attributes and their values associated with these sketches. Most importantly, it allows us to develop some initial understanding of the entire class of known design concepts. We begin to see similarities and differences between various concepts, not only in terms of attributes but also in terms of their various engineering features. For example, two design concepts may be identical in terms of their configurations (and several symbolic attributes associated with the configurations) but one may use concrete and the other one steel; that is, in the first case the symbolic attribute "Material" will have value "Concrete," and in the second case this value will be "Steel."

The second objective (finding unknown design concepts) requires all our knowledge and creative/abductive talents and skills. When all the sketches of known design concepts are known, we need to study them again in order to find out what changes in values of their attributes can be made to create new concepts, which will be still feasible but different from the known concepts. For example, if the known concept of a local bracing is in the form of an X-bracing, we could consider a bracing in the form of two independent diagonals, which are not connected by a joint in the middle of the braced cell in our structure (Figure 6.10).

The search for unknown design concepts requires a lot of time and excellent working conditions. It results in a class of unknown combinations of symbolic attributes and their values. Some of them may be infeasible but usually many of them are feasible. When a knowledgeable and well-prepared

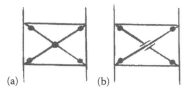

(a)   (b)

Figure 6.10 (a) X-bracing and (b) diagonal bracing.

inventive engineer does it, it may even result in patentable design concepts. In any case, these new design concepts, even if not patentable, will become a foundation for the next stage in our process of knowledge acquisition.

Instead of carrying out a *manual* search for unknown design concepts, this could be done using a random generation of various combinations of symbolic attributes and their values, as it is in morphological analysis. Based on the Author's experience, this approach is recommended, although there is another challenge. When this generation is done using a computer program, in no time thousands of combinations will be generated. Unfortunately, their feasibility analysis must be done by a human expert, and this is extremely time consuming and difficult (Arciszewski 1985).

6.6.4.4 Construction of a morphological tree

When a collection of sketches of various design concepts and the associated combinations of symbolic attributes and their values are both available, we assume that they constitute our *closed world*. The term means that this collection contains the entire body of knowledge about a specific domain represented by the identified design concepts. We transform the knowledge hidden in our design concepts (implicit knowledge) and into explicit knowledge, that is, into organized and transparent knowledge.

The construction of a morphological tree is a truly creative part of the entire process of knowledge acquisition. It takes time, but it is also fascinating to realize in the process how our understanding of a given domain gradually emerges.

First, we put each sketch and its formal identification on a separate sheet of paper. Next, we place all these papers on a large table. We look at them and compare them in various combinations. Our goal is to find a sketch representing in our collection the most basic, or the earliest, type of an engineering system within this collection. Such a type is called a *root type* and becomes the first piece of our morphological tree, the base of its main *trunk*. Obviously, it is placed on the bottom of our tree in its central part. Next, we are looking for similar types that differ from the root type only in one value of symbolical attributes. Such types will be in the morphological distance of 1 from the root type. These types are placed on the same level as the root type. Which two are placed next to the root type is a matter of engineering judgment and some experience in building morphological trees.

Table 6.5 Morphological table for diesel engines

Symbolic attribute	Symbolic values		
A_1 Number of turbochargers	0	1	2
A_2 Type of fuel	Regular	Low emission	

For example, let us consider learning about diesel engines. For clarity and simplicity we will use only two descriptors—two symbolic attributes. These attributes and their values are provided in Table 6.5.

We should bear in mind that experienced engineers always begin designing with the simplest design concept available and move to a more complicated one only when their initial concept fails. This strategy can be called "Keep it simple" or "Simple is beautiful." It is similar to the strategy we would use in our everyday life while renting a moving truck. We would never rent a 2.5-ton truck if a 1.5-ton truck were sufficient for our moving purposes, a truck that is definitely smaller, lighter, and most likely less complicated to operate than a 2.5-ton truck.

In the case of our example of design knowledge acquisition, the objective is to acquire knowledge in the form of decision rules guiding the selection of the next design concept to be considered when our present one (including the root type) fails.

In our example, the root type is the simplest type of a diesel engine, that is, one without any turbocharging and that uses regular diesel fuel. In this case, the value of the symbolic attribute A_1, "Number of Turbochargers," will be equal "0" ($A_1 = 0$) and the value of the symbolic attribute A_2, "Fuel Type," will be equal to "Regular" ($A_2 =$ Regular).

There are, however, types of diesel engines with one and two turbochargers, that is, with values of the symbolic attribute A_1, "Number of Turbochargers," being equal to 1 and 2, appropriately ($A_1 = 1$ and $A_1 = 2$). It makes a lot of sense to put both design concepts on the same side of the root type, for example on the right, with the design concept with a single turbocharger next to the root type (closer to the root type because it is simpler than the one with two turbochargers). Obviously, the design concept with two turbochargers will be placed further right. In this way, when we move from the root type to the right, first we will find a design concept with the value of A_1, "Number of turbochargers," being equal to 1 and next a design concept with the value of this attribute equal to 2 (see Figure 6.11).

Let us consider a diesel engine with a single turbocharger using low-emission diesel fuel. It will differ from the root type in the values of two symbolic attributes. First, the value of the attribute A_1, "Number of turbochargers," will be 1 and not 0 as for the root type. Next, the value of the attribute A_2, "Type of Fuel," will be "Low Emission" instead of "Regular" as for the root type. In this case, the morphological distance from the root type is 2. Therefore, this design concept should not be placed on the same

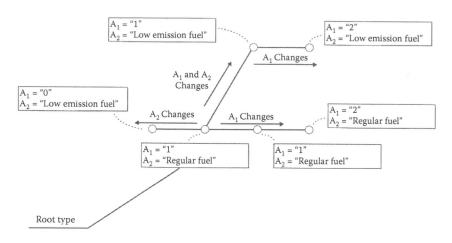

Figure 6.11 Morphological tree for diesel engines (grossly oversimplified).

level as design concepts that have a distance of only 1. We should place this design concept *above* the root type but on the trunk (See Figure 6.11.). Obviously, a design concept of an engine using "Low Emission Fuel" but with two turbochargers should be placed above the root type and on the right of the trunk, above the design concept situation on the bottom level with the value of A_1 "Number of turbochargers" being equal to 2. In this way, we have created some order, a reflection of our growing understanding of relationships existing among the considered design concepts. Also, we have just constructed a part of a morphological tree (Figure 6.11.).

For reasons of clarity and to avoid lengthy domain-specific explanations, we have used an example that may be oversimplified for mechanical engineers. In the case of real applications, we would definitely have a larger number of descriptors, and the process of morphological learning would lead to a much larger morphological tree. Real morphological trees are usually quite complicated and difficult to explain for engineers from outside their domains. Such detailed examples are available in Arciszewski (1985); Arciszewski and Uduma (1988), and in Arciszewski (1984). In the first case, a morphological tree is provided and discussed for patented joints in steel space structures. In the second case, such trees are developed and analyzed for various classes of wind bracings in skeleton structures of tall buildings. Both publications provide the depth of the domain-specific knowledge and general patterns of structural shaping that have much more universal nature.

6.6.4.5 Identification of decision rules

When our entire morphological tree is constructed, the final stage—acquiring formal knowledge in the form of decision rules—begins. We know changes in the values of symbolic attributes occurring horizontally on the individual levels of the tree. Also, we know such changes occurring vertically between levels along the trunk of our tree. Now, we need to determine decision rules describing, or governing, these changes. Sought-after decision rules will become the core of our understanding of how the individual design concepts are interrelated and, more importantly, how we can develop an entire class of design concepts beginning with our root type.

Let us reveal four decision rules that can be deduced from our decision tree. We could formulate such decision rules as

1. If $A_1 = 0$ and $A_2 =$ Regular Fuel are insufficient then make $A_1 = 1$.
   The engineering interpretation of this decision rule will be

   If the root type of a diesel engine is insufficient, then evolve the root type by adding to the engine a single turbocharger.

2. If $A_1 = 0$ and $A_2 =$ Regular Fuel are insufficient then make $A_2 =$ Low Emission Fuel.
   The engineering interpretation of this decision rule will be

   If the root type of a diesel engine is insufficient, then evolve the root type by adopting the engine to low emission fuel.

3. If $A_1 = 0$ and $A_1 = 1$ and $A_2 =$ Regular Fuel are insufficient, then make $A_1 = 2$.

   If the root type of a diesel engine and a concept of a diesel engine with a single turbocharger working on regular fuel are both insufficient, then evolve the root type by adding to the engine two turbochargers.

4. If $A_1 = 0$ and $A_1 = 1$ and $A_1 = 2$ and $A_2 =$ Regular Fuel are insufficient then make $A_1 = 1$ and $A_2 =$ Low Emission Fuel.

   If the root type of a diesel engine, a concept of a diesel engine with a single turbocharger, and a concept of a diesel engine with two turbochargers, all using regular fuel, are insufficient, then evolve the root type by adding to the engine a single turbocharger and modify it to work on low-emission fuel, which improves the efficiency of the engine (this may not be entirely true—the Author).

5. If $A_1 = 0$ and $A_1 = 1$ and $A2 =$ Regular Fuel and $A_2 =$ Low Emission Fuel are insufficient then make $A_1 = 2$.

   If the root type of a diesel engine, a concept of a diesel engine with a single turbocharger, and a concept of a diesel engine with two turbochargers, all using regular fuel, are insufficient, and a concept of a diesel engine using low-emission fuel without a turbocharger and with a single turbocharger are all insufficient, then evolve the root type by adding to the engine two turbochargers and modify it to work on low-emission fuel.

In the case of practical applications, our morphological trees will be much more complicated, and the decision rules describing them will reflect this complexity. Also, their number can be easily on the level of tens or even hundreds.

6.6.4.6 *Learning abstract thinking*

In the past, future engineers worked for years with their more experienced colleagues and gradually acquired factual knowledge and an intuitive understanding of engineering (such a system is called *master–disciple*). Their intuitive understanding was simply an abstract body of knowledge in the form of heuristics acquired through induction. Today's engineering students cannot study for 10 or 20 years, and their education is formal and strongly utilizes computing. Students also need to study the use and programming of computers, and that requires thinking in terms of mathematical models, programming languages, and both numerical and symbolic variables. Additionally, during the last several years the importance of Artificial Intelligence (AI), and particularly of knowledge-based systems and intelligent agents, has finally been recognized. Today, it is expected that engineering students will be prepared not only to use but also to develop various computing systems with roots in AI. However, learning how to understand and use symbolic attributes, which is crucial in this area, is usually at best a byproduct of learning mathematical modeling and simulation. This is a difficult situation that has a negative impact on progress in engineering.

In the described situation, learning about morphological analysis, and particularly practicing building various morphological tables, naturally allows inventive engineers to develop their ability to deal with symbolic attributes and their values. For many engineers, this ability may be crucial in their careers.

6.7 BLACK AND WHITE

6.7.1 White: Positive features of morphological analysis

- Fritz Zwicky was a modern Renaissance man who lived in several cultures, spoke several languages, and had many professional interests, both scientific and engineering. His life, and particularly his various contributions, are truly inspiring. For all these reasons, he should be considered as an outstanding example of an inventive engineer—an example to follow.

- It is not a coincidence that morphological analysis has been introduced as the first method in our book. This method combines rigors of traditional and deductive engineering thinking with simple "mechanistic" engineering creativity driven by a random generation of design concepts. Therefore, it constitutes a "bridge" closing a gap between engineering and heuristics, the science of creative problem solving.

- All students love the method. It is significantly different from the known to them analytical methods, but it is still possible for them to comprehend and use the method. The shock of inventive engineering is not too big, as it is sometimes with the other inventive methods.

- The method is quasi algorithmic (Arciszewski and Kisielnicka 1977); that is, at least we have a nondeterministic algorithm to follow. This algorithm is probabilistic, and following it only increases the probability of finding desired results but never guarantees it. However, the existence of this algorithm has a tremendous positive impact on engineers, who feel that the method is systematic and thus possible for engineers to use.

- The morphological table can be interpreted as a design knowledge representation space and can be also used for various AI applications. We could say that learning morphological analysis is a natural introduction to AI in engineering.

- The method is the first attempt to acquire and use knowledge in order to produce novel design concepts. When it is properly introduced, it provides a knowledge-based framework for engineering creativity. Therefore, it is an excellent introduction to other more difficult methods that are also knowledge based. Thus, it teaches inventive engineers how to approach inventive problems from the knowledge perspective.

6.7.2 Black: Negative features of morphological analysis

- When the method is well presented, it seems to be simple and easy to use. However, real engineering applications require a lot of effort and time commitment to create a *complete* morphological table, which is actually never complete. For example, when the Author worked on the morphological table for joints in steel space structures (Arciszewski 1984), it took him about five weeks of full-time work with the assistance of several domain experts. Also, when a given morphological table is developed and verified by experts, usually it needs modifications and improvements after some period of time. In fact, such changes should be expected and a morphological table should be maintained with regular updates over a long time period.

- No matter how well pegging has been done, we always feel that more could be done and much more knowledge should be acquired. It is associated with the fact that we do not have any strong and deterministic stopping mechanisms.
- When we use the method, the results are in the form of sequence of symbolic attributes and combinations of their values. They need to be transformed into sketches or verbal descriptions. This is not a trivial task in the case of complex domains, for example in the area of steel skeleton structures, although today it can be done by a computer (Murawski et al. 2000). Literally weeks may be necessary to analyze and visualize several hundred combinations, as the Author experienced when he worked with the method in Nigeria.
- The method is best used by truly interdisciplinary people who are able to look for knowledge in several domains. Also, they should be mentally willing to "atomize," or to divide, the acquired knowledge and to recombine its various pieces into a new quality—into new knowledge in the form of unknown design concepts. These are high requirements for the majority of engineers.

Chapter 7

Brainstorming

7.1 CREATOR

Alexander Faickney Osborn (Figure 7.1) was in many respects an unusual man. He could have easily qualified as a modern Renaissance man in the American world of advertising and business during the traditional 1940s and 1950s, when he was mostly active and made his everlasting impact.

Osborn was born in New York City in 1888 and spent most of his professional years in that city and in Buffalo, New York. In his day, both places were booming multicultural centers, encouraging people to understand and accept various cultures. In this way, these two environments contributed to the development of his rare ability to change context, which is essential to human creativity. This ability is critical to practicing brainstorming, but it is also absolutely necessary for users of Synectics (see Chapter 8 on Synectics). There was another key to Osborn's unusual success: his studies at Hamilton College in Clinton, New York. This is one of the oldest and finest colleges in the United States, offering excellent liberal arts education with a focus on interdisciplinary studies and the integration of knowledge. It also stresses the importance of writing and public speaking to a degree that is unusual among American colleges (Hamilton College website, visited on December 1, 2014).

After graduation, Osborn worked as a reporter in Buffalo for two local newspapers, the *Buffalo Times* and the *Buffalo Express*. There is a legend saying that when he was fired from the first one, he approached the editor of the second one. He was hired immediately because his clippings "showed ideas." Supposedly, Osborn suddenly realized the importance of ideas and decided to build his career, if not his life, around them. "If ideas are that valuable, why don't I try to turn out more of them?" is his famous saying, reflecting this fundamental philosophical change in his life.

After working as a newspaper reporter, Osborn worked as an administrator (assistant secretary) for the Buffalo Chamber of Commerce and as a sales manager for a manufacturing company. Only after all these jobs did Osborn begin working for an advertising company. It was a clear evolution of interests from the position of a passive observer to an executive but

Figure 7.1 Alexander Faickney Osborn. (With permission. Drawn by Joy E. Tartter.)

creatively passive position, and then to another executive position but this time with high expectations regarding initiative and action. In fact, this last position was most likely the key to his success in the advertising industry, since it forced, or at least encouraged, Osborn to think creatively and to develop new ideas that would improve the sales of his manufacturing company. In the process, he gained a priceless understanding or knowledge of how society operated on the level of individual people (as a reporter), on the level of local government (when he worked for the Buffalo Chamber of Commerce), and on the level of industry (when he was the sales manager for a manufacturing company). He acquired a large body of differentiated knowledge, which, when integrated, became his stepping stone to success in the advertising industry.

Osborn joined the marketing company Burton & Durstine in 1919 and worked for them until he retired in 1960. During these years, the company evolved to become one of major advertising companies in the United States, and Osborn played an important role in this success, serving in various executive capacities, including the position of the chairman of the board of directors.

Osborn was so much more than just a successful business executive. Early in his career, he began his studies of advertising and human creativity. His first book on the principles of advertising was published in 1921 (Osborn 1921), and his first books on creativity were published in 1952: *Wake Up Your Mind: 101 Ways to Develop Creativeness* (Osborn 1952a) and *Your Creative Power. How to Use Imagination* (Osborn 1952b). In fact, he published six books on the subject.

In addition to studying creativity and publishing books on the subject, Osborn was also concerned about creativity education in the United States and published articles and a book on *The Creative Education Movement* (Osborn 1962). Finally, he used to give talks to faculty and educators to promote his ideas. Moreover, in 1954 he established the Creative Education Foundation, supported by the royalties earned from his books. All these activities clearly demonstrate that Osborn, like Zwicky, was a renaissance person with many interests who also balanced his professional and social work and who made an impact in many ways. Creating brainstorming is only one of his achievements, although it is his best-known contribution.

7.2 HISTORY

We have already learned from the history of morphological analysis that inventive problem-solving methods do not simply emerge from an intellectual vacuum. They are usually a product of a long evolution of human thought, which at a specific time becomes articulated by a brilliant scholar and begins its new existence as a method.

In the case of brainstorming, its emergence may be explained best in the context of cognitive psychology and studies of human thinking styles. Sternberg (1997) distinguished three major thinking styles: legislative, judicial, and executive. The legislative style means thinking focused on the creation of new ideas, while the judicial style identifies thinking concentrated on the analysis and assessment of existing ideas. The first kind of thinking is mostly abductive, and the second one is mostly deductive. The third style, the executive style, describes thinking that is focused on the execution of a process leading to a predefined goal, and this kind of thinking is also mostly deductive.

From this psychological perspective, people can be roughly divided into legislative, judicial, and executive thinkers. Legislative thinkers are known for their creativity and ability to generate ideas but also as people who are simply not interested in a solid analysis of their ideas, not to mention their implementation. Judicial thinkers always criticize other people and their ideas but never have their own ideas. They are the first to criticize and they are masters of that. Finally, executive thinkers constantly look for something to do, for ideas to implement, and they are not concerned about the quality or novelty of ideas they are trying to implement. In fact, we have here three extreme examples of various thinkers. In real life, thinking styles are usually mixed with one dominant style.

For the last 11,352 years, legislative thinkers have been generating ideas and judicial thinkers have been criticizing them. This was a slow but effective mechanism that resulted in the gradual evolution of our civilization and its technological progress. In the twentieth century, as a result of the industrial revolution, our civilization entered the stage of accelerated progress that can be interpreted as the *growth period* on the S-curve (see Chapter 9). The demand for new ideas was rapidly growing and the traditional mechanism that was driving progress became grossly insufficient. Suddenly (in historical terms), a strong need emerged to develop a better way to generate new ideas. This need was particularly visible in the advertising industry, where the demand for new ideas was much stronger than in other industries. In the 1950s, after the Second World War, the United States was the only superpower. The country was undergoing a period of unprecedented growth and enthusiasm about the future, and the advertising industry was gaining respect while rapidly growing.

Osborn was the right person in the right place. He was a brilliant and a highly educated man working in a leadership position in the advertising industry. Since his early years he had an interest in studying and in professional writing. He had already published his first book in 1921 and knew the impact a single book could make. He also knew that his industry was on the brink of becoming a major industry, but there was a problem; there was not a single method that could be systematically used to produce ideas, the main product of the advertising industry. Therefore, he became a man on a mission to find such a method.

Surprisingly, in the early 1950s Osborn had already begun writing about the separation of legislative and judicial thinking in the same way as we talk today about right-hemisphere creative thinking versus left-hemisphere analytical thinking (see Section 3.2). He also claimed that delayed analytical thinking leading to judgment could result in improved ideas generated by creative thinking. "Deferment-of-judgment during ideative effort is crucial to keep the critical faculty from jamming the creative faculty" (Osborn 1962; Wheeler 2014) is his famous saying and the key to understanding brainstorming.

Osborn claims (1953) that in 1938 he began organizing group sessions to develop ideas. The name "brainstorming" spontaneously emerged as the best one for activities that could be described as "using the brain to storm a problem." There is no question that Osborn was the American pioneer in this area. We should bear in mind, however, that a similar practice was used several hundred years ago in India. It was known as Prai-Barshana and was used by Hindu teachers.

Brainstorming changed the advertising industry in the United States and it is still widely used. However, its impact has been much more important than that. The introduction of brainstorming created a revolution that brought human, emotional, and psychological perspectives to problem solving and to creative problem solving in particular. Also, brainstorming led to the development of Synectics, which is the most powerful heuristic method available (this is discussed in the next chapter). Most importantly, the introduction of brainstorming changed the public perception of human creativity in the context of problem solving. Before the brainstorming revolution, the ability to solve creative problems was considered a mysterious ability given by the universe to very few people, an ability that could not be learned or improved. After the brainstorming revolution, this ability, particularly in engineering, came to be considered as one of many abilities people can learn and master, although it is highly dependent on people's personalities and their attitudes to life and its challenges.

7.3 ASSUMPTIONS

The method has been based on the fundamental principle of *delayed judgment*, which may be related to the seventeenth-century Hindu teachers (see Section 7.2). This principle and 16 other more specific and operational assumptions create the methodological core of the method. The total number of assumptions, 17, is magic, and many people consider the method as a magical tool for solving all kinds of problems. Unfortunately, this is not true, but there is no question that brainstorming is a powerful method if used properly with a good understanding of its assumptions and procedure.

Osborn's original assumptions have been expanded, interpreted, and adapted for inventive designing by the Author as a result of many years of teaching inventive engineering. These assumptions are that

1. It is a heuristic method.
2. It is for group use.
3. It is intended only for the development of design concepts.
4. It is a closed-world approach; that is, the problem domain knowledge as well as any other relevant knowledge has been acquired by the group and it is exclusively used to produce design concepts.
5. A design concept of an engineering system (which may be actual or abstract) is described by a finite number of its features. Such description should be at least necessary and sufficient to identify all known concepts and to distinguish between them.
6. It involves delayed judgment.
7. It involves the complete separation of concept generation and evaluation.
8. "More is better."
9. No criticism is allowed.
10. Neither quality nor novelty matter.
11. Ridiculous, meaningless, and incomplete concepts are sought.
12. There is no individual ownership of design concepts.
13. All participants are equal.
14. The group composition must maximize its creativity.
15. Cooperation is a must.
16. Chain thinking is required.
17. The physical environment must relax and stimulate both individual and group creativity.

All assumptions must be understood in the context of the powerful, although hidden, goal of a brainstorming session:

> The ultimate goal is to stimulate creativity of the individual group members.

Johannes Müller (1970) introduced a simple criterion to distinguish between heuristic and deterministic methods. If an individual or a group using a method is considered as a system, then this system may be modeled as a black box transforming a problem into a solution. If the probability of this transformation is 1.0, we have a deterministic method that repeatedly produces the same results with the probability of 1.0. Obviously, this is the case with all analytical engineering methods for the analysis of stresses, deformations, and so on. When brainstorming is considered, the probability of the transformation will never be 1.0. We can only expect that a solution, or solutions, are produced, although each time the method is used a different solution may emerge. This characteristic of brainstorming is at the same time its greatest weakness and strength. It is a weakness because engineers always seek the certainty and predictability associated with deterministic methods. It is a strength because inventive engineers always seek novelty and always want to have a second chance if initial results are unsatisfactory.

Osborn developed brainstorming as a group method, that is, a method that is intended for use by several people working together. This assumption may be considered revolutionary. All earlier methods, for example morphological analysis, were intended for a single person. As we know (see previous section), early in his creativity studies Osborn discovered that a group is so much more than a collection of several individual people. There are at least two powerful explanations of this fact.

First, when several people begin interacting on both the personal and professional levels, they become a *system*, and a system always provides different and much more sophisticated results than its individual elements. Second, from the psychological point of view, when an isolated group of people begins working together, it activates a number of complex psychological mechanisms, all supporting human creativity. First, people begin, consciously but mostly unconsciously, to compete with the other group members. That leads to their emotional involvement, which activates their entire brains, leading to both deductive and abductive thinking. This in itself multiplies their ability to produce new ideas. Also, a complex combination of feedbacks on the person-to-person and person-to-group levels emerges. That leads to the emergence of synergy, a unique phenomenon in which interpersonal intellectual and creative stimulation takes place, producing the best group dynamics possible under given circumstances. A brainstorming group is like a symphonic orchestra—so much more than several violinists, trumpet players, and other musicians only playing together. Unfortunately, as in the case of a symphonic orchestra, the best and most unforgettable performance requires many preparations, a carefully arranged physical environment, and a "magic moment," that is, the right selection of people who at any given time are ready to work together and are at the peak of their mental abilities.

A brainstorming session may be also conducted by a single person, but this is a very rare ability and it is difficult to learn without exceptional talents. In this case, a gifted and knowledgeable person focuses on a problem, and after a moment of concentration, of entering the proper "mood," he or she begins producing a stream of interrelated ideas that gradually evolves as the process progresses. Such an instantaneous generation of ideas is called *kaleidoscopic thinking*. Edison is known as a master of kaleidoscopic thinking (Gelb and Miller Caldicott 2007), but very few people could be compared with him as far as the transdisciplinary understanding of engineering and creative talents are concerned.

Today, an individual may conduct a pseudo-brainstorming session working with a computer tool like IdeaFisher (a free download is available at ideafisher.soft112.com). This particular program uses a large database of interrelated terms or concepts and guides the user from the first concept entered by the user to all other concepts that are directly or indirectly related to the first one. It is a purely mechanistic approach, but it is sometimes surprisingly effective, as the Author discovered with his students several times. Unfortunately, many of today's students prefer to work alone with a computer tool instead of working with other humans. For such students, programs like IdeaFisher may be an alternative. However, there is no question that the best results will be produced by a group of humans, not by a computer program, which may only be a weak substitute for a group of people.

Assumption No. 3 claims that the method is exclusively intended for the development of design concepts, for only a single step in the process of inventive problem solving (e.g., morphological analysis is intended for the entire process). Brainstorming is a highly specialized method that is not intended for knowledge acquisition or for problem formulation. This assumption means that a brainstorming session must be carefully prepared in terms of bringing appropriate knowledge to the table through the proper selection of participants and consideration of their knowledge. Also, the problem must be formulated in such a way that it will be at the same time sufficiently precise and fuzzy. It must be precise in order to make sure that the right problem will be solved. It must be also fuzzy so as to stimulate creative thinking and not to overconstrain the solution space.

Assumptions No. 4 and 5 are interrelated. They are crucial for the ultimate success of a brainstorming session. "Judgment is delayed" is the mantra of the brainstorming community. It describes our mental state during a session when we know that only the generation of ideas counts and that our judgment will come much later during an entirely different stage of the inventive problem-solving process. Most likely, it will not be even our concern, since a different group of people will much later conduct the feasibility analysis or judgment of our ideas. Assumption No. 5 addresses the same issue but on a more operational level. It underlines the fact that we clearly distinguish between the generation and evaluation of design concepts and that these two entirely different activities are two completely independent processes, separated not only from the methodological point of view but also separated in time and possibly also done by different people.

"More is better" is an assumption usually considered counterintuitive by students. After taking many engineering courses, students wrongly believe that engineering is about finding a single solution, that is, the best solution in a given situation; and that finding this single solution should be achieved with minimum effort. Therefore, trying to find as many as possible solutions is not only unnecessary but it is also simply wasteful and counters the principles of best engineering practice. This common sense engineering belief could not be further from truth in the case of inventive engineering and of brainstorming in particular. There are at least two reasons why this belief is wrong in the case of brainstorming. The first one comes from knowledge engineering and the second one from psychology.

The development of design concepts is a process of knowledge acquisition, a process of acquiring knowledge about our future design concepts. It is obvious that when conducting knowledge acquisition, we want to learn as much as possible, that is, to acquire knowledge from all available sources. Similarly, in the case of conceptual designing, before we develop our final concepts, we want to learn from as many concepts as possible. Therefore, our goal should be to maximize their number (and differentiation) so as to learn as much as possible about the domain and to create a large body of knowledge in order to develop novel final concepts.

A brainstorming session should *not* be like a group of gloomy adults with stone faces who are sitting motionlessly and reading in steady and bored voices their ideas, two ideas per person, ideas that were all prepared individually long time before the session. A brainstorming session should be like a scene in a preschool learning center in a colorful room full of art and flowers, and with a group of happy children playing a game and learning new words in the process. The children are visibly excited, they walk around and talk, they are entirely mentally and intellectually engaged, they are in the mood, and they have already become part of the game. The number of new words is rapidly growing, and each new word is met with shouting and screaming and nearly immediately leads to more and more new words.

In the first case, the participants could be sitting in their offices in various towns and faxing ideas to each other. There is no cross-stimulation, no learning from others—simply a group of people slowly and mechanistically sharing ideas. The second case is entirely different. We have here a complex human system rapidly learning and using all kinds of stimulation to improve and accelerate the process. There are many obvious differences between these two extreme situations, but one is particularly striking. In the first case we have a very small and constrained number of ideas, while in the second their number is practically unlimited. Our conclusion should be simple. It is the flood of ideas that changes the dynamics of a group session and creates all the positive psychological effects that lead to a true explosion of human creativity.

When a session is in progress, "Criticism is ruled out," as Osborn (1953) wrote. This assumption can be explained from the perspectives of an individual participant and of the entire group. In the case of a single participant, we want to switch the dominant thinking from the left to the right hemisphere of the brain; that is, we want to fully activate the right, creative brain hemisphere, responsible for abductive thinking and generating creative ideas. At the same time, we want to deactivate the left hemisphere, which specializes in deductive thinking; it immediately quantitatively evaluates all ideas coming from the other hemisphere and in the process eliminates the majority, if not all, of new ideas.

When the entire group is considered, we want to temporarily transform its members, a group of responsible and sophisticated professionals, into people who act like small children, feel happy, and simply enjoy the moment. Moreover, they feel free to say whatever words come into their heads without taking any responsibility for their actions. Such a situation stimulates them to propose ideas that they would never even consider when somber—that is, under normal professional conditions. Moreover, the situation may fire them up to start kaleidoscopic thinking. It would not be a typical part of a brainstorming session, but it has happened to the Author several times and it is much more common than is believed.

Knowledge engineering provides another justification for the elimination of any criticism during brainstorming. A session is a learning process. Human learning is mostly inductive, that is, it is learning from both positive and negative examples. Both classes of examples are equally necessary to acquire knowledge, and none can be disregarded if we want to acquire knowledge. If criticism is allowed, it will immediately eliminate all negative examples: all ideas that are "crazy," infeasible, too complicated, too simple, or simply have "too many notes," as Austrian Kaiser France Joseph I once said about one of Mozart's compositions. As a result of such an approach, nothing will be learned; no new ideas will emerge.

Assumption No. 10 is "Neither quality nor novelty matter." This is a surprising assumption for engineers, who always believe that they are supposed to produce results that are at least of good quality and eventually even novel. The purpose of this assumption is to convince participants that any ideas are sought without any concern for their quality or novelty as perceived by the session's participants. It is like a college recruiter being paid by head, no matter what the qualifications of the recruited students.

The next assumption, No. 11, may be considered as an extension of the previous one. However, it is much more general and represents a fundamental departure from traditional engineering philosophy. The core of this philosophy is the use of rational or deductive thinking, which usually results in feasible solutions that are everything but ridiculous or meaningless and are usually complete.

If we consider conceptual designing as a search through the solution space, following traditional engineering philosophy means that only small regions of this space will be searched—regions that satisfy all imposed requirements and constraints as well as all known traditional heuristics. In effect, only a very small fraction of the entire space will be searched, and the real chance to find anything potentially interesting is truly miniscule. Eliminating constraints and requirements, not to mention traditional heuristics, radically changes the situation and opens the entire solution space for search. In this way, the probability of finding potentially interesting pieces of information is significantly increased. These pieces of information may initially look meaningless to a traditional eye but may stimulate the inventive engineer or direct him or her to novel solutions. A brainstorming session should be the equivalent of searching the entire solution space, and therefore seeking nontraditional solutions is very important.

If we consider a brainstorming session from the human point of view, it can be seen that feeling free from any constraints is not only liberating but also activates complex psychological mechanisms. Respectable and accomplished adults behave like happy playing children, begin talking "nonsense," and create chains of interrelated but meaningless ideas, and the phenomenon of emergence takes over. Suddenly, from the flood of ridiculous ideas, one or two interesting ideas emerge. They would never emerge without going through the phase of generating exclusively ridiculous, meaningless, or incomplete solutions. We might even say that these "useless" ideas were our key to inventions.

The assumption No. 12, "No individual ownership of design concepts," is critical in transforming a group of people into a team and in creating a team spirit. On the level of individual participants, their attitudes must change and that is difficult. Humans are possessive by nature, and sharing ideas with others is not usual and requires strong motivation and that we break existing habits. Even more, engineers are educated about the importance of intellectual property and about their ownership of the ideas that are developed, which may be patented, bringing fame and fortune. For all these reasons, all participants must be reeducated about the ownership of ideas in the context of a brainstorming session. Each participant needs to be convinced that cooperating with others, and particularly helping them, ultimately means serving his or her own interests. Doing so will contribute to the team's success, which potentially may be on a different order of magnitude in terms of novelty than would be possible in the case of a single person.

Again, we could come back to our example of children playing a game and learning together. Children are no as possessive as adults, and they simply enjoy the moment. Such behavior gives them much more than they would learn individually. Their behavior is entirely intuitive, but they subconsciously employ the optimal strategy to maximize their learning.

"All participants are equal" is our Assumption No. 13. It is also counterintuitive for engineers and needs to be explained. Engineers have deductive minds, and they always classify everything and everybody. They then use their classifications to determine hierarchies and use them in their activities. It is natural and usually serves them well. However, in the context of a brainstorming session all participants must be equal, like children playing together, and no emotional, not to mention formal, dependencies must take place. Particularly dangerous is the participation in a session of both a superior and his or her subordinates. Such a situation usually creates all kinds of problems and should be avoided at any cost. It is usually difficult to convince the boss that his or her participation will be counterproductive, but it must be done. Otherwise, it will freeze subordinates, practically destroying the session.

Assumption No. 14 is about the group composition and says that "The group composition maximizes its creativity." In the 1950s, Osborn recommended about 12 people as being the optimal size for a brainstorming group. He even claimed good results produced by a large group of more than 200 people (Osborn 1953). In any case, he recommended an odd number of participants so as to avoid situations when the group is suddenly divided into two opposing and equally sized factions and a confrontation emerges. Today, however, the best practice is to use much smaller groups of from five to nine people. The Author recommends seven people as the optimal group size. It is a compromise between knowledge and creativity requirements on one side and group manageability on the other side. In the case of a larger group, the interactions are impersonal; people do not even remember the names of other participants and may feel lost in a crowd. A larger group develops a complex dynamic and may become divided into competing or even fighting factions, and in general its internal inertia strongly affects its performance.

A group of several randomly selected people would never result in a productive brainstorming team. To create the true creative magic of a brainstorming session, we need the perfect blend of various professional backgrounds and abilities. The best explanation of this issue comes from the system's perspective. In this case, all participants must become not only a group but also a system that uses the available knowledge and transforms it into a class of creative solutions. Therefore, we need knowledge available within the system as well as various abilities necessary for the creative transformation of this knowledge into the desired solutions.

We have here a closed-world system, and the knowledge must come from the participants. Therefore, we need two or three participants who are very familiar with the problem domain. Interestingly, however, we do not need experts, for a most unusual reason: They would know simply too much and would be convinced that they know everything about the problem domain. As a result, they would dominate the session and, even worse, would be able to indirectly impose on others their frozen-in-time understanding of the problem domain. We need access to a sufficient body of knowledge but not much more. The professional background of the remaining four or five participants should be strongly differentiated and definitely not related to the problem domain. There is no magic formula, but two or three practicing artists (painters, artists specializing in sculpture, or musicians) and two or three other professionals (like economists, mathematicians, doctors, etc.) is a good guess.

The participants should be also selected for their thinking styles. The majority of participants should be legislative thinkers, because the development of ideas is most natural for them. However, executive thinkers might also develop ideas, although their ideas would be more pragmatic than those of legislative thinkers. Even judicial thinkers should be invited, but as a minority, and they should not be the dominant part of the team.

When brainstorming is regularly used in a given company, organizers tend to use the same tested people again and again. Such a policy is convenient for the administrators, but it may lead to making brainstorming another routine activity with people repeating their behaviors, arguments, questions, and so on. For this reason, it is a good practice to rotate participants, particularly guests, so that their input will not become predictable.

"Cooperation is a must" is our powerful assumption No. 15. There is no brainstorming without cooperation; it is absolutely critical for the success of a session, and it should be understood from several perspectives.

From the most fundamental and important systems perspective, cooperation changes a group into a system that acquires, integrates, and transforms knowledge into new ideas. There is a fundamental and qualitative difference between a group and a system in terms of its performance, particularly that related to creativity. Brainstorming requires the emergence of a system, and the decisive factor behind this phenomenon is cooperation. All participants must clearly understand that cooperation is not a matter of a personal decision but the fundamental requirement for a successful session. A decision to participate in a session is simply a decision to cooperate with the other participants. Unfortunately, even a single uncooperative participant may change the dynamics of the entire group and practically destroy any chances of success.

Our two previous assumptions, Nos. 12 and 13, have addressed the issues of the individual ownership of design concepts and of the equality of all the participants during a session. They are directly related to the "cooperation" assumption (No. 14) and can be seen as necessary although insufficient cooperation conditions.

From the human point of view, the missing link to achieving cooperation is the motivation of all participants. The most effective motivation must be both intellectual and emotional, with each component being equally important. In the first case, all participants must understand that cooperation is in their best interests if they want their session to succeed. The systems explanation provided above is not only rational but also powerful and is more than sufficient to intellectually motivate all engineers.

The emotional motivation is a little more complex since it has several interrelated parts. First, engineers usually tend to be introverts; that is, they prefer to work individually rather than in teams. They become team members only when there is intellectual motivation, as has been explained, and when they "feel" various benefits coming from becoming team members. These emotional benefits may include feeling being appreciated, having fun, going back to early childhood and to the happiness associated with it, feeling like they are contributing to making an impact, and so on. Second, a brainstorming session has its own dynamics, which gradually engages even the most passive or reluctant participants. As a result, participants become emotionally engaged and identify themselves with the group and its final success. Finally, even engineers have dreams about fame and fortune. It is possible to relate these dreams to the gradually growing momentum of a session when all the participants are becoming more and more excited, a good "energy" is flowing, and all are becoming very happy about the entire situation. That in turn increases even more the engagement of all the participants, both intellectual and emotional. Again, the entire complex psychological process, crucial for the final session's success, would not take place without cooperation.

Assumption No. 16 is about the required form of thinking, called *chain thinking*. In the context of brainstorming, this is a thinking process that involves several participants who become subsequently involved and propose ideas, or design concepts in our case, which are inspired or stimulated by the ideas presented earlier. Chain thinking is the essence of brainstorming and is absolutely critical for the success of a session. All participants must be aware of this fact and must actively engage in chain thinking. It requires some practice to master, but it comes naturally during a good session when a momentum is building and participants become truly emotionally involved. The process is described in more detail in Section 7.5, "Tool Box."

The last assumption, No. 17, is about the importance of the physical environment for the success of a session. Engineers are not known for their appreciation of art and beauty. Their working environments are usually simple and utilitarian, deprived of any paintings, sculptures, or even small pieces of art. The lights are bright, the wall colors gray, the furniture is functional but not necessarily esthetically appealing. Such uninspiring surroundings represent the bare minimum sufficient for doing traditional and routine engineering work. Unfortunately, they are grossly insufficient for a brainstorming session, which is creative work and requires an entirely different and a very unique physical environment with a sophisticated ambience.

It is not a matter of the personal comfort of the session's participants but a matter of the session's success or failure.

Brainstorming is about the development of new ideas. Therefore, it mostly depends on abductive thinking, which is mainly conducted by the right hemisphere of the human brain. This hemisphere is also responsible for feelings, emotions, and all kinds of sophisticated and qualitative reasoning that are definitely neither quantitative nor deductive. Routine engineering work is mostly deductive in nature and requires a very active left hemisphere, which becomes dominant and forces the right hemisphere to idle in order not to interfere with the rational and systematic activities of the left hemisphere. Often we hear experienced engineers saying to their young colleagues to forget about all their crazy ideas and simply do their work, that is, to shut down their right hemispheres and use only their left ones when they are on company time. In the case of brainstorming, we have the entirely opposite situation. We need to shut down our left hemispheres and activate our right ones when we are on company time. This is a fundamental shift, which is difficult, and the right ambience in terms of the surroundings makes this shift easier or even possible in many cases.

7.4 PROCEDURE

Osborn was a true pioneer. He proposed brainstorming as a method at the time when design science simply did not exist. He correctly assumed that creative problem solving may be seen as a three-stage process. This simple model reflects the state of the art in the early 1950s, but it was probably also a result of his underlying assumption that spontaneity is a must in brainstorming and that a more detailed procedure might hurt it. He distinguished the following stages:

1. Fact finding
2. Idea finding
3. Solution finding

These three stages may be described in casual language as "Preparations," "Production," and "Selection." There are described with some level of detail below.

The first stage, "Fact finding," may be described as a three-step process, including

1. Picking out
2. Pointing up
3. Preparation

As Osborn presented it, *picking out* may be understood as opportunity seeking in a given domain, looking for problems waiting for solutions and using your knowledge and curiosity in order to find actual problems, which have not been discovered yet by the other people but that constitute problems whose solutions may lead to new products. It is a very modern understanding of problem solving from the entrepreneurial perspective, which only recently became widely accepted as a part of innovation engineering. Today, we teach students how to look for opportunities, but nearly 70 years ago, including opportunity seeking as a part of problem solving was a major advancement in our thinking about problem solving.

Pointing up is formulating a problem. Osborn considered this step to be critically important for the final results and underlined his belief with Einstein's classical theory, saying that

> The formulation of a problem is far more often essential than its solution, which may be merely a matter of mathematical or experimental skill. (Osborn 1953)

Preparation may be understood as two fold. First, it is knowledge acquisition about the problem and the preparation of key facts for the session's participants. Second, it is the preparation of the participants and of the physical environment where the session will take place.

"Idea finding" is the second stage in the creative problem-solving stage. Within this stage, the ideas are produced and developed. First comes the *production of ideas*, and during this step the initial ideas or parts of ideas are produced. This is the most important and difficult part of brainstorming. When the products of this step become available, they become "input" to the next stage, called the *development of ideas*. During this step, the initial ideas are developed, that is, refined, changed, combined, simplified, or expanded; and a class of ideas emerges.

Solution finding is the last third stage in the process. The previous stage usually results in a large number of ideas, which have been developed without any concern about their quality and feasibility. During this previous stage, judgment was delayed, and this is the key aspect of brainstorming. Now, in the third stage, the judgment finally comes. It is time to use engineering, or quantitative means, to verify the feasibility of the produced ideas and to determine their novelty. Obviously, only a few ideas survive this stage, and they become the final products of brainstorming.

7.5 TOOL BOX

7.5.1 Introduction

On the surface, brainstorming looks from the outside like an entirely unstructured process with individuals randomly generating various ideas. Up to a certain point, such a picture is correct, but it is much more complicated. In the case of a well-prepared session, its participants have learned earlier all kinds of thinking and various mental operations that could be effectively used during a session. The big picture still shows chaos, but behind this chaos are many operations conducted in a disciplined and thoughtful manner, which are expected to produce the desired results in the form of creative ideas or links to such ideas. The method is heuristic and the entire process is probabilistic, with no certainty that desired results will be produced. However, brainstorming usually works, and learning how to produce ideas will definitely increase the probability that a session will be successful. Below we have an overview of the methodological knowledge that it is most important to know in order to become an effective participant in a brainstorming session.

7.5.2 Chain thinking

Osborn (1953) has identified chain thinking as the major mechanism driving idea production. It is often called *ping-ponging* or *hike jacking*. The concept of chain thinking has already been introduced in Section 7.3, Assumptions, and is described as

> a thinking process that involves several participants who become subsequently involved and propose ideas, or design concepts in our case, which are inspired, or stimulated by the ideas presented earlier.

The process begins when one participant proposes an idea that is immediately transformed by another participant and leads to the next idea; that idea consequently inspires another participant to present his or her idea, and so on. This is a complex process in which participants must feel comfortable feeding on other people's ideas and using all kinds of associations to develop their ideas, which in fact belong to the group, not to the individual participants.

Chain thinking produces a *chain of ideas*, or design concepts in our case. They are interrelated or associated, and they result from associations taking place during a session. The initial idea may be randomly generated or proposed by any session participant, and it leads to the next one, and so on. The first idea starting the entire process may be ridiculous and may be seen by those participants with knowledge of the problem domain as ridiculous and entirely useless. It is *not* a bad beginning, because such a crazy idea has so much more stimulation power than useful and feasible ideas, which lead to similarly useful and feasible ideas, and the chain of ideas soon ends without producing any truly novel ideas. Also, we should understand that the participants of a successful session ought to search for ideas in a solution space that is as large as possible. Therefore, beginning this process far away from known and feasible ideas will only improve the probability that truly novel ideas will be found.

As an example, let us consider a session of brainstorming. The group is looking for creative ideas to reduce the noise in a tall office building. Traditional noise insulation is heavy and takes away priceless space within the building. The first participant, Jack, proposes, "Let use a big heavy ax to kill the noise"; everybody laughs, and Janet says, "Let us kill the noise gently, without any axes." Tomek says, "Let us kill the noise with a counter noise"; everybody laughs again, and Jack says, "We could use a white noise that will be matched to the original noise for each room separately." Finally, everybody becomes excited and starts talking about a system of microphones and speakers that would monitor the noise in all rooms and adjust the level of white noise produced by speakers in individual rooms depending on the need. Mark recalls that similar systems already exist in various luxury cars where they reduce the noise and play music. Immediately, the next idea is born: Use the public address and sound system in the building to produce white noise wherever it is necessary. Everybody is very happy because they do not know about any tall office building in the world that has an active acoustic noise-suppression system. They all feel excited about finding an idea that may change the way the noise is handled in tall buildings and particularly that the idea may result in increased space utilization and reduced costs, and, most importantly, may improve working conditions in tall buildings.

7.5.3 Associations

Osborn (1953) claims that the ancient Greeks were first to recognize the importance of associations in human creativity. They introduced three classes of association: *contiguity, similarity*, and *contrast*. The first class, contiguity, contains cases of associations in which both ideas or design concepts are close. For example,

- a "large span heavy steel truss" is associated with a "railway bridge," or
- a "three-bay heavy concrete frame" is associated with a "parking structure."

In the case of the similarity class, ideas are somehow similar. For example,

- a "rigid frame in a two-story office building" reminds us of or is associated with a "100-story rigid frame in the Empire State Building in New York, the United States."

Finally, the contrast class contains associations in which ideas represent opposite ends of spectrum or are strongly contrasting and for this reason often come together. For example, a "hollow box column" reminds us of or is associated with a "solid column," or

a "high strength steel" is associated with "iron," or the "construction" of a building is associated with its "collapse."

All associations are intended to initiate or expand chains of thought, sequences of ideas, or lines of evolution of ideas (using the terminology of the theory of evolution of engineering systems [Zlotin and Zusman 2006]). They are like catalysts in the process of chain thinking.

There are magic words activating all kinds of associations: "That reminds me ...." These words should be memorized by all participants and used whenever a need for a new idea becomes clear. Even when we do not know what to say, using these magic words will bring up associations that will surprise us. This will be a result of our subconscious thinking, which is always working; we need only to get access to its results. Obviously, using our magic words is one of the ways to do this.

The other magic words are "How about this ...?" We can also use these words when we have no idea what to say, and they will usually bring up all kinds of interesting, if not creative, ideas. The best way to learn how to use our magic words is to play with children. The winning team produces the longest chain of ideas; their feasibility and novelty do not matter, only their number. We have already learned that and why "more is better."

7.6 SEVEN SIMPLE ACTIVITIES

7.6.1 No. 1: Analyzing the problem

Analysis is a process leading to an improved understanding of the problem. It can be conducted at any time during the entire brainstorming session whenever questions about the nature of the problem emerge. However, in this case, the goal of analysis is limited: It is not to develop a full understanding of the problem but one that is only sufficient for the progress of brainstorming. Osborn believed that having too many previous examples, or simply knowing too much, is harmful for human creativity. When many previous examples are known, they may attract an inventive engineer who will try to adapt them instead of developing anything new. Therefore only a few fundamental facts should be acquired, and those facts should be inherent in our problem. Eventually, a few facts having some bearing on the problem should be also acquired.

Let us assume that we are looking for new ideas regarding power generators with diesel engines, and all our examples will be related to this problem. The most important magic question is why? This should be asked in various forms and many times until sufficient, but only sufficient answers come. In the case of our example, we could formulate such questions as

Why do we really need this device?
Why do we need this type of energy?

We could also use a little more inspiring questions like "why so?" and "what if?" In the first case, we could formulate a question like

Why is it that only diesel fuel must be used?

In the second case, such a question could be formulated:

What will happen if we replace diesel fuel with vegetable oil?

7.6.2 No. 2: Targeting

This activity is most effective at the beginning of a session, but it can be used later, when we realize that the ideas coming are too vague and we need more specific or better targeted ideas. Usually, the problems given to solve are too complex to understand and solve during a short session. Targeting helps to improve our problem understanding and to focus on its most important aspects.

As we have a chain of thoughts, in brainstorming we also have a chain of predefined *basic questions*, which are interrelated and lead to a better problem definition. We have such questions as

1. *Why* is it necessary?
2. *Where* should it be done?
3. *When* should it be done?
4. *Who* should do it?
5. *What* should be done?
6. *How* should it be done?

The key words are "why," "where," "when," "who," "what," and "how." They should be provided to all participants before a session with a heuristic

Use the six basic questions when losing focus or looking for a better target.

The basic questions bring an element of systematic analysis to brainstorming and usually help to refine the search for creative ideas. Also, when our target is already specified, we may try to convert it into questions we want to answer during our session.

7.6.3 No. 3: Breaking down the problem

The concept of division is fundamental for understanding morphological analysis (Chapter 6), but it is also associated with brainstorming. It is recommended by Osborn as one of many activities that are potentially useful early in the brainstorming process for developing a better understanding of the problem. It often happens that instead of working on the entire problem, a group focuses on a subproblem and produces worthwhile ideas that later become the key to solving the entire problem.

7.6.4 No. 4: Asking stimulating questions

Osborn believed in the extraordinary power of asking questions during a brainstorming session. It was a revolutionary idea more than half a century ago, but today many cognitive psychologists agree with Osborn that asking questions stimulates the human brain and allows attention to be shifted in a new direction. Not all questions are equal; the most effective questions are general, if not vague, because they activate the entire brain. Simple technical questions can be easily answered using only deductive reasoning—that is, using the left brain hemisphere—and they do not help us in creative thinking. On the other hand, imprecise questions cannot be answered in a precise way by the left hemisphere only; they engage the entire brain, including its right hemisphere, which immediately brings into action emotions and abductive thinking and thus initiates creative thinking.

Osborn (1953) provides tens of stimulating questions that can be used in various situations during a session. Many of these questions are provided in the following sections of this chapter. They can be used directly by inventive engineers or modified by them to better match specific engineering domains. It is a good practice to put all these questions on a single sheet of paper. They do not need to be memorized, but studying these questions before a session usually helps inventive engineers to play an active role during a session and to surprise other participants by their unusual ability to analyze the problem and to lead the process of solving it.

7.6.5 No. 5: Creating combinations

We have already learned about morphological analysis and about combining subsolutions, or subconcepts, as its major mechanism for generating design concepts. In this case, the creation of various combinations is random and is conducted in a purely mechanistic way without any human involvement.

Brainstorming may use the same operation, but in this case humans—the participants—produce combinations. We have yet another appropriate magic phrase related to this operation: What if we combine ideas X and Y? Both ideas may come from the same person or from two different participants.

For example, if a group were brainstorming ideas for new structural material ideas, one would say, What if we combine concrete and bamboo bars? The answer would be, "We will create reinforced concrete with reinforcement in the form of bamboo bars." At first sight, this combination seems absurd, but the Author remembers an experimental study conducted at the University of Zaria in Northern Nigeria in the 1970s. It was focused on the feasibility of using bamboo bars as reinforcement in temporary concrete structures. At that time steel, including r-bars, was extremely expensive in Nigeria, while locally available bamboo was practically free, and using bamboo as reinforcement seemed to be an idea worth at least exploring. Unfortunately, the Author does not know the results of this study.

Another good example of combining ideas comes from the area of mechanical engineering and is related to designing high-performance brakes in sports cars. In the 1970s, the idea of disc brakes was introduced, and about 20 years later, the idea of high-strength ceramic products was introduced. When these two ideas were combined, the new idea of ceramic disc brakes emerged. Today, nearly all high-end sports cars, like Ferraris or racing models of Porsche, have ceramic disc brakes, which have many advantages with respect to the regular disc brakes, although price is not one of them.

7.6.6 No. 6: Adapting, modifying, and substituting solutions

We can distinguish two types of adaptation: functional and conceptual. Both are important and both can be used during a brainstorming session. The term *adaptation* comes from biology, where it describes physical and behavioral changes animals undergo as a result of environmental changes as they are trying to fit into their changing environment.

In the first case, we consider the fact that each engineering system has been developed to provide a specific function. However, many, if not all, engineering systems can be used for other than their originally intended purposes. Changing a system's function is called *functional adaptation*, that is, the adaptation or use of an existing system to provide a new function.

For example, farmers use hammers with long handles and heavy heads in the form of two perpendicular wedges to split logs, and such hammers are appropriately called *log splitters*. However, a log splitter could be also used as a powerful weapon when a farmer is attacked in the woods by a charging bear. He uses his hammer as a defensive weapon and changes its original function, adapting it to the changing circumstances.

We can easily produce many questions to stimulate functional adaptation. For example, we are looking for various applications of a new material, which has been originally developed for structural purposes. We could formulate such questions as

Could we use it in medicine to create artificial bones?
Could we use it as a sound insulation?
Could we use it as a shock absorber?

Could we use it in building a rocket?
Could we use it in building a car?
Could we use it in building a boat?

Conceptual adaptation is an adaptation of ideas. It is a well-known practice in literature. For example, a beautiful love story takes place in Greece in the 1930s. Another writer uses the same plot or adapts this plot for his novel. Names will be different, the location will be changed from Greece to Portugal, and the time period will be different—not the 1930s but the 1940s—but the plot will be the same. This often happens in engineering. After the first tall buildings with steel skeleton structures in the form of rigid frames were constructed in Chicago in the nineteenth century, the tall buildings explosion began. Over the last century, literally thousands of tall buildings with steel rigid frames were constructed in many countries; in each case, the idea of a three-bay rigid steel frame was adapted to the local conditions and successfully used.

Asking appropriate questions can also stimulate conceptual adaptation. For example, we are working on the development of a new type of a vacuum cleaner. One of the participants has brought to our brainstorming session a recently developed Roomba 880 vacuum cleaner, which is a robotic device and is considered to be one of the most advanced vacuum cleaners on the market. Several US patents protect its design, but we still may adapt some of its ideas without violating US patent laws. Here we have examples of questions:

What laws of nature have been used to develop this design?
Which law of nature is the most important?
How have these laws of nature been used?
Is there any other way to use the same laws of nature?

Adaptation of ideas is a powerful activity during a brainstorming session and should proceed unrestricted by any concerns about the intellectual property of the creators of ideas we would like to adapt for our project (no judgment during the session). However, we have to be aware that the issue of intellectual property must be addressed later, when our ideas are evaluated not only from the perspective of their feasibility and novelty but also taking into consideration patent law. Inventive engineers must be particularly sensitive to the issue of intellectual property, not only for ethical and legal reasons but also because of the simple fact that they are knowledge workers; knowledge, and especially ideas, constitute the core of their professional life and must be protected at any cost, and this applies not only to their ideas but also the ideas of other people.

Modification is a similar activity, but in this case we not only adapt an idea, we modify it; that is, we change some of its components to create our own idea. Modification may be explained best using our formal definition of a design concept as a unique sequence of symbolic attributes and a combination of their values. Let us use a simple example of a design concept (an idea) of the steel Howe truss, which is widely used in roof structures (Figure 7.2a). Attributes describing such a truss are provided in Table 7.1. The combination of these attributes and their values defining the Howe truss is

$$A_1 = \text{Steel}, A_2 = \text{Triangle}, A_3 = \text{Hinges}, A_4 = 3, A_5 = 2, \text{ and } A_6 = 4$$

We can easily modify the concept of the Howe truss and develop the concept of the double Howe truss simply changing the values of attributes A_4, A_5, and A_6. After this modification, we have a combination describing the double Howe truss, as shown in Figure 7.2b:

$$A_1 = \text{Steel}, A_2 = \text{Triangle}, A_3 = \text{Hinges}, A_4 = 5, A_5 = 4, \text{ and } A_6 = 6$$

Osborn has recommended various universal questions that should help in the process of modifying existing ideas. For example,

What if this were somewhat changed?
How can this be altered for the better?
How about a new twist?
What change can we make in the process?
How about changing the shape?

All these questions could be easily adapted for engineering applications and modified for our specific problem. We could add questions about

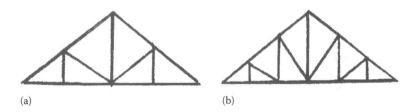

(a) (b)

Figure 7.2 (a) Howe roof truss and (b) double Howe roof truss.

Table 7.1 Howe and double Howe roof trusses

| | | Attribute Values | |
Attribute No.	Attribute	1	2
A_1	Material	Steel	
A_2	General shape	Triangle	
A_3	Joints	Hinges	
A_4	Number of vertical members	3	5
A_5	Number of diagonals	2	4
A_6	Number of bottom chord members	4	6

changing the materials, questions about changing sequence of operations, and so on.

Modifying ideas developed by others is a well-known engineering practice. However, when we come to evaluate our final ideas, we should be always aware of the intellectual property issue. Even if changing one or two features of a patented concept allows us to avoid any legal responsibility, not to mention paying any licensing fees, we should at least recognize the fact that we have just benefited from somebody's else idea and that that person should be somehow "rewarded" for his or her contribution to our success. A nice letter, a call, or simply mentioning the source of our inspiration in the description of our product would be at least what is expected from a conscientious inventive engineer.

Substitution is another important way to develop new ideas from existing ones. In this case, we simply change the values of one or more attributes in such a way that a concept of a different engineering system suddenly emerges. In a more casual engineering language, substitution is a process of developing a new engineering system from an existing one by replacing its elements without changing the function of the entire system.

Osborn (1953) calls substitution "The Technique of This-for-That" and recommends several universal questions to help us do substitution during a brainstorming session:

What can I substitute?
What else instead?
What other ingredient?
What other process?
What other power might work better?

For example, when leg power was substituted by electric power in sewing machines, it led to revolutionary changes in the garment industry.

7.6.7 No. 7: Using mathematical operations

Addition and multiplication can be jointly called *magnification*. Both operators are surprisingly effective in the development of new ideas from existing ones.

In the case of engineering brainstorming, addition can be described as a magnification operator in which two or more design concepts are simply added. For example, the concept of a motorbike was developed by adding the concept of a bike to the concept of a motor. More specifically, in the 1860s, Pierre Michaux, a Frenchman, established "Michaux et Cie" ("Michaux and company"), the first company to construct bicycles with pedals, which at that time were called velocipedes. In 1867, his son Ernest added a small steam engine to a velocipede and created the first motorbike or motorcycle (Wikipedia, retrieved on January 3, 2015).

Multiplication operator means using the same design concept many times by simply multiplying it. The concept of a wheel is well known and it has led to the development of a class of vehicles based on various numbers of wheels used in a single vehicle. A single wheel can be used, and such a vehicle is called a single-wheel bike; two wheels are used in a regular bike; and three wheels are used in a three wheeler (Figure 7.3). We have even vehicles with 18 wheels, which are appropriately called 18-wheelers.

Maximizing is a form of magnification in which we purposefully consider extreme cases of using components of huge size or using a very large number of components in the system under development. For example, we are working on a new concept for a short-span bridge. Our initial idea is a truss bridge with two parallel chords (see Figure 7.4a). A participant in our session proposes to increase by the factor of 100 the depth of the top chord, and a tall, simply supported beam (Figure 7.4b) is created as a new concept for our bridge.

In the case of the second form of maximizing, that is, using a huge number of components, as an example we may consider a group working on a

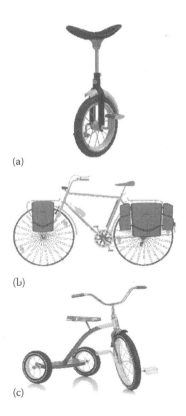

(a)

(b)

(c)

Figure 7.3 Multiplication in action: (a) single-wheel bike, (b) regular bike, and (c) triple-wheel bike.

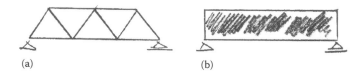

Figure 7.4 Size maximization: (a) truss bridge, (b) beam bridge.

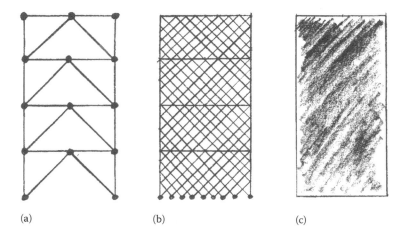

Figure 7.5 Number of components maximization: (a) truss bracing, 24 members; (b) truss bracing, 600 members; and (c) shear wall.

new concept of a wind bracing in a five-story skeleton structure. The first idea is a truss bracing with 24 members (Figure 7.5a), but a participant suggests using 600 members (Figure 7.5b), and this idea immediately leads to the concept of a shear wall (Figure 7.5c).

The process of magnification may be stimulated by using various universal questions (Osborn 1953), which can be read from our secret list of magic questions during a session of brainstorming. Examples of such questions are provided below; obviously they can be adapted for engineering applications and make much more specific.

What to add?
Should it be stronger?
Should it be bigger?
How about more time?
How can we add more value?
How about maximizing the size?
How about doubling the system?
How about using a jumbo size?
How about using a swarm?

Magnification can be described as doing "more-so"; obviously opposite operations of doing "less-so" can be also used. Osborn (1953) called such operations "minifying," but the more appropriate term "minimizing," taken from engineering optimization, should be used today. There are at least three basic categories of minimizing, including reduction, elimination, and division.

Reduction is widely used in engineering and is focused on reducing dimensions, time, cost, noise, and so on. The most important is the reduction of dimensions, often called *miniaturization*, which can be observed in all areas of engineering. Car engines are getting smaller, TV sets thinner, and so on. In fact, according to the theory of the evolution of engineering systems (see Chapter 9), engineering systems evolve by moving from macro to micro level, that is, they evolve in the direction of miniaturization.

Elimination is also known in engineering, mostly in the context of designing simplified systems with eliminated redundant subsystems or subsystems providing noncritical functions. For example, recently Volkswagen introduced in the United States a new model of Passat that is less expensive than its predecessor but does not have all the features of the earlier model.

Finally, division can be understood as a process of minimizing by dividing a system along functional lines into two or more independent systems, each providing a separate function, previously a subfunction of the original complex system. Division is often used when portability is important or when the separation of subsystems may lead to their easier operations or less costly maintenance by specialized crews.

Osborn (1953) has provided many stimulating questions that can be used to stimulate the magnification or minimizing processes, several of which are provided here:

What if it were smaller?
How about miniatures?
How about condensing?
What if this were lower?
What could I omit?
What can we eliminate?
Why not fewer parts?
How about dividing?
How about separating this from that?

7.7 SEVEN HEURISTICS

7.7.1 Introduction

When Osborn studied human creativity, he was not only interested in the development of a universal method of creative problem solving; he was also searching for all physical and social factors contributing to increased human creativity as well as for heuristics that might help in this area. In this section, we have an overview of seven such powerful heuristics, which are important for inventive engineers, not only when using brainstorming but also for their "everyday creativity." In many cases, following only one or two of these heuristics may not be enough, but their various combinations should definitely work.

7.7.2 No. I: Work harder

Before Osborn, an act of human creativity was considered an emerging phenomenon that could not be predicted, stimulated, organized, or planned. Osborn fundamentally changed the situation. He "proved" that it is a "creative work," that is, it is a process requiring time and effort as do all other kinds of work. Yes, talent and knowledge matter, but they are usually insufficient without a lot of effort—without a great deal of systematic work. The final idea may emerge in a moment of enlightenment, but it is usually preceded by days or even by months or years of studies, experiments, and continuous thinking about the problem. For all these reasons, inventive engineers should understand well that effort matters and should always be a part of their creative work.

7.7.3 No. 2: Get motivated

We usually use only a small fraction of our available brainpower. It is generally sufficient for doing routine work. However, generating new ideas is much more difficult than following repetitive and simple instructions. It is so much more than deductive reasoning. It requires activation of the entire brain: both its hemispheres, including the right one, which is responsible for abduction and for creative thinking but also for human emotions. For this reason, the best way to activate the entire brain, the key to inventions, is to become motivated.

All inventors are motivated by the fame and fortune waiting for them, but this is usually insufficient. Motivation should be more specific and should relate their inventive challenges to various emotional issues like winning a war or protecting people from the Ebola virus. For example, Nigeria is not particularly known as a source of inventions. However, during the Nigerian Civil War (1967–1970), Igbo engineers invented various weapons, including "flying mines," a.k.a. *Ogbunigwe* (one that kills a multitude), which significantly delayed the progress of the federal army and the fall of Biafra. Flying mines were land mines modified in such a way that they could be placed in the air about the narrow roads in the crowns of leafy tropical trees that stand on both sides of the roads in southern Nigeria. Next, they were activated when a column of advancing federal troops entered the killing zone.

The development of Ogbunigwe was not simple. Its evolution began with regular land mines, but it required using several creative operations to come to the final idea. First, the operation of *multiplication* was used to move from a single mine to many mines. Second, the operation of *division in time* was used to change the usual immediate reaction of a mine to a delayed reaction. Third, the operation *division in space* was used to move ignition from the location of a mine to the location of a soldier controlling Ogbunigwe. Next, the analogy *like a bird* (see Chapter 8, "Synectics") was used to convert a land mine into a flying object located in the air.

Ogbunigwe was a powerful weapon, developed by a group of engineering professors working at the University of Nigeria, where the Author also worked in the late 1970s and early 1980s. It is a strong example of the power of motivation but also of the creativity of the Igbo people, who were capable of creating such a sophisticated invention.

7.7.4 No. 3: Make notes

Leonardo Da Vinci, Thomas Edison, Marie Curie-Sklodowska, and many other scholars of global reputation have been known for always walking around with their notebooks and writing down all their observations. Such a practice not only releases the brain from remembering all mundane facts so it can better use its memory for reasoning; it also serves two important interrelated purposes. First, writing down thoughts and facts means restructuring and reformulating the pieces of knowledge being acquired, and that improves their integration and ultimately their understanding. Second, the process of writing down our thoughts activates subconscious thinking, the continuation of our conscious thinking, which is so critical for human creativity.

7.7.5 No. 4: Plan to invent

"Inventing is a spontaneous activity, new ideas suddenly emerge and we obviously cannot plan them." This is a traditional perception of inventing that is fortunately inconsistent with the state of the art in cognitive psychology. We need to reject this perception through changing our frame of reference.

If we assume that inventing is a kind of work, it obviously can be planned. Moreover, planning it increases the probability that inventions will be developed. In fact, it is an excellent idea to plan on a regular basis a period of time to work exclusively on inventions. It may be an hour or two a day, or 2–3 hours on Mondays, Wednesdays, and Fridays. When we have in our daily planner specific days and times scheduled exclusively for inventing, it matters. Our life changes and inventing becomes a part of our life in the same way that various routine activities are. This is a profound change with a huge impact on our ability to develop inventions.

During your "inventive hours," only inventing exists; all other activities are strictly forbidden. It is only you, your inventive problem, and 2–3 hours of time. Even your phone is disconnected, email does not exist, and you are exclusively focused on your inventive problem. At the beginning, you feel uncomfortable with this new situation and you do not know what to do with your new "free" time, but gradually it becomes a part of your professional routine, and you discover that this time is priceless for learning and for gradually developing an improved understanding of the inventive problem. Later, various ideas start coming, and very soon you are working on their development, it came so easily.

The described implementation of the heuristic "plan to invent" is not a part of brainstorming, but it is definitely a step in a long process of preparations for a successful session.

7.7.6 No. 5: Get into the "inventive mood"

There are few exceptional people who simply close the door, turn off their iPhones, and start kaleidoscopic thinking, producing a long stream of ideas. Unfortunately, there are very few people like that. All other people need to work hard to get into the "inventive mood," that is, to be ready for inventing.

There is no single prescription for getting into the inventive mood; it is a highly individual matter. For some people, a day in the woods, a good workout, or splitting logs for an hour or two are enough. Other people need several hours of silence, or music, or good food, or watching a comedy. In practical terms, each inventive engineer needs to determine for himself or herself what really works.

The most universal advice on how to create the inventive mood, but also the most difficult, comes from a new science called the "Science of Well-Being" (Dieter 2009). This is a branch of cognitive psychology focused on human well-being. The notion of well-being is difficult to explain, but it may be best understood as a feeling: a balance between dreams and reality, a feeling of deep happiness and mindfulness. One recent finding is that the perception of well-being has a positive impact on human creativity and most likely is even a requirement for any creative activities. This perception may be temporary or lifelong, with the first kind being much easier to achieve and already sufficient to satisfy a precondition for human creativity. From this perspective, the perception of well-being has to be created first, before inventing begins. How this can be accomplished is still an open question, but initial research results strongly suggest that regular meditation and practicing yoga are very effective.

7.7.7 No. 6: Know only enough

We always want to learn as much as possible. Surprisingly, in the case of brainstorming, this is not the right attitude. In fact, we must be aware that learning "too much" creates a powerful psychological mechanism that will stifle our creativity.

Our "excessive" knowledge will allow us to criticize all new ideas; we will see problems with all new ideas and, what is even worse, we will believe that we are right. From this perspective, we should acquire only major facts, understand only the fundamental aspects of our problem, and leave all the details to the experts, who will later thoroughly criticize our new ideas and will find all real and imaginative problems with them.

Usually, inventors have 5–10 years of professional experience. Less experienced engineers simply do not have enough understanding of their domains to develop inventions. On the other hand, more experienced engineers know too much, and that creates a powerful fixation on the existing solutions and does not allow them to explore and develop new solutions.

7.7.8 No. 7: Establish "creative space"

This heuristic is directly related to the Heuristic No. 5, since an inventive space will directly contribute to building a temporary perception of well-being.

It was Leonardo da Vinci who formulated one of his rules for creative people, *sensazione*. This rule deals with the importance of all senses for human creativity. Da Vinci considered synesthesia (see Chapter 2) as the highest form of mood that creative people may achieve, and he believed that this mood requires the activation of all the senses. Therefore, the creative space must positively affect all the senses.

According to Da Vinci, the most important sense for artists is vision. Therefore, the creative space should be painted blue (which stimulates creativity) or in other warm and inviting colors. Obviously, it should also provide all kinds of stimulating art, including paintings, watercolors, graphics, or even computer-based visual art. The Author has discovered that sculptures, particularly in soft and warm wood, always attract attention and positive reactions. In general, the art should be abstract but not aggressive or evoking feelings of stress, depression, or fear. Lights should be appropriately selected to generate a nice feeling of natural light but at the same time should not be too bright.

The second most important sense for artists is hearing. Therefore, in the United States or in Europe, light spiritual music or classical piano music played in a discreet way from a good sound system may not be even noticed but it will be definitely felt and will contribute to building the perception of well-being. There is not one single kind of music that should be played. In Asia or in Africa, other kinds of music should be used after consulting with psychologists and local musicians.

The other senses are not as important as the two already discussed but should not be neglected. Everything in the creative space should be soft and nice to the touch, including a lot of wood, preferably pine, which is particularly healthy for humans. Food will be provided and it will be light, sophisticated, tasty, and beautifully served so as to contribute to the perception of temporary well-being. Plenty of water should be available, but strong tea or coffee should not be served as they make many sensitive people overexcited and prone to arguments.

Finally, the sense of smell may also help us to enjoy the moment, but smells must be not overwhelming; preferably burning incense or using very light fragrances are the best options.

7.8 SESSION ORGANIZATION

A good and productive brainstorming session may be very short, about 15 minutes; or a little longer, about 45 minutes. The Author's experience indicates that about 30 minutes is the optimal length; after that the momentum is usually lost, participants loose enthusiasm and attention, and a break is necessary or the session needs to be continued the next day. We need to remember, however, that the length of a session cannot be predetermined with military precision. It depends on so many poorly understood factors, including the group dynamics, on losing or gaining momentum, on the nature of the problem and, most importantly, on the emotional dimension of the problem. Problems causing emotional responses usually lead to longer and better sessions. Such problems are associated with health issues, social issues, safety, security, the environment—all kinds of issues that are important and emotional for participants. Even knowing that the session will be most likely over in 30 or 40 minutes, it is prudent to plan it for half a day, leaving the remaining half a day free; in this way we create a time comfort zone so the participants will know that on this day the brainstorming is king and nothing else matters; iPhones are turned off, access to the Internet is cut off, there is no watching TV, and so on.

Preparations to a session are two pronged: They are focused on the participants and on the preparation of the "creative space," on the physical environment for the session. Both are equally important and require a lot of time and careful attention. The session may take 23 minutes, but the preparations may last for weeks. In the United States, it is possible to hire consultants who specialize in organizing and running brainstorming sessions. The Creative Education Foundation in Buffalo, New York, can provide their addresses and should also confirm their credentials. This is an effective although expensive way to organize a brainstorming session.

If we decide not to hire brainstorming professionals, we need to identify the potential participants first. The best practice is to announce our session so as to attract a number of volunteers. Volunteers will be motivated to succeed, willing to prepare for a session, and truly interested in becoming members of an elite group of "creative thinkers" in a given organization who might participate in a number of brainstorming sessions. Usually a successful session leads to many more sessions, and organizing them becomes a practice in a given company.

In Section 7.3, "Assumptions," we formulated and discussed 17 major assumptions of brainstorming. Assumption No. 14, "The group composition maximizes its creativity," discusses the optimal size and composition of a brainstorming group and provides details about the selection of its members. Let us assume that we want to use a group with seven participants, that is, we need about nine or ten potential participants, taking into consideration last-minute resignations, sudden deaths, and so on. Even in a small company, we will have many candidates to choose from who will be familiar with the problem domain. Therefore, it will be easy to choose only two or three of them. They should *not* be experts for reasons discussed earlier: Experts simply know too much and have many strong opinions, also they are usually more senior than other candidates and their participation might freeze the enthusiasm of the other participants. In the case of a large company like Google, the remaining candidates (four or five) can be easily found within the company. A small company will need to hire consultants, and contacting a local university or college is the best practice. Most likely, its departments of history, literature, or art will have many highly qualified and eager faculty members who will be interested in consulting and in participating in brainstorming sessions.

All potential participants need to be prepared for a session; that is, they should learn about brainstorming and particularly about its assumptions and major heuristics. It is a good practice to inform the group members about the problem in advance and to provide major facts from the problem domain. The Author has found that preparing a short document, at most two or three pages, about brainstorming and the problem itself is very helpful. It should be a printed booklet and should be distributed several days before the session. It will encourage the participants to start thinking about the problem and in some cases may even lead to individual Internet searches—to acquiring additional knowledge, which is always useful. The participants should also know that their individual preparations for the day of the session should be like those of a sportsman getting ready for an important game: a light meal the night before, a walk, enjoyable reading, and a good and long sleep.

All group members are equal, but we need a leader (the chair), who will know more about brainstorming than the other participants and will have a secret list with all the questions to be used to initiate the session and to restart it or redirect it later. Osborn (1953) also recommended the selection of an associate chair who would act as an "idea collector." Osborn's second recommendation was absolutely critical in the 1950s, but today the entire session may be digitally recorded and obviously this is a far superior way to record ideas than to use a human recorder, who will often disrupt the session with clarifying questions and may destroy the momentum of a rapidly evolving session. More details about the selection of group members are provided in Section 5.2.3. in the paragraph discussing Assumption No. 14 regarding group composition.

Before a session begins, the chair should prepare a number of solutions to the problem. His or her solutions could be realistic or absolutely crazy; they will be used only to begin the session (crazy solutions will work best) or to restart it when the session significantly slows down or the chair wants to change the direction. In any case, they should not be imposed on the group.

Our Heuristic No. 7 (see Section 7.7.8), "Establish Creative Space," provides many details on how such a space should be created. The most important goal is to create a temporary perception of well-being when we enter this space. It should not be a working space that is only temporarily converted into a brainstorming haven. From the psychological point of view, it is quite important that we have a dedicated space that is exclusively used for special purposes and has a "proper energy," not a "regular heavy energy" associated with boring and routine work. Everybody entering our brainstorming heaven should immediately feel a difference, a breeze of new "energy" boosting his or her enthusiasm and helping to transform an engineer or an invited professional to become, at least temporarily, a very happy and creative person. The entire interior should send a message that organizers have done everything possible to make the brainstorming group members happy, relaxed, and ready for an enjoyable (and productive) session.

A successful session begins with informal introductions, and using first names is strongly recommended. Even in the case of a group of military officers, ranks should be forgotten for a moment so as to underline the fact that all are equal and should act like a group of friends getting together to have fun. Next, the ice should be broken by several jokes, playing an exciting piece of music, and having everybody dancing or jumping. Also distributing dark chocolate or cookies is a good idea to get everybody relaxed and happy. The Author still remembers a brainstorming session chaired by him in 1974 or 1975. He brought to the meeting a large box with pastries from the best pastry shop in Warsaw, and that entirely changed the mood of the session from a formal meeting of serious scholars from competing universities into a get-together of happy friends. In fact, all kinds of "light" activities are suggested simply to relax the participants and help them to identify themselves with the group and to make sure that they are both intellectually and emotionally engaged.

When everybody feels comfortable and relaxed, a short mini presentation should be given by the chair to review the fundamentals of brainstorming and to provide the basic facts associated with the problem. The chair or experts could answer several questions but not more than two or three. Everybody should be aware that the presentation's goal is to provide the "big picture," not all the details that could be more harmful than helpful for the final session's success. At the end of the introduction, the chair should make an unexpected announcement about numerous awards waiting for the participants: awards for the craziest idea, for the most practical idea, for the most active participation, for the best tie, for making faces while generating ideas, and so on. These awards may be purely symbolic and with no material value but they will still have a positive impact on the participants and make them even more emotionally involved. As we know, all people, like small children, love awards and will invest huge amounts of energy to get a worthless award purely for the adrenaline boost associated with getting an award. It is another simple psychological mechanism that should be used for the benefit of a session.

After the chair's introduction, a mini- "warm-up" session, lasting 5–10 minutes, could take place. The problem should be very easy and casual and used only to help participants to discover on their own that they are capable of generating ideas. Unfortunately, many engineering students, not to mention practicing engineers, are convinced that they are incapable of being creative, and changing this deep-rooted but false belief is essential. Good examples of warm-up problems are provided here:

How to make driving more enjoyable?

How to sleep better in a hotel room?

How to fight jet lag when you arrive in Australia from Washington, DC?

After the warm-up session, the group is ready for real action. The first step is to make sure that the problem is appropriate for brainstorming. It happens sometimes that administrators naively believe that all problems may be solved using brainstorming. The truth is that only open-ended problems will benefit from brainstorming because only such problems have multiple solutions, which will be developed using abduction. Traditional analytical problems, or formal decision-making problems with a limited number of well-known potential solutions, are simply inappropriate for brainstorming; they require deduction, not abduction. For example, the problem "How to build a safe car seat?" is perfect for brainstorming, while a problem like "What is the normal stress value at point A in the steel frame shown in a drawing under a given system of loads?" is inappropriate because it is strictly analytical and has only a single answer. A decision-making problem like "Select the best decision from the four given alternatives using the provided decision tree" is not recommended for brainstorming either; it can be solved analytically and also has a single answer.

Discussing the appropriateness of the problem has two goals. First, inappropriate problems should be immediately eliminated before the group invests a lot of time and energy only to discover that these problems should not have been given to them. Such a situation is always frustrating and has a negative impact on the group's enthusiasm. Second, if a group agrees that the problem is good for them and they accept the problem, they will become more emotionally involved, and that obviously means better results.

The actual development of creative ideas, the essence of brainstorming, begins with the chair initiating a discussion of the problem with the goal being to define and understand it better. This is the first of the "Seven Simple Activities" discussed in Section 7.5, "Tool Box," and is appropriately called "Analyzing the Problem." After that, "Targeting" begins, which is the second activity discussed within the group of the seven simple activities. If necessary, the third activity in this group can be conducted, called "Breaking Down the Problem." The problem should not be too broad; Osborn even recommends working on simple rather than on complex problems. Complex problems may be overwhelming, their solution space huge, and the chance of finding creative solutions very slim.

When the participants seem to feel comfortable with the problem, the chair offers one of the ideas that he or she developed earlier; the craziest one will work best. When this is done, usually everybody laughs and chain thinking begins. Obviously, it does not necessarily need to be the chair who starts the process—any participant could do this—but the chair must be prepared to begin if nobody else volunteers.

Chain thinking is the essence of brainstorming. Sometimes there is an excellent group dynamic; everybody is involved and excited and has many ideas, which are developed from the ideas already presented by others or are entirely original (and begin new lines of action). In such a situation, the best strategy is to allow a long and unrestricted chain of thinking that will continue without any intervention until the end of the session. Obviously, associations (discussed in Section 7.5, "Tool Box") mainly drive the chain thinking, but entirely original ideas emerge parallel to the dominating line of chain thinking. They are the result of individual creativity stimulated by the creative behavior of the entire group, and they are a product of "free-wheeling," as Osborn (1953) calls such spontaneous generations of ideas; or of "kaleidoscopic thinking" as it is called by Gelb (and Miller Caldicott 2007).

Unfortunately, in many cases the unsupported chain of thinking dies after several minutes and after producing only a few ideas. In this case, the chair needs to help the group to regain the momentum and to restart chain thinking. There are many ways to do this. First, the chair may inject one or two of his or her preprepared ideas to initiate another chain of thinking. Second, the chair may continue on an already existing and promising line of thinking but begins asking stimulating questions from his or her secret list (see Section 7.5, "Tool Box," Activity No. 4 from the collection of "Seven Simple Activities"). Usually, such questions greatly help to recover the lost momentum, but if they fail or result only in temporary progress, the chair may propose creating combinations (see details in Section 7.5, Activity No. 5 from the collection of "Seven Simple Activities"). Again, if creating combinations is not sufficient to create and maintain brainstorming, the chair still has two classes of activities that may be initiated using the already developed ideas as an input. Each class of activities is associated with many stimulating questions, which may be read by the chair from his or her secret list of questions. The first class, "Adapting, Modifying, and Substituting Solutions," is usually very effective and is relatively simple to use. The second one, "Using Mathematical Operations," is particularly attractive for engineers because it directly relates to their mathematical skills and deductive thinking. All details of these two classes of activities are provided in Section 7.5, being the last two activities from the collection of "Seven Simple Activities."

No matter how productive and exciting a brainstorming session is, its pace gradually slows down. It is a system and naturally follows the S-curve law of a system's evolution. When this happens, it is time to review the most important ideas developed, to distribute awards, and to thank all the participants. The last task of a good chair is to ask the participants to continue working on the problem. On the surface, this is an absurd request. brainstorming is supposed to drain all ideas from them and no unexplored idea should be left. It is an extreme statement; it may be even correct at the moment when a session ends. However, when the human brain is activated, its subconscious part will not stop working on an exciting problem just because the session is over. It may continue for at least several days. Therefore, it is prudent to contact the participants about four or five days after a session and to ask them if they have had any additional ideas. The results may be surprising; a brainstorming echo may actually produce valuable results.

7.9 EXAMPLE

7.9.1 Introduction

The example to be discussed involves an application of brainstorming that took place about 10 years ago at George Mason University in the Department of Civil, Environmental, and Infrastructure Engineering, where the Author used to work. It describes a brainstorming process organized and run by a group of undergraduate students taking the Author's class Introduction to Design and Inventive Engineering. The session was prepared in cooperation with the industry, namely with the Chrysler Technology Center in Auburn Hills, Michigan. Dr. Kalu Uduma, senior technical specialist in the Vehicle Safety Engineering Department, was directly responsible for this cooperation. Dr. Uduma's expertise is in vehicle crash safety development. He is also an inventor in the area of vehicle crash safety, and his involvement guaranteed the industrial relevance of the problem and access to the state-of-the-art problem domain knowledge. A brief description of the entire process follows.

7.9.2 Fact finding

7.9.2.1 No. 1: Picking out

At the time of the brainstorming reported here, crash safety was a focus of public opinion, and all car companies were working on the development of new safety measures. Air bags were the new state-of-the-art safety restraints in the automotive industry. They were initially considered a great success in saving human lives, particularly in the case of front collisions. However, after the period of initial euphoria, there was great concern that deploying air bags posed potential dangers to pregnant women and small children sitting in the front seat. Front air bags were also ineffective in the case of side collisions and rollovers. The safety concerns posed by deploying air bags led to a wave of criticism from the press and lots of regulatory pressures from the federal government. Understandably, the automotive industry was in an urgent search for new safety measures that would eliminate the limitations of front air bags. It should be mentioned that, at that time, the next generation of "intelligent" air bags with less powerful exploding bags had not yet been developed. In the described situation, our expert saw an extraordinary opportunity and the need to develop safety restraints that would not have the negative consequences of air bags. His lecture on the state of the art in the area of automotive crash safety engineering helped to clearly formulate an emerging opportunity for future inventive engineers in my class:

> Front air bags are important invention, saving human lives. However, front air bags are *not* the ideal solution and need improvements or a replacement by a better solution. You have a rare opportunity to save human lives and to make a fortune if you find a better safety measure than an air bag. It should work in all kinds of collisions.

Students reacted with real interest, showing even emotional involvement (many had had relatives or friends killed in car accidents) and had many questions. After that they were ready to move to the next step in the brainstorming process.

7.9.2.2 No. 2: Pointing out

This step is also known as *formulating the problem*. It can be understood as a process of transforming or translating an inventive opportunity into an inventive challenge, that is, into an inventive problem.

In our case, the students had a brief discussion about the presented inventive situation and of the state of the art in the problem domain. They also carefully analyzed the provided inventive opportunity. After that, the students went through several drafts and finally formulated the following inventive problem:

Develop an inventive design concept of an engineering system intended for passenger cars. Its function will be to provide protection for front-seat passengers during all kinds of collisions and without the unintended consequences and limitations of the front air bags.

7.9.2.3 No. 3: Preparation

The first part of preparation is acquiring knowledge relevant to the inventive problem. In the reported case, the expert provided this knowledge during a special lecture. Also, he gave students access to several key publications and answered all their questions. After that, the students felt strongly confident that they had already acquired sufficient basic problem domain knowledge and were ready for brainstorming.

The second part of preparation is focused on preparing participants and the physical space for the session. In the reported case, the preparation of participants took literally years, while the idea-generation period lasted only about 20 minutes. Participants were mostly seniors, and that meant that they had already accumulated a large body of engineering knowledge. Next, they were in the Author's class for about seven weeks before the brainstorming took place. During that time, the students had already developed a broad understanding of human and engineering creativity. Finally, they had a three-hour lecture on brainstorming and a lecture of about two hours on crash safety and air bags. After all this studying, students were well prepared for the session. It was a relatively small class of about 12 students, and the decision was made that the group would not be divided. Only a volunteer chair was chosen, and the session was not recorded because its nature was academic rather than industrial.

George Mason University does not have any space specifically designed and prepared for brainstorming, and the session took place in a small students' laboratory that was at that time used mostly by graduate students and teaching assistants. The room was definitely not optimal for brainstorming; the interior was uninspiring, the lights were too strong, the chairs were uncomfortable, there were too many visual distractions, and so on. To mitigate the problem and to raise the spirits of the participants, the Author provided cookies and soft drinks, eliciting a very positive response from the students. Fortunately, the limitations of the space were more than compensated for by the enthusiasm of students and by their strong motivation combined with their relatively good intellectual preparation for the session.

7.9.3 Idea finding

The idea-finding stage of brainstorming, the heart of the process, began with the chair talking for several minutes. He reiterated the major assumptions of the method and focused on the delayed judgment principle and on the liberating freedom to say anything and to propose even the most absurd ideas in the moment, which could later prove to be the most inspiring and important. He also mentioned that the goal was simply to produce a flood of ideas, not a very few good ideas. "Be like small children playing a game. Remember that if you say something, you will get a cookie, but if you say something stupid, you will get two cookies." Everybody laughed and seemed ready for action.

Finally, the chair reminded his fellow students of the inventive problem definition (see the previous paragraph). He also mentioned that the original equipment manufacturers (OEMs) in the automotive industry were averse to the aggressive way that state-of-the-art air bags deployed at that time because they posed potential dangers to pregnant women and small children in the front seats of vehicles. Injuries attributed to aggressively deploying air bags were creating terrible publicity and all kinds of legal problems. Yes, air bags contributed significantly to the safety of vehicle occupants during a crash; however, the unintended consequences of their introduction were horrible, and OEMs wanted the inventive group to come up with a better solution to save lives during vehicle collisions. "We have a mission to save human lives and in the process to make a fortune. Think about exploding air bags and all the fatalities attributed to those deploying air bags. Do something about it. It is your chance to change the world."

The last remarks created the intended emotional response from the students, who nodded their heads and showed real engagement. The session began.

The following is the record of the dialog that subsequently took place. It is based on the Author's recollection; therefore it may not be exact, but it reflects well the nature of the session and its major findings.

STUDENT NO. 1: If air bags kill, why not get rid of them?

STUDENT NO. 2: But they represent progress; they do not always kill, they kill only weak bodies.

STUDENT NO. 3: We need to make them better or to introduce something to replace them.

STUDENT NO. 4: But how to improve them?

STUDENT NO. 5: They should recognize pregnant women and small children.

STUDENT NO. 6: But how? Should we weigh them?

STUDENT NO. 5: But pregnant women are as heavy as men; weighing will not help.

STUDENT NO. 7: Should we measure their volume and shape?

STUDENT NO. 8: Yes, let us ask all women and children to take a bath before riding in a car and measure the volume of water displaced by their bodies and later weigh the water.

STUDENT NO. 9: I would suggest hormonal tests and an on-board computer that would analyze results and tell the air bag to reduce the explosive force.

STUDENT NO. 10: The computer would also control the acceleration and braking rates when pregnant women and small children are on board. When I was a teenager and drove my pregnant sister, she nearly delivered after my driving.

STUDENT NO. 7: I guess that your parents were not very happy about your driving.

STUDENT NO. 11: Let us come back to safety. I like the idea of air bags. It is beautiful. Air bags exist when we need them. and they do not exist when we simply drive.

STUDENT NO. 12: It is like in TRIZ, the principle of separation in time.

STUDENT NO. 9: Exactly, let us get rid of the air bags but leave the principle.

STUDENT NO. 7: Something else emerging during the crash?

STUDENT NO. 4: But what? My grandparents living in Florida have rolling shutters in their house and the shutters protect the house only when a hurricane is coming.

STUDENT NO. 2: Rolling shutters? In Florida they need to be hard, but in a car they need to be soft.

STUDENT NO. 1: Soft like an air bag? What about using two air-bag curtains behind the windshield?

STUDENT NO. 6: What about a single air-bag curtain behind the windshield?

STUDENT NO. 8: What about several air-bag curtains surrounding all passengers?

STUDENT NO. 4: But we need to protect only sensitive parts of the body. What about a system of curtains with holes to reduce the volume of gas we need to fill them?

STUDENT NO. 2: Do we need many separate curtains with holes? We may connect them.

STUDENT NO. 4: In this way, we will have something like a cage with members filled with gas.

STUDENT NO. 3: But we know of racing cages with steel members. They work wonderfully. A cage with plastic tubes filled with gas will provide universal protection.

STUDENT NO. 7: But we wanted to protect only pregnant women and children.

STUDENT NO. 8: I am not a pregnant woman, not at all, but I would not mind being protected during a crash.

STUDENT NO. 6: We are using the *separation principle*. Could we separate pregnant women from the car during a crash in a different way?

STUDENT NO. 8: They could levitate during a crash.

STUDENT NO. 12: Not all women know how to levitate—definitely not all pregnant women.

STUDENT NO. 6: But OEMs could organize levitation courses for all pregnant women drivers or front-seat passengers of moving vehicles.

STUDENT NO. 7: And there should also be such courses for children.

STUDENT NO. 10: What if all passenger vehicles were sold with instructions on how to levitate during crashes?

STUDENT NO. 12: Instructions and videos could be available on the OEM's website.

STUDENT NO. 3: The issue is not levitation but separation. Antigravity devices could also temporarily separate pregnant women from the car.

STUDENT NO. 4: What about a portable antigravity device for pregnant women and small children to wear while driving?

STUDENT NO. 11: What about separation used in the fighter planes? Only pregnant women and small children would sit in the ejection seats.

STUDENT NO. 7: What about ejection seats for mothers-in-law? I would definitely buy one.

STUDENT NO. 4: My mother-in-law is a true angel. I would not want such a seat in my car.

STUDENT NO. 7: But your mother-in-law died two years ago.

STUDENT NO. 6: Enough about mothers-in-law, but you have inspired me. What about using angels to provide separation services?

STUDENT NO. 3: They occasionally provide such services.

STUDENT NO. 4: But only to those who believe in angels.

STUDENT NO. 2: It is debatable. Dr. Uduma has not provided any detailed data on the subject. During a crash everybody believes in angels.

CHAIR (SILENT TO THIS MOMENT): We are at George Mason University (GMU), political correctness rules; we simply cannot even debate the use of angels during the crash. It implies that angels exist and that thought might offend all agnostic students. Let us review our ideas and develop them.

Considering the serious and real dangers of political incorrectness at GMU, the students immediately moved to the *development of ideas* stage. They discussed all the ideas developed during the session and almost immediately identified several key ideas, including

1. Sensors to weigh passengers
2. Sensors conducting hormonal tests to identify pregnant women and small children
3. Intelligent air bags that would adapt the explosion force to the weight of passengers
4. Safety devices based on the principle of separation in time
5. Air bag–type curtains behind the windshield
6. A system of several perpendicularly positioned air bag–type curtains surrounding passengers
7. A system of several perpendicularly positioned air bag–type curtains surrounding passengers with holes
8. A cage with plastic gas-filled tubes
9. Using levitation and levitation training
10. Levitation-based safety devices
11. Antigravitation safety devices

12. Ejection seats
13. Help of individual custodial angels
14. A service organized in cooperation with angels and providing crash protection

Short descriptions of all the ideas were prepared, and the idea-finding part of the brainstorming process ended with everybody congratulating the other participants on their success.

7.9.4 Solution finding

This part of the brainstorming took place immediately after the idea-finding stage ended. The judgment time came. Surprisingly, only one idea survived: No. 8, "A cage with plastic gas-filled tubes." Students argued that the concept had already been proven by its wide use in racing cars in the form of metal cages. Their invention should work equally well but with many advantages with respect to metal cages: they would invisible before the crash, weigh less, allow easier access to the seats, and represent tremendous public relations value.

Unfortunately, this brainstorming session took place about 10 years ago when the car industry was struggling with bad publicity and huge legal costs associated with deaths caused by air bags. In this situation, the idea of using in a car a complex system of interconnected plastic tubes filled with high-pressure gas raised all kinds of concerns with our expert and was not implemented. The concept was simply premature; a year later, BMW introduced the air bags in the form of side curtains and today each car has at least four to six air bags.

7.10 BLACK AND WHITE

7.10.1 White

Brainstorming is so much more than just one of many heuristic methods. Its introduction in the early 1950s created a revolution in the popular perception of human creativity. Moreover, it suddenly democratized creativity. Before Osborn, creativity was a secret domain of the very few who were given unusual talents by God. After Osborn, creativity was simply one of many human activities that can be simply learned and practiced by everybody. No more mystery—just a lot of systematic effort, time, and learning of various specific creativity-boosting procedures.

Brainstorming may be practiced by anyone, and that means that even by engineers. For them, brainstorming is much more difficult than for anybody else because engineers are educated to be rational and to exclusively practice deductive thinking. On the other hand, the rewards of learning brainstorming are so much more meaningful for engineers than for other people. For engineers, learning brainstorming is a life-changing and transformational experience. It changes not only their professional lives, creating in the process entirely new professional opportunities, but it also changes their private lives. This aspect is likely the most important, because an ability to be creative is associated with mindfulness, which ultimately leads to well-being.

Learning and practicing brainstorming is a must for inventive engineers. First, learning it provides a body of fundamental inventive engineering knowledge. Second, practicing it leads to the development of useful practical creative skills and, much more importantly, to building the belief that creative activities can be effectively learned, and that belief is absolutely critical for inventive engineers. Finally, understanding brainstorming, particularly its revolutionary assumptions and heuristics, is the stepping stone to learning Synectics, the most powerful inventive designing method, which is discussed in the next chapter.

Learning brainstorming together with other inventive design methods gives students not only various pieces of specialized knowledge related to the individual methods but leads to the integration of this knowledge. This integration is probably the most important outcome of learning about various methods, because it results in acquiring transdisciplinary knowledge. This knowledge is in the form of metarules or methodological heuristics that guide inventive engineers through the process of selecting appropriate methods for various types of inventive problems. Having this transdisciplinary knowledge gives an inventive engineer a tremendous advantage with respect to a traditional engineer who has learned only a single method and tries to use this method in a mechanistic way, no matter whether it is appropriate for a given problem or not.

Learning the principles of brainstorming is relatively easy and does not require long preparation, particularly in the case of "regular" participants (chairs need more studies, more motivation, and more preparation). It is also enjoyable and reminds many people of the happy days of their childhood. Also, each successful brainstorming session makes all participants happy, at least for a short time, and creates a strong boost in their energy levels.

Each session is different, even if it is on the same topic and features the same group of people. For this reason, an unsuccessful session may be always repeated with a good chance that the results will be much better, as they usually are (the subconscious in action). A brainstorming group is like a complex system: Its behavior cannot be simply predicted; each session has its own trajectories and attractors, and the results of two sessions may be fundamentally different. Therefore, there is always hope that the next session will bring the desired results.

Today, after more than 60 years of development and evolution, a huge body of knowledge exists about brainstorming, and finding the necessary information or instruction is not difficult, particularly as various courses are easy to find and professional help is available.

7.10.2 Black

Brainstorming means too many things to many people, and its public and even its professional perception is confusing. For many people, brainstorming is a magic tool, but for many others it is a suspicious, poorly understood, and ineffective way of spending money for a group of randomly selected people who know nothing about the problem concerned but try to find inventive solutions to it. This perception hurts the image of people using brainstorming, even if they are very successful. We need to understand the causes of this situation:

- Coming from the advertising industry, which is not the most respectable segment of the American industry
- Too much popularity too soon
- Often too many promises unsupported by results
- Too many pseudo-experts, who are simply unprepared for the job and are focused on making money, not on solving problems
- Lack of good academic programs and courses teaching the method in a balanced academic way
- Too many poorly prepared and executed sessions
- Too casual use, often without any understanding of the method and without any preparation
- Underestimation of the importance of establishing the "creative space" and motivating participants

There is another essential reason why so very few professionals, and particularly engineers, have a good and balanced understanding of the method. Usually, brainstorming is taught as an independent entity and a universal method, without any broad background knowledge. Such a presentation of brainstorming gives students a false and at best only a very limited understanding that is simply insufficient for the effective use of the method. Brainstorming should be taught as a part of a system called inventive engineering. In this case, students learn about the method only after they have at least gained some understanding of human creativity from various perspectives (historical, social, psychological, economic, etc.). They should also learn about other inventive designing methods so that they will see brainstorming as only one of several "tools" in their "inventive toolbox." In this situation, they will be using brainstorming only when it is appropriate, not always.

Chapter 8

Synectics

8.1 CREATOR

William J. J. Gordon (Figure 8.1) (1919–2003) was in many respects an unusual man, like many creative people and particularly inventors. Most of his life was spent in the state of Massachusetts (a coastal state in the northeastern United States), and without doubt his various interests and intellectual attitudes well reflected the culture and the intellectual richness of his *heimat* (a German word that perfectly identifies a region that has a great impact on an individual). In the late 1930s, he studied at the University of Pennsylvania, but supposedly he never graduated. This may be the key to understanding his passion for continuous learning and in that he reminds us of Leonardo Da Vinci, for whom the lack of a traditional academic education also led to lifelong individual experiential studies, ultimately giving him a unique transdisciplinary knowledge and understanding of the world.

During the time period from 1950 to 1960, Gordon (1961) was in charge of the *Invention Group* at Arthur D. Little (ADL). This was, and still is, a consulting company focused on innovation. ADL claims on its website that it is the world's first consultancy, and its 125 years of existence obviously support this claim. It was the perfect environment for a Renaissance man like Gordon, who thrived in his position and was very successful, not only creating many inventions but also doing a lot of pioneering research on the inventive process. It was a systematic and rigorous research with a strong psychological component, which was well ahead of its time.

At the time his groundbreaking book on Synectics was published (Gordon 1961), he was the president of the Invention Research Group in Cambridge, Massachusetts. He was also lecturing at Harvard. In the same year, Gordon established, with three friends and associates from ADL, a company called Synectics Inc. Its mission was to continue and expand their research on the inventive process. After several years, Gordon left Synectics Inc. to focus on the potential applications of the method of Synectics in education and founded the company Synectics Education Systems. His new company was focused on creativity and education and on the promotion of

Figure 8.1 William J. J. Gordon. (With permission. Drawn by Joy E. Tartter.)

his educational ideas. In many respects, his professional evolution and the various activities he carried out later in his career were similar to those of Alexander Osborn, the "father" of *brainstorming*, in his final years. It is probably unsurprising that Gordon also established a foundation to promote creativity in education.

Gordon has been known as a psychologist, most likely because of his deep psychological knowledge and his many contributions to this area, particularly those related to human creativity. Only in recent years have neuropsychologists tried to integrate psychological knowledge with neuroscience in order to explain various creativity-related phenomena, but Gordon, with his team, discovered them a long time ago in the late 1950s and early 1960s.

Gordon has also been seen as a talented inventor with a huge body of transdisciplinary knowledge, which has allowed him to develop inventions in many supposedly unrelated areas. For example, he has patented inventions as diverse as a rotatable multiple tool support unit (1956), artificial snow (1962), a dispenser (1962), a boat fender (1962), and a lipstick container (1973).

Gordon's method, Synectics, is an excellent reflection of its creator's transdisciplinary knowledge, appreciation of psychology, and understanding of the emotional dimension of human creativity. Most importantly, his method reflects his fundamental understanding—which is, however, never explicitly revealed—that human creativity works best in a group setting when a group behaves like a complex system; that is, it undergoes constant adaptive changes and is unpredictable but multiplies the creative power of the individual people.

8.2 HISTORY

The Boston area of the United States has a long tradition of rich intellectual life going back to the beginnings of the country. This area has also a unique combination of the best universities (e.g., Harvard, the Massachusetts Institute of Technology [MIT]), research companies, and art institutions. Boston acts like an attractor to a large number of creative people of all kinds who interact and contribute to the local liberal culture. That culture further promotes crossing of disciplinary boundaries and has ultimately led to the emergence of transdisciplinary knowledge, the foundation of human creativity. It is the best American example of the Medici effect taking place. Obviously, it is not a coincidence that Synectics emerged in the Boston area.

Gordon (1961) claims that the experiments began in 1944, but they grew from earlier studies of human creativity and from the area of esthetics. From the beginning, the focus was on discovering mechanisms that drive the creative process but which were not known to the creativity scholars. Interestingly, at the beginning, the subject of observations was only a single inventor working on a problem in instrumentation. The inventor's goal was to develop a new kind of aircraft instrument, which would eliminate aircraft accidents caused by misreading gauges with dial faces. The inventor played a dual role: He was inventing a new instrument, and at the same time he was conducting psychoanalysis of his inventive thinking. During the entire process, he was talking and describing his thoughts, which were revealing his creative process.

The study revealed four psychological states that seem to be associated with the inventive process:

1. *Detachment:* The inventor feels as if he is looking at the object from a certain distance and says, "I've got to take a real look ... from way out."
2. *Involvement:* The inventor identifies himself with the object and says "How would I feel if I were a spring?"
3. *Deferment:* The inventor realizes that understanding must come first and solutions later and says "Solutions are the payoff! But to hell with them. ... Otherwise I'll invent the same thing all over again."
4. *Speculation:* The inventor speculates about a seemingly impossible form of the object and says "If there was an enormous spring?"
5. *Autonomy of object:* The inventor feels that the object is an entire independent entity and says "I have the feeling that this thing is on its own, completely outside of me. ..."

These five psychological states were observed at times when the inventor was making breakthroughs, and the frequency of their occurrence was growing at the end of the inventive process. The researchers verified the existence of these five states through a series of interviews with creative individuals in art and science. All the interviewees recognized the psychological states previously identified but also admitted that they had never explicitly articulated them or were even aware of their existence. Therefore, Gordon and his team deemed their discovery important but still insufficient for practical purposes because they did not know how to create these states. They decided to pursue this line of research and in 1945 began working with groups solving problems in hydrodynamics and acoustics at Harvard Underwater Sound Laboratory. Unfortunately, this research did not bring any improved understanding of psychological states during the process of creative thinking, mostly because these states were not sufficiently known to study them experimentally.

The next step in the search for the holy grail of creative problem solving was to shift the research focus from the individual psychological states associated with creativity to the much more general psychological conditions related to the entire creative process. The goal was to acquire a sufficient body of knowledge about human creativity that would enable people to increase the probability of a breakthrough leading to creative solutions.

For all these reasons, in 1948 the Rock Pool experiment was conducted. Groups of various artists were invited to live together (usually 12–20 people at a time) and to interact during the summer months in Lisbon, New Hampshire. The participants were expected not only to continue their artistic activities but also to do jointly some physical work to accelerate the process of group integration. Only artists were invited because earlier the researchers had found that artists were much better than engineers or accountants at articulating their insights and feelings. On the artistic side, the experiment was extremely successful, and its participants won a number of awards for their art. Unfortunately, the experiment did not offer any improved understanding of human creativity, although it resulted in an unexpected and important conclusion: *People become much more creative when they are members of a group.* In this way, one of the fundamental assumptions of Synectics was born.

Analysis of several of these meetings revealed that communication took place through shorthand free associations and, equally importantly, that a single person working on a problem does not have to talk out loud, while in a group setting, people talk out loud all the time. This seemingly small difference has a significant psychological impact on all participants and is a powerful factor in creativity stimulation.

In 1944, in addition to continued experimental studies, a traditional academic analysis of the *state of the art* of human creativity began. It covered the classical works in this area and autobiographical records of such creativity giants as Einstein, Goethe, or Edison. The analysis brought interesting results. First, it was found that the majority of contemporary studies were focused on finding test methods for identifying creative people, not on the creative process itself and the psychological mechanisms behind it.

Moreover, these studies dealt mostly with creativity in art, where creative results are difficult, if not impossible, to measure; their evaluation is therefore much more subjective than in engineering.

Earlier, Gordon and his team had concluded that the creative process is basically the same in art and engineering. Therefore, they decided to continue their research with a focus on engineering creativity, that is, on the development of patentable inventions. Obviously, the results of engineering creativity are much easier to quantify and objectively evaluate, and Gordon's decision greatly accelerated the research's progress. The goal was to discover, or to develop, a *scheme* that people could comprehend and use to increase the probability of their creative success. More specifically, the search was for psychological mechanisms and psychological states, which could be repetitively used and created, respectively.

In 1952, an *operating group* was established at Arthur D. Little Inc. in Cambridge, Massachusetts. Its mission was to produce patentable inventions. The group had five members, and their professional backgrounds and interests reflected the Gordon's emerging understanding that a group of *Synectors*, that is, people practicing Synectics, should possess a rich body of knowledge coming from various domains. The initial group included

- A physicist with an interest in psychology
- An electromechanical engineer
- An anthropologist with an interest in electronics
- A graphic artist with a background in industrial engineering
- A sculptor with some background in chemistry

The group developed its own style and worked on inventions through conversations and discussions. As a result of monitoring the group during many sessions, various meaningful results were gradually produced. First, an improved understanding of the role of the leader was developed. It was discovered that a single strong leader was not the optimal solution and that the best results were produced when the role of the leader shifted from person to person. Second, enjoyment, or fun, was recognized as a very desirable psychological phenomenon stimulating creativity and an important indicator that the inventing process is moving in the right direction. Third, it was observed that an associated phenomenon was a spontaneous child-like play, which also stimulates human creativity and helps people to deal with seemingly irrelevant information. Next, and surprisingly, engineering elegance was similarly identified as being directly related to the enjoyment of inventing.

Engineering elegance may be described as the esthetics of the design, its symmetry, the smooth flow of internal forces or stresses, the desirable nature of deformations, and so on, but beauty is obviously in the eye of the beholder. Most likely, a given person recognizes elegance using his or her experience—that is, his or her acquired knowledge in the form of metaheuristics. Based on the author's experience, he strongly believes that elegance, or beauty, is the best indicator of the optimality and simplicity of engineering designs.

Finally, in 1957, an important finding took place. It was discovered that two abilities were necessary for inventing: *an ability to tolerate irrelevant information and an ability to use it.* Both abilities are associated with the ability to change the context of a given piece of information or knowledge and, in this way, to create a new understanding of or new uses for a given object, which are often the key to an invention.

Only in 1958 did a major breakthrough take place, and an important activity was discovered to be repeatedly used during inventing: *making the familiar strange,* that is, changing the context of a well-known concept (this is discussed with examples in Section 8.4). This activity can be used consciously by simply asking a question: How can we make the familiar strange here?

Around the year 1958, the research focus was also on dealing with irrelevant information as having great inventive potential and being related to making the familiar strange. Irrelevant information is easily accepted and used in children's play, but it was not known how to create the psychological conditions of such play for a group of adult inventors working on a problem. Finally, the psychological mechanisms were found (discussed in Section 8.5) that actually do not only create but also stimulate children's play conditions:

1. Playing with known words, with their various meanings and definitions
2. Playing with various laws of nature and trying to expand or change their usual application domains
3. Playing with metaphors

Before the method was fully developed, the first Synectics groups at industrial companies were organized in 1955. The critical issue was to find appropriate people, that is, people able to use the various psychological mechanisms associated with Synectics and who would also be emotionally as well as intellectually ready to entirely change their operational paradigm. This shift required moving from the mostly *rational* and deductive traditional engineering work to mostly irrational playing with irrelevant information and the frequent use of abduction. The issue of selection of proper people for a group of Synectors is addressed in Section 8.7.

Learning the history of Synectics is particularly meaningful because it shows the complex process behind its gradual emergence. It also explains how this sophisticated method has been gradually developed from several simple, nearly casual observations with the proper use of experimentation, literature studies and, first of all, years of consistent and persistent effort, which always make a difference.

8.3 ASSUMPTIONS

The presented assumptions have been partially explicitly articulated by Gordon (1961) and partially formulated by the author, based on Gordon's writings:

1. Synectics is a heuristic method for the development of creative solutions in art and engineering.
2. The creative process is the same in art and engineering.
3. The creative process is a mental activity resulting in artistic or engineering inventions.
4. The creative process is about joining together different and apparently irrelevant elements.
5. The creative process is a process of intertwined knowledge acquisition and integration.
6. The creative process should be a group activity.
7. A Synectics group can compress into a few hours the kind of semiconscious activity that might take months of incubation for a single person.
8. The creativity of individuals is stimulated by the presence of and interactions with other people.
9. An individual's creativity can be significantly improved if he/she understands the psychological process by which he/she operates.
10. In the creative process, the emotional component is more important than the intellectual, the irrational more important than the rational.
11. Understanding emotional, irrational elements is the key to increasing the probability of success in problem solving.
12. There are five psychological states:
    a. Involvement
    b. Detachment
    c. Deferment
    d. Play
    e. Autonomy
13. There are two key activities:
    a. Making the familiar strange
    b. Making the strange familiar
14. There are four psychological mechanisms:
    a. Personal analogy
    b. Direct analogy
    c. Symbolic analogy
    d. Fantasy analogy
15. Problem solving is a two-stage process: problem definition and the formulation and development of solutions.

8.4 PSYCHOLOGICAL STATES

All five psychological states in the process of human creativity can be described as states in which the human brain uses mostly its right hemisphere; that is, thinking is creative, the dominant reasoning is abduction, and emotions play the dominant role. Obviously, not all five states take place at the same time, but during a synectics session all states could occur and therefore should be understood to become operational.

Detachment is the state in which the inventor attempts, usually successfully, to see the object from various perspectives. Relatively simple is seeing the object from a physical distance, from a bird's view or from a mouse's view (Figure 8.2). Such different perspectives lead together to a better understanding of the object and to the discovery of new aspects that had not been seen before.

For example, a group is working on a new concept for a family SUV. The latest emerging concept is of a vehicle with two decks, like a London bus. The lower deck is for parents and for two other adults, while the upper deck is for children and provides some open space for playing. A view from a distance may reveal that the vehicle looks huge and massive. The bird's perspective may show that the vehicle looks like a truck. Finally, the mouse's view may tell us that the ground clearance is huge and that the vehicle may be very attractive not only for families with children but also for hunters and fishermen. All these three perspectives simply provide unnoticed earlier aspects of the object of our interest, and they may definitely help the group to continue developing the concept of a new family SUV. We need to remember, however, that acquired knowledge should not be judged but considered as objective (facts of life) and used for inspiration.

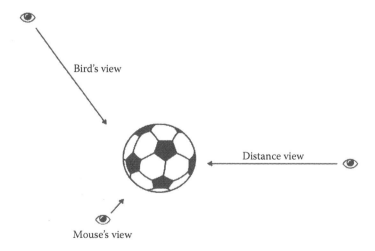

Figure 8.2 Detachment: Seeing an object from different physical perspectives.

Much more difficult is seeing the object from various personal or professional perspectives. For example, a Synector may see the object from the perspectives both of a child (who knows very little) and of an expert (who knows "everything"). Similarly, a Synector may see the object from various professional perspectives. A building may be seen from an architectural perspective (elevations), from a structural engineering perspective (structural system), or from a heating perspective (heating system).

The emergence of the state of detachment may be stimulated through asking appropriate questions, for example,

- How would this object look from the distance of a mile? What would be seen?
- How would a mouse see the object?
- How would a flying eagle see the object?
- How would a child see the object?
- How would an expert see the object?
- How would a chemical engineer see the object?
- How would a mechanical engineer see the object?

Detachment is relatively easy to comprehend and to use. *Involvement,* also called *empathy,* is initially difficult to accept for many engineering students because it involves using emotions, such a foreign notion to so many of us. However, we want to become inventors. That clearly means that we need to learn how to use the entire power of our brain, not only its deductive power (absolutely essential for all engineers, including inventors), but also its creative or abductive power. We will be able to use our entire brainpower when we are in the psychological state called involvement or empathy. It is a state when our emotions are activated and we are becoming involved, not only on the rational or intellectual level but also on the emotional level. Our involvement means that we are able to identify ourselves with the object of our interest and literally *feel* like this object. Such a situation gives us a great insight into our problem and may lead to inventive solutions.

For example, you are working on the development of a new concept for a temporary crash barrier on a highway. You have already learned that such a crash barrier must absorb and/or dissipate a large amount of kinetic energy brought by the car hitting the barrier. You are the barrier. You see the car coming but you are unable to move. You feel the car hitting your stomach. First, your stomach deforms (and absorbs energy); next, it literally explodes (and dissipates energy). It is not the end of your experience. Your bones are broken (and more energy is dissipated). Sorry, your ordeal is not over yet. You feel terrible, terrible pain, but suddenly your body begins to move. To your sheer terror, you realize that the car is dragging your body on the rough surface of the highway, but fortunately the car is slowing down. You can feel rapidly growing heat associated with the friction between your body and the concrete (and more energy is dissipated). Finally, the movement ends and you realize that your badly injured body has stopped the car and saved the lives of several maintenance workers who were behind you/the barrier. You are in terrible pain, but you also feel happy about the saved human lives. You are gravely hurt but try to think rationally, from your engineering perspective, about what has actually happened. You realize that you have discovered all the secrets of a temporary crash barrier on a highway. Now you know that it absorbs energy through large elastoplastic deformations and dissipates energy through plastic collapse and breaking apart as well as through friction associated with movement. At last, you clearly see the problem. You feel shaken and exhausted but ready to find a means to translate your experience into an invention. You know that you will save human lives.

At the beginning of their inventive careers many engineers find it difficult to identify themselves with the objects of their interests. Obviously, gaining practical experience will gradually help, but asking appropriate questions may bring immediate results. Coming back to our example, we could ask various questions helping us to develop empathy for our object, or to get into "the mood." Exemplary questions:

> How would you feel being a crash barrier?
> What would you see standing in the middle of a lane on a highway?
> Would you see or hear the oncoming car?
> What is the most terrifying part of its image? Lights or bumpers?
> What will you hit first? The bumper?
> Will you see the driver's face?
> Will it terrify you?
> Where will you be hit first?
> What will happen to your stomach?
> Will you feel pain when your bones are being broken?
> Will you think about your burned jacket when you are being dragged by the car?

Deferment is an interesting psychological state that helps us to move away from our usual constant focus on results. We, as engineers, consciously or subconsciously always want to be productive, whatever we do, and we always have our final results, our products, on our mind. We could say that we are obsessed with obtaining results, and unfortunately that is simply harmful when we want to be creative; it actually prevents us from producing results. For all these reasons we need to defer, or *postpone*, our focus on results and concentrate on the *process* of creating them. The psychological state of deferment helps us not to think all the time about the final results but to emotionally engage in creating them, and that is the big difference.

One of the most important assumptions of Synectics is assumption No. 4 (see Section 8.3). It says that the creative process is about joining together different and apparently irrelevant elements. This cannot be done in a rational way because it would not make any sense. It must be a result of unrestricted thinking without any concern about the rationality of our actions or about the final results. It must be like children's play, when children simply play for the enjoyment of playing. This pure enjoyment of playing is the key to human creativity, and we need to create a psychological state, deferment, which allows us to temporarily forget about the expected results. It is never easy, but it can be done using various spontaneous or prepared questions and statements.

For example, we are working on a new concept for drilling a tunnel. The problem has been clearly presented, including the known drilling technologies. Unfortunately, the first proposed concept, involving using a system of nuclear explosive devices, was met with a surprising reaction of gloomy faces and visible disapproval, although nobody said anything critical, all members of the group being in the Synectical mood. The leader immediately recognized the problem and felt that the group was not ready for action. Her first reaction was an attempt to create deferment. Let us look at her several questions and statements.

> "OK, let us forget about drilling the tunnel and have some fun. I had a long day yesterday and want to forget about it. I want to be happy like a child again."
> "Who could sing us a song? Three cookies for a volunteer!"
> "You want to dance for us? A snake dance? A drilling snake dance? No, not a drilling snake dance. Perform a cobra dance. We need to be fresh before we start inventing."
> "Let us play a simple game of words."
> "What is the opposite word to drilling?"

What followed was a funny dance performed by a horizontally and gravitationally challenged Joan, and *Oh My Darling Clementine* performed by Peter (who had no clue about singing but loved to perform). Suddenly everybody was smiling and relaxed; we could see the changing body language. Obviously, nobody was concerned anymore about the radioactive by-products of serial nuclear explosions and everybody seemed to be focused more on playing with the concept than on considering its final form and its consequences. All participants were in the state of deferment.

The psychological state of *play* is also called *speculation* or *use of irrelevance*. All three names correctly reflect various aspects of this state. The discovery of play was the major breakthrough in the development of Synectics and became its essence. It is also difficult to understand, and its explanation requires introduction of several concepts that may be strange for engineers, but we will make them familiar in the spirit of Synectics.

The first concept is of a human activity called play. It is an activity whose purpose is not to produce any physical or abstract objects but simply to create an emotion, a feeling of pleasure or satisfaction. For centuries, people were fascinated by the question of why humans would be interested in such an apparently useless and pointless activity. For a long time, philosophers and psychologists were looking for the answer. Finally, only late last century, they found it in our second new concept: *hedonic response*.

Hedonic response is a rapidly building feeling being a mixture of intensive pleasure and satisfaction, in many cases nearly orgasmic. It is also associated with a sudden energy boost and a flood of positive thoughts. Considering the importance of the hedonic response for Synectors, and obviously for all inventors, we will have two examples explaining this concept.

8.4.1 First example: Performance driving

Inventors, and particularly young male future inventors, like driving sports cars, so our example will be easy for them. Let us assume that you drive a car, a powerful sports car with a turbocharged 456 HP engine. You are just at the central apex of a tight curve, and you floor the accelerator to leave the exit apex under full power. Suddenly, both turbochargers are spooling at the maximum speed and you physically feel a powerful kick in the butt, the essence of high-performance driving, not to mention the sound of a screaming engine. Life is beautiful. Your heart starts pounding and you literally feel the boost of your own energy, of your ego, and an incredible shot of optimism. You are experiencing unbelievable emotions, and you feel that you have just unleashed them. You are the king or queen of the world and everything is possible; suddenly your mood has changed. You feel that you are ready for all coming battles. You have just experienced a hedonic response.

8.4.2 Second example: Playing chess

Let us assume that you are a chess player. You have talent, you have been playing chess since you were three, and now you are an international chess master. You participate in an international tournament and you have already won all but your final game. Your last opponent is a famous Russian chess master, who is considered invincible but also has a bad reputation for intimidating his opponents. You are concerned about your challenge but focused, and you play the game of your life. At first, you are losing, but gradually you recover and finally win. You get a standing ovation, everybody applauds you with sheer disbelief; you were destined to lose, but you have just won, and now you are being given a standing ovation by hundreds of people in the audience. You cannot believe what has just happened and suddenly you are very happy, nearly ecstatic. You are overwhelmed by emotions you never had before. Suddenly, the world belongs to you, everything is possible, and you are ready to accept any challenge. It was only a game, but it has triggered a hedonic response.

Both examples describe different forms of play leading to the hedonic response. In fact, all kinds of play may produce a hedonic response. Under certain circumstances, our brain simply does not distinguish between real-life situations and play. This fact gives us a tremendous opportunity to create the hedonic response whenever we need a boost in our energy, in our optimism and, most importantly, in our ability to use our entire brain, to activate our emotions and to conduct abductive thinking, which is absolutely necessary in order to develop creative ideas. Our two examples also explain why the psychological state of play is so important for inventors. It also confirms the well-known fact that for centuries, creative people—great artists and inventors—have considered the process of creation as a wonderful experience bringing happiness and fun. It was Edison (Gelb and Miller 2007) who said "I never did a day's work in my life, it was all fun." (This should be also our mantra.) Now we know that our journey to engineering creativity and inventions must lead through the gate called play.

When we convince ourselves that we are *only* playing, anything may happen, and that is good. Our brain subconsciously strives to have another hedonic response and is ready for action and for doing truly imaginative and speculative things. We forget about our engineering common sense and begin talking nonsense; we start speculating about *impossible* forms or behavior of the object of our interest. We immediately notice that the more we talk, the more nonsense is coming, to our surprise, and we have fun, we are playing.

For example, we are working on a new type of transport plane. We have already learned about various types of planes, and members of our group become relaxed and the play gradually begins.

MEMBER NO. 2: Do we need a plane or do we need to transport the goods?
MEMBER NO. 1: What if our plane were flying goods?
MEMBER NO. 3: Container with wings?
MEMBER NO. 5: Do we need wings to fly?
MEMBER NO. 3: What if there were invisible wings?
MEMBER NO. 4: What if each container could fly independently?
MEMBER NO. 1: What if there were big birds carrying individual containers?
MEMBER NO. 3: What if there were drones carrying individual containers?
MEMBER NO. 4: Do we need containers to transport goods?
MEMBER NO. 2: What if there were drones carrying individual goods?
MEMBER NO. 1: We have just invented a delivery system recently successfully tested by Amazon.com. Congratulations to everybody.

We all suddenly feel the rush of the hedonic response and we simply feel that we are making progress, we are moving in the right direction.

The last psychological state identified by Gordon and his team is the state of *autonomy of object*. In this case, the inventor suddenly feels that the object is an entirely independent entity with its own life—that it is simply an adaptive system. The inventor begins watching the object moving, responding to the input from its environment, and it is like a movie slowly developing in front of his eyes, like virtual reality in action.

For example, we are working again on a transport plane. We are playing, and Member No. 3, Eva, unexpectedly says,

> "I see our object flying. It constantly changes its shape and speed, it moves up and down. It is carrying three packages, but it descends and leaves one package on the ground, it changes its shape, becomes smaller and now it is ascending, now it is flying like an arrow. Oh, it is slowing down, descending again, becomes huge, it is landing, it releases a package and it shrinks. What an incredible video, a flying object changing its shape, changing its size. Are we seeing a modular flying object? A modular plane? A modular drone? I love it, I love it."

She definitely feels a hedonic response. It is infectious, and unexpectedly all the group members become excited and begin rapidly talking about their own ideas. The session moves forward and everybody is happy. That brings more and more ideas.

All five psychological states are obviously interrelated and desired during a Synectics session. We already know about their importance and have at least some tools to stimulate their emergence. The next two sections will bring more specific knowledge about the creation of these states on demand.

8.5 KEY ACTIVITIES

When the problem is sufficiently understood, the stage of development of solutions (design concepts, in our case) begins. Two activities are particularly important: making the familiar strange and making the strange familiar as listed in "Assumptions." Both can be understood as knowledge integration, that is, transforming the available knowledge from various domains into transdisciplinary knowledge, which is the knowledge foundation for engineering creativity (see Section 4.4.). In this case, knowledge integration means changing the context of a given piece of knowledge and discovering in the process a new meaning of this knowledge, which may become the key to the invention. This process is not always easy, requires preparation, knowing just enough, being in an appropriate psychological state, and having the operational ability to use various psychological mechanisms that are discussed in the next section.

The concept of making the familiar strange is best explained considering a simple example.

Everybody knows that the function of a freestanding brick wall is to carry its own weight, or exactly *vertical forces* associated with weight. However, one day about seventy years ago, a structural engineer working on a skeleton structure in a tall building saw a freestanding brick wall (Figure 8.3). He looked again and asked himself a profound question: Why not use it as a wind bracing carrying *horizontal* wind forces? He thought about it again for a while and could not find any reason why a wall could not be used together with a skeleton structure in a tall building, but he determined that its function would be different than usually assumed: carrying horizontal forces instead of the usual vertical forces. The engineer used the familiar concept of a wall under vertical forces and made it strange; by transforming it into a new concept of a wall under horizontal forces, he made the familiar strange. His invention has made a huge impact on designing skeleton structures in tall buildings. Today, walls are widely used in tall buildings as the so-called *shear walls*, that is, walls carrying horizontal wind forces that produce internal shear forces.

The concept of making the strange familiar is much more difficult to understand and to use, and thus it is not as popular as making the familiar strange. In this case, we are imposing a concept known to us on a strange unknown concept, making it understandable and useful for us. Again, a good example will explain this concept.

All engineers understand well the concept of a water distribution system in a chemical plant: a system of interconnected pipes of various diameters (Figure 8.4). A typical engineer has no clue what a shark's blood circulation system is like; it is an unknown and strange concept to him/her. However, when our typical engineer is told that the shark's blood circulation system is like a water distribution system, he/she will smile and respond, "But of course, now I understand it." The strange concept of a shark's blood distribution system was made familiar through the imposition on it of the known concept of a water distribution system.

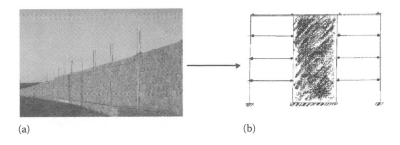

(a)   (b)

Figure 8.3 Making the familiar strange: (a) a brick wall and (b) a shear wall (skeleton structure).

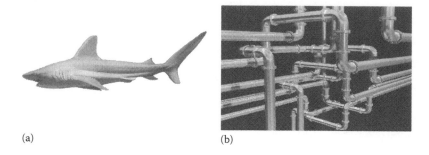

(a)   (b)

Figure 8.4 Making the strange familiar: (a) shark's blood distribution system and (b) a water distribution system.

8.6 PSYCHOLOGICAL MECHANISMS

A *metaphor* is an expression with which an object is described using notions usually associated with an entirely different object. In this way, these two objects are equated, and meaningful intellectual insight plus emotional response are created. Metaphors enrich human language and allow the expression of complex notions in very few words. They also have huge stimulating and emotional power. As a matter of fact, *stimulating power* and *emotional power* are themselves metaphors. More examples follow:

> "Love is a journey"
> "Love is a hell"
> "Sister Lucy is a volcano"
> "Star artist"
> "Skeleton structure"

An analogy is a kind of a metaphor in which two objects are not equated but compared, usually using expressions such as *like* or *as*. Examples:

> "A steel structure is like a skeleton of a building"
> "A beam is used as a bracing"

Metaphors must become a part of the inventor's language in order not only to communicate complex and difficult notions but also to engage his/her emotions, which are the key to human creativity and ultimately to invention. Gordon (1961) identified four types of analogies that are the most important for inventors. He described these analogies as *operational mechanisms*, which can be used when necessary during a session of Synectics. They were extensively tested and verified during the many years of Gordon's research, but obviously their use does not guarantee getting any desired results, because their nature is heuristic. Three analogies are usually used when the group is in the psychological state of play. They often begin as a joke but sometimes end as inventions. The fourth one, the symbolic analogy, is different since it is mostly quantitative and does not require any activation of emotions.

8.6.1 No. 1: Personal analogy

Personal analogy is also called *wishing* or *fantasizing*. In this case, a Synector identifies himself or herself with the object of interest and activates his or her emotions and subconscious. In this way, his/her perspective rapidly changes, rational thinking disappears, and he/she is in a fantasy land where everything is possible, even seeing an atom of blood trying to swim through fat obstacles in the human brain during a stroke. Results may be amazing and groundbreaking. Many famous scientists, like Faraday or Einstein, admitted using this analogy to make fundamental scientific discoveries, although they did not know the name and were doing it on their own. The author also admits using the personal analogy in teaching structural analysis and helping students to identify themselves with flowing internal forces through members and joints in a frame under analysis. (It always works!)

Personal analogy practically requires the state of play. Within this state, or even before this state begins, it may be activated asking appropriate questions or template statements, such as

> "What would I feel being XXX?"
> "How would I feel being XXX?"
> "What would I see being XXX?"
> "What if I were XXX?"
> "I wish I were XXX."

Using words like *feel* or *wish* is nearly magical; it is like switching our brain from the process of rational and deductive thinking to that of creative and abductive thinking. Suddenly we are in the world of our object; in fact for our brain we *are* our object, and our eyes are now open to see things impossible to notice through the eyes of an engineer. We are like XXX, at least for our brain.

A simple example will explain the concept of personal analogy.

Let us assume that our Synectics group is working for the famous Italian producer of light and sweet wines, Asti. The company wants to introduce a new line of sparkling wine, Ariana, and wants to use a new kind of cork, which could be advertised as a cork with a spirit—whatever this word means in the context of sparkling wines. We are in the middle of our session, and we have already acquired enough knowledge about sparkling wines to begin the development of new corks. We are in southern Italy; last night we had outstanding Italian cuisine for dinner, with some wine (it should be admitted), and now we are all happy and ready to have some fun. The group leader serves a fantastic tiramisu with a touch of Asti. We all are smiling, relaxed, and happy, and Anna, our group member, begins talking. What follows is an abbreviated version of her monologue while enjoying tiramisu and sipping Asti Villa Jolanda in the best Italian tradition.

I feel like a cork. Sorry, I *am* a cork. It is so dark. I am squeezed by cold and rigid glass walls surrounding me. I cannot move, I can't breathe. Suddenly it is getting even colder. I must be in a refrigerator. I realize that there is liquid below me but it must be compressed, under pressure. I feel my bottom constantly attacked by particles of gas trying to escape from their prison. This pressure is growing fast, it is pain, terrible pain. Out of the blue, I hear a human voice talking about pulling me out, but ... by using a bottle opener. I do not have eyes, but I can see the big sharp tip of this bottle opener coming to my top and ... I need to find a solution, I need this solution NOW. I am pressured from the bottom and I will be punctured from the top. How to avoid both dangers at the same time? What if I open my mouth and let the gas out and fly out of the bottle? But I do not have any mouth. Do I need a mouth or a way to release gas and at the same time to avoid being punctured when the bottle opener touches me? No, no mouth, please, but a secret passage for gas nearly to the top, a very narrow channel filled with gas. And a *mouth* on the top, a piece of plastic sitting on the top of the channel. This piece looks like lips, my beautiful lips, because it is slightly pre-cut. Now I have a solution. When the tip of the bottle opener comes, it will not puncture me but it will touch my lips and create a small hole between my lips, my lips will be open and the gas will be released and I will be able to leave this terrible hole in cold glass (Figure 8.5).

Figure 8.5 Personal analogy: Anna and a cork. (With permission. Drawn by Joy E. Tartter.)

Everybody listened with astonishment. She was a promising electrical engineer and now she was talking like ... a squeezed cork, but it was apparently working. It became clear to everybody that a solution had just emerged: a cork with a centrally located channel and a built-in plastic valve on the top of this channel. This valve could be easily opened using a bottle opener or a fork; in this way the pressure would be released and pulling out the cork would be easy. Everybody has also realized that the proposed solution would significantly reduce the effort needed to open a bottle of sparkling wine. At this stage nobody was concerned about the feasibility of the solution and or its cost. It would come later. Everybody knew, however, that the play had only begun and the best was to come.

8.6.2 No. 2: Direct analogy

Engineers love the *direct analogy*. It is simple and easy to understand and use. It is also quasi-deductive, and that is greatly appreciated by all engineers. The direct analogy is a psychological mechanism, a process in which two objects are compared on the high level of abstraction. Alexander Graham Bell (MacKenzie 1928) recalled:

> It struck me that the bones of the human ear were very massive, indeed, as compared with the delicate membrane that operated them, and the thought occurred that if a membrane so delicate could move bones relatively so massive, why should not a thicker and stouter piece of membrane move my piece of steel. And the telephone was conceived. (Figure 8.6)

In the context of knowledge, using a direct analogy may be understood as discovering a previously unknown relationship between two objects. For example, if we consider Isaac Newton's discovery of gravity while watching a falling apple, his analogy may be formulated as

> If an apple is attracted by the earth, why can't any two objects be attracted to each other in the same way? (Figure 8.7)

For engineering applications, Newton's analogy can be generalized and presented as

> If X works in a desired way under given circumstances, why can't Y work in a similar way under the same circumstances?

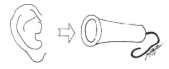

Figure 8.6 Direct analogy: Human ear and Bell's telephone receiver. (With permission. Drawn by Joy E. Tartter.)

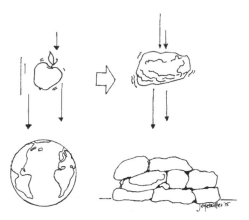

Figure 8.7 Direct analogy: Newton and the law of gravity. (With permission. Drawn by Joy E. Tartter.)

When a direct analogy is presented in such a general form, it becomes a heuristic and can be used for many engineering and creative purposes. This heuristic helps us to reason that if steel can be used to build a car chassis, aluminum could be also used, as Ford did recently in its 2015 F150 light truck.

The use of the direct analogy requires a sufficiently large and differentiated body of knowledge coming from various fields. The French inventor Pasteur wrote (Valler-Radot 1902) that his world-changing inventions were based "on varied notions borrowed from diverse branches of science," and that clearly demonstrates the importance of this analogy. In engineering, knowledge from biology is particularly attractive and should be always considered when inventing new engineering systems. For this reason, Chapter 10, "Bio-inspiration," specifically discusses how this could be done.

8.6.3 No. 3: Symbolic analogy

The *symbolic analogy* is also called *compressed conflict,* and this second name probably better reflects its nature. It is a difficult analogy, and a lot of training is necessary to develop the ability to use it well during a session. However, the effort is justified, particularly because this analogy can be used independently in our engineering practice since it also has intrinsic value outside Synectics. The name refers to a mostly deductive process resulting in a short and abstract statement identifying the essence of the problem. It is like making the strange familiar and ending with a piece of poetry reflecting our newly acquired understanding of the problem. The statement may contain contradictory words or notions and can be understood as a contradiction, which is the heart of the problem and must be eliminated to solve it. In Chapter 9 on TRIZ ("theory of inventive problem solving" in Russian), the concept of a contradiction will be discussed in detail, since this concept is absolutely crucial for understanding and using TRIZ and probably for understanding inventive engineering.

The best-known classical examples of the symbolic analogy provided by Gordon (1961) are Shakespeare's analogy of the *captive victor* and Pasteur's analogy of *safe attack*. The analogy of the captive victor can be interpreted in at least three ways. First, it is about losing a battle and becoming a prisoner of war but remaining victorious. Second, it is about winning a battle but still remaining surrounded by the enemy. Finally, it may be about losing a battle in order to capture the opponent and to win the war later. None of these interpretations are definite, and that is their power; they force us to look for other interpretations and rapidly expand our search space.

Pasteur's analogy was the key to the development of his vaccines, which are generally based on the concept of creating viral attacks on the human body, but attacks that are safe because the viruses used were weakened earlier. The body is able to defend itself and in the process becomes immune to attacks by truly vicious viruses. This example is also particularly appropriate for engineers, because it shows how an aggressive and potentially dangerous action may result in a definitely positive outcome.

For example, applying compressive forces to a structural member is potentially dangerous and may lead to its destruction through compression or buckling. However, if a compressive force is appropriately applied to a concrete beam under bending in the beam's tension zone, it reduces tensile stresses that are particularly dangerous for concrete. In this way, attacking a beam under bending with compressive stresses significantly increases the beam's load-carrying capacity. It is a practical engineering application of the symbolic analogy of safe attack. In fact, structural engineers are using an entire class of structures, called *prestressed structures*, which were developed using the principle of the safe attack. The best known are prestressed concrete beams (widely used in various multistory garage structures), but prestressing may be also used in steel structures (Figure 8.8).

There is another and easier interpretation of the symbolic analogy, which is that it is simply a statement that synthetizes the entire problem using several abstract terms or describing all the essential elements of the system under consideration. Such use of the symbolic analogy can be best explained by an example involving the author.

When the author was in his early twenties, he worked at the Department of Metal Structures at the Warsaw University of Technology, Poland. Once, he was invited by a senior faculty member, a world-class expert in the area of steel structures, to determine with him the mechanism by which a large steel structure had collapsed in a small town in northern Poland. This determination had various legal implications and was quite important for all parties involved in the incident.

The structure to be investigated was a large steel gas tank with a floating roof. The roof structure was in the form of several steel arches, which

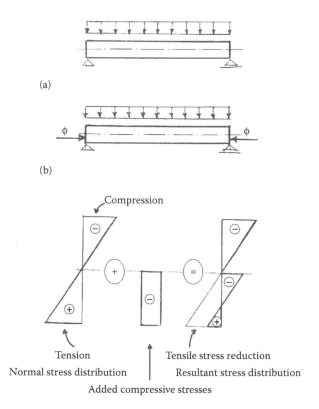

Figure 8.8 Symbolic analogy: "safe attack." (a) A beam under bending (transverse load). (b) A beam under bending and compression.

supported parallel beams on which steel sheets were placed. During a severe winter with a lot of snow accumulation, a period of thaw took place, and after that the water became frozen; after more periods of thaw and freeze, the roof collapsed. It was overloaded by a huge weight of snow and ice, exceeding the design load by a factor of three or four. Nobody was killed as, fortunately, it happened during the night, but the financial losses were substantial.

First, the author spent several days reading the entire technical documentation of the structure, including analysis of loads, analysis of internal forces, structural design, detailing, maintenance records, and so on. After that, the expert and the author traveled to see the collapsed structure. They saw a mountain of twisted steel, still covered by snow and ice, as well as the faces of the concerned engineers from the gas company, which owned the tank. The expert and the author examined the arches particularly carefully. One of them was deformed in a very specific way, and plastic hinges were found; this was entirely consistent with plastic collapse rather than with buckling, which had been suspected. Suddenly, long hours of analysis based on deductive thinking, the thorough examination of many structural members—all this systematic and purely quantitative effort was transformed into two words of our opinion: a plastic collapse. We synthesized the mountain of data using our background knowledge, and finally the symbolic analogy simply emerged. This symbolic analogy became obviously the equivalent of all the documentation and of all this twisted steel and in very few words perfectly described the situation.

The symbolic analogy is similar to the personal analogy in that that both produce abstract results. The fundamental difference is in the emotional engagement. Practically, the precondition for using the personal analogy is the occurrence of the psychological state of play, while the symbolic analogy is mostly the product of high-level abstract thinking, which is rational but unusually complex and therefore difficult for engineers. In fact, the symbolic analogy is the only one that is not dependent on human emotions and may be used by all engineers, not only by inventive engineers.

8.6.4 No. 4: Fantasy analogy

This term refers to a powerful psychological mechanism based on the assumption of suspended judgment and critical thinking, and even of our self-consciousness. It is like the rational world does not exist and we can freely develop and consider all kinds of ideas that are generated by our creative mind. When engineers are doing a simple routine analysis, the left deductive hemisphere of their brain is fully engaged and controls their thinking, while the right abductive hemisphere is idling. When using the fantasy analogy, we have a reversed situation; the left hemisphere is entirely deactivated, while the right hemisphere is completely in charge. At the time this deactivation happens, even very serious and often narrow-minded engineers begin thinking like creative people or like inventors, producing, to their great surprise, unexpected and exciting results. These results may not be directly useful, but they may initiate new lines of thinking or may simply stimulate other group members to propose their ideas, related or not. Children often use the fantasy analogy as a learning mechanism, helping them to explore the world and acquire knowledge through a series of statements, which provoke feedback from adults providing the meaning and context of these statements.

Sigmund Freud, the Austrian father of psychoanalysis, developed a theory of human creativity (Freud 1930). The fundamental assumption of his theory is that human creativity is the fulfillment of a wish or a fantasy. He was particularly interested in creativity in art, but Gordon practically proved that creativity in art and in engineering are the same in terms of the psychological mechanisms that are engaged. Therefore, Freud's theory provides a good explanation of why the fantasy analogy is so important and powerful, and why it is the essence of Synectics. Freud studied also human dreams: daydreams, and particularly night dreams. He strongly believed that our night dreams in particular are a subconscious reflection of our wishes, wishes that we are afraid even to articulate but which are secretly controlling our lives. In this context, the fantasy analogy is like a window onto our subconscious, revealing our wishes but also helping us to utilize the hidden mechanisms of subconscious in order to advance our inventive agenda.

The fantasy analogy is a short conditional statement like

"If I were rich and famous"
"If the beam were rigid and soft at the same time"
"If the cable were short and long at the same time"

The analogy is relatively easy to understand and to use, but it requires some prior training and a proper psychological state. It is usually used during the state of play and at a time when all participants are already relaxed, their emotions are activated, and they are ready to play like children. They subconsciously desire the hedonic response, and that strongly motivates them to forget about all the limitations of the real world and to become creative.

The use of the fantasy analogy will be illustrated by a simple example of a group of Synectors working on a challenging problem of cleaning and washing windows in a tall building. The author recorded a part of the session and the transcription of his recording is provided here.

PARTICIPANT NO. 1: We have this huge glass wall covered with dirt and dust, ugly. I wish I were able to clean it quickly and safely, like having a magic wand; I wave it and the glass is perfectly clean.

PARTICIPANT NO. 3: Yes, you are right; no cranes, no cables, no falling carts. Pure magic and clean glass.

PARTICIPANT NO. 5: Guys, guys, I have a dream. I am dreaming about having a rare ability to be suspended in the air with a water hose and cleaning materials.

PARTICIPANT NO. 2: So you are saying, "If I were able to fly and be suspended in the air."

PARTICIPANT NO. 5: Exactly.

PARTICIPANT NO. 4: Magic means to me that I do not need to be suspended in the air and do the cleaning. But if there were a system capable of being suspended in the air and doing the cleaning job, it would be a fulfillment of *my* dream.

PARTICIPANT NO. 3: You mean a helicopter?

PARTICIPANT NO. 5: Yes and no. A helicopter is large and heavy; it is making a lot of noise and it cannot fly too close to a building. Its airflows are long and it must stay at least 50 ft. from a building. Something like a helicopter but small.

PARTICIPANT NO. 3: I wish there were small helicopters capable of flying very close to the building.

PARTICIPANT NO. 1: But there are. They are called "helicopter drones," they cost several hundred dollars and are quite capable.

PARTICIPANT NO. 2: But a drone can carry so little cleaning materials and will stay in the air for only 15 minutes.

PARTICIPANT NO. 3: But if I were a building owner, I could buy hundreds of them and create waves of drones bringing water, cleaning "stuff," and soft cotton towels to polish glass. They would be acting as a kind of intelligent robotic flying swarm. Some of these drones could also play music. People hate the cleaning window days but now they would enjoy them.

PARTICIPANT NO. 4: I am not sure about the music. Who would decide about the kind of music to play? The owner of the building or the trade union?

PARTICIPANT NO. 5: Do not worry about that. The federal government will immediately regulate this area and recommend playing only politically correct music, which would not offend anybody.

PARTICIPANT NO. 3: Let us leave the music alone. It is a serious political issue. Only politicians can solve it, but we are good people, we are engineers and we have already found the solution: *a swarm of drones*.

8.6 PROCEDURE

Gordon (1961) saw Synectics as a two-stage process. He suggested a more detailed sequence of events when solving the problem, but it was a mere suggestion without the power of imposing this sequence on Synectors. The essence of Synectics is the full utilization of the human creative potential through the use of various sophisticated psychological states and mechanisms. For this reason, there is no need for a detailed procedure. Moreover, such a procedure would be followed literally by the engineers as all analytical procedures are, and that would hurt their ability to relax and to properly engage in a session focusing more on the procedure than on the creative process. This is the classical example of an old engineering heuristic: "Less is better." The procedure is simple:

1. Problem identification and formulation
2. Problem solving

Problem identification is usually provided by an outside engineering body, and it is developed in the traditional engineering and deductive way. If it is not provided, Synectors can develop it using, for example, the methods discussed in Chapter 5, Section 5.3.

The *problem formulation* process is unique for Synectics and it is called *springboarding*. It can be understood as a process of making the strange familiar. Problem definition is given from outside the group and initially it is strange for Synectors. However, as a result of springboarding, the group develops an understanding of the problem definition and reformulates it into a springboard, which is their unique product; obviously it is understandable and familiar to them.

Springboarding is a powerful process initiating the core part of a Synectics session. It is like a bridge connecting the real world and its problems, reflected in the *problem identification*, with the Synectics world of fantasy, suspended judgment, and so on. As a transitional system, it has integrated elements of both worlds. The product of this process is called a springboard. It is a short statement, the goal of which is to communicate the essence of the problem and at the same time to inspire Synectors. A good springboard should have several interrelated features:

- It should contain all key words associated with our problem.
- It should present our initial understanding of the problem.
- It should show the context of our session.
- It might contain our initial solutions in very general terms.
- It should contain our main initial thoughts about the problem, along with all kinds of associations and even images.

All key words associated with our problem should be identified but preferably substituted in the springboard by words that are alike but more abstract, following the simple heuristic "Use the most abstract equivalent words." The objective is to use words with the most power to trigger other words or ideas through the use of abstract key words. Our understanding of the problem should be also presented in abstract terms in order to avoid implying specific and obvious solutions. The springboard should reveal whose point of view is accepted or rejected by the team. In this way the context of the session is provided. A Synectics session is a situated activity (Gero 2007), and knowing the context is absolutely necessary to communicate with others and simply to be a productive Synector. The springboard might even contain an initial solution or a class of solutions, but again they should be on a very high level of abstraction and should suggest a large class of solutions. For example, it might introduce the concept of a bridge (vs. a ferry) but not a much more specific concept of a steel bridge or a concrete bridge. Finally, including initial thoughts, associations, and visions, all in a very abstract form, might help the team to move from the analytical activities (associated with the problem formulation) to the creative activities (associated with the Synectics session).

A springboard should start with "I wish ..." or "How to ..." followed by a short statement. Such a formulation is consistent with the fantasy analogy, and this is not coincidental. Other formulations may be also used, like "I would like to know how to fly without a plane," but such formulations are longer and their stimulating power is smaller than of the two recommended. In this way, Synectors slowly enter the psychological state of play and are getting ready for the development of new ideas. The following examples will aid a better understanding of the concepts of springboarding and of a springboard.

> "I wish that this column were lighter, stronger and longer at the same time."
> "I wish I could stay in the air above the bridge and observe traffic."
> "I wish that my car were faster and safer at the same time."
> "I wish that my vacuum cleaner were like a rolling stone."
> "I would like to know how to build a safe plane."
> "Making this door fireproof is the question."

Developing a springboard is a process involving all the participants in a session. Initially, it is a trial-and-error process, but gradually, as a result of discussion and improving understanding of the problem, a springboard emerges that is clear and acceptable for all group members. By *clear* should be understood not a black-and-white definition but a statement that correctly reflects the understanding of all group members, although formulated in an abstract and purposefully vague language stimulating the human imagination.

The *problem solving* stage begins when the springboard is ready and everybody feels comfortable with it. At that time, the group is most likely in the state of play, and it is the time to begin using various psychological mechanisms. The sequence in which the analogies are used is highly subjective and depends on the group. Usually, direct and personal analogies are used first, since they are easier to use than the other analogies. Symbolic and fantasy analogies are generally used later when the group is truly engaged in the development of solutions and has reached the state in which all the participants are at the peak of their creative performance. When this peak is over and all the group members begin showing signs of exhaustion, or, even worse, signs of boredom, it is time to employ and method called *excursion*, which is also unique for Synectics but could be effectively used outside the Synectics session.

Excursion is an interesting method that is used when the progress of a session stalls—when we hit the wall and need an additional stimulus, or we could say that we have encountered a roadblock and need a breakthrough. The other situation when excursion could help is when the group wants to move in an entirely different direction. The casual understanding of excursion is that it is a way to get fresh and stimulating ideas when we need them and that they will come from outside the traditional understanding of the problem. It is like taking a mental vacation from the problem in order to see it from a different perspective and to find the missing solution (Gordon 1961).

There are at least two interpretations of this method when we assume that excursion is a random search for new ideas through a certain body of knowledge. First, we may assume that this body of knowledge is the total knowledge of all session participants. When only this body of knowledge is searched, it is *exploitation* (Figure 8.9) in accordance to (Gero and Schnier 1995). In the second case, excursion can be interpreted as moving the search outside the total knowledge of all participants (Figure 8.9) to an entirely different and unexplored point outside this knowledge. The term *exploration* can be used in this case (Gero and Schnier 1995).

Excursion is a powerful method that usually brings surprising results. First, it enhances the *fun factor* and helps the participants to fight their reluctance to become entirely relaxed and start playing like children. Second, and probably more important, it actually works. It reformulates the problem and its understanding and brings a new start to the deadlocked session.

From the psychological point of view, excursion changes the context of the discussion and moves it, for instance, from chemical engineering to mechanical engineering. The same problem in these two domains may be understood entirely differently, and bringing the second perspective may greatly help. For example, in chemical engineering, excessive heat is understood as a by-product of a runaway chemical reaction, while in mechanical engineering, it may be understood as a by-product of heavy breaking.

Excursion has a six-step heuristic procedure, as shown and explained here:

1. Temporarily forget about the problem.
2. Go to the springboard.

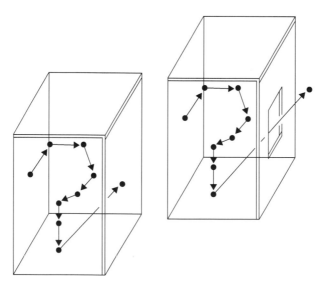

Figure 8.9 Exploitation and exploration.

3. Randomly select one of the key words.
4. Randomly select an irrelevant word/concept (*concept X*) from the surrounding outside world.
5. Concentrate on this irrelevant concept X and study it.
6. Begin thinking again about your problem and the selected key word and formulate a second-generation concept or concepts.

In the first step, we do everything we can to forget about our problem; for example, we begin talking about our last vacation on the Cayman Islands, about our children and dogs, even about our mothers-in-law (very effective because it creates so many highly emotionally charged responses). We simply need to change the focus of our brains. When we have an impression that it has already happened, we ask one of the participants to read our springboard out loudly once or twice. Next, we ask the youngest, or the oldest, member of our group to select his or her favorite key word from the springboard and write it down. The following step, No. 4, "Randomly select an irrelevant concept X from the surrounding outside world" is crucial, but it is also a lot of fun moving in an unpredictable direction. There are at least twenty-nine ways to find our irrelevant concept X, but we will briefly overview only four of them.

8.6.1 First way

Ask one of the group members to describe his/her favorite activity, for example playing a flute or yodeling. Next, use one of the words used in this description as a fresh concept X.

8.6.2 Second way

Send one of the group members to a nearby public library to fetch an old copy of *Time* magazine. Next, close your eyes, open the magazine, and put your finger on the text. Your finger will show you the concept X to be used.

8.6.3 Third way

Open the eighth volume of the *British Encyclopedia* on page 273, go to line 34, and find the 14th word from the left margin. It is your new inspiring concept X.

8.6.4 Fourth way

Ask one of the female members of your group for permission to open her purse and to pull out a single object, for example a car key, a pen, an iPhone, or a screwdriver. Next, use this object as your new imagination stimulating concept X.

These four examples show how many various random forms of search for the irrelevant concept X can be used, all equally effective and all providing the desired new concept X that seems to be entirely irrelevant to our problem solving, at least for the moment. When the concept X is finally known, it needs to be studied. First, everybody complains that a wrong concept X was found (which is simply impossible) and that it is useless. The group members talk more and more about the concept X, and their discussion gradually becomes stormy. Suddenly, one of the group members begins talking about using it to address the problem and proposes the first second-generation idea rooted in the concept X. The excursion is over, and the group comes back to the main flow of ideas.

The procedure can be best explained using the simple example of finding a new type/concept of an energy-efficient residential ceiling fan. Your company, Happy Fans, a medium-size residential fan manufacturer in Virginia, the United States, has just learned about the incredible success story behind BigAss fan, which has had an increase in sales of 86% over the last quarter. Your company has decided to develop a new concept for a residential ceiling fan, potentially patentable, which could help it to restore its position as an innovation leader and to increase stagnant sales.

The Synectics session is in progress, but great ideas are simply not coming, and the group decides to conduct an excursion. The group leader reads twice the simple springboard developed by the group: We wish to develop a butterfly of ceiling fans, whose smooth and gracious movements will create an air of happiness and comfort in medium-size rooms while improving the ambience through its majestic and calming looks.

Milena, the youngest member of the group and the rising star in the accounting department, is send to the local public library to bring her favorite book, and she comes back with an old nineteenth-century romance novel. Everybody is surprised that an accountant would select a romance but also impressed by the exceptional beauty of the illustrations and the poetic language used to describe the adventures of two young lovers. John is asked to find randomly a word from the book, and he opens it on a page with a description of the lovers galloping on a beautiful black stallion. He closes his eyes and his finger lands on a paragraph describing the horse and the smooth moves of the animal's left ear while he is galloping. The horse ear becomes the concept X, and the fun begins. As expected, at the beginning nobody thinks that the horse ear has anything to do with the design of residential fans. Mark even suggests that another book is found, one more focused on engineering than on love stories, but several group members want to try this concept X.

Next, a heated discussion begins, with Milena and another group member arguing that the horse ear is obviously used by the animal for hearing but also for cooling the head through complex movements, not to mention for fighting flies. This perspective attracts the attention of the domain expert, who is a mechanical engineer. He states that the function of a residential fan is to move air, not to rotate the airflows, and that an air-moving device that replicates movements of the horse ear could be actually designed and built. Everybody is surprised by this unexpected change of focus. Now, the group is moving from searching for a concept of a fan with rotating airflows to a class of air-moving device, and everybody feels happy about it. All the group members start clapping hands, laughing, and the group dynamics changes in the most positive way. Also, everybody thinks that a significant breakthrough has taken place, and the group may come back to the development of the problem-related ideas but with a new direction and enthusiasm.

8.7 SESSION ORGANIZATION

There are obvious similarities and differences between the organization of brainstorming and Synectics sessions. Similarities result from the analogous roots of both methods in psychology. The use of these two methods is supposed to engage human emotions and in this way to maximize the creative power not only of the individual group members but also the power of the entire group. In fact, authors of both methods claim that individual members greatly benefit from interactions with the entire group and that its feedback and stimulation make them more creative.

As in the case of brainstorming, preparations for a Synectics session are in two parts. First, the *creative space*, or the physical environment for the session, must be created. Chapter 7 provides detailed guidelines on how the creative space for a brainstorming session should be prepared. All these guidelines are also valid for a Synectics session with only small modifications.

Considering the analogical and abstract thinking involved in Synectics, art might be slightly different with the use of more abstract and allegorical art to create the feeling of *sfumato* and complexity and to prepare Synectors for action. Salvador Dali's art seems to reflect well the spirit of Synectics through the surrealistic and random combinations of various objects. These objects, taken together, suddenly create an emerging and surprising meaning, a hidden message; just like in Synectics, playing with irrelevant objects, physical or abstract, may create inventions. Also "engineering" drawings by Maurits Cornelis Escher (M.C. Escher) generate an atmosphere of abstract creativity with roots in "solid" traditional science like mathematics.

The second part of preparations regards Synectors. They must be selected and properly trained. As in the case of brainstorming, both parts are just as important and need a lot of effort. In the United States, an entire Synectics industry exists. Therefore, it is costly but prudent to use consultants specializing in organizing and running Synectics sessions, particularly considering the complex and sophisticated nature of the method but also the possible huge payoffs.

A Synectics session may be relatively long, even lasting several hours (a productive brainstorming session may be as short as 15 minutes), and many experts suggest scheduling it for an entire day. The length of the session is simply a reflection of two facts. First, it is a psychological fact that future session participants are familiar with the *bad reputation* of the method, that is, its relative difficulty, and are simply afraid that very little will be accomplished in a short time period. Next, the creation of psychological states, the use of analogies, high-level abstract discussion, and so on actually need a lot of time without any constraints and expectations that anything meaningful would be accomplished in a half an hour or so. We need a mood conducive to play, not that of a race. We should also remember that the formulation of the springboard may take at least 32–67 minutes, and it is obviously only the beginning. Realistically, a time frame of at least 3–5 hours should be expected, and in this case *longer* may actually mean *better*. It is a good practice to have short breaks with sweets and coffee, but these breaks should not be longer than 10 minutes; longer breaks change the focus of participants and they are simply counterproductive. The author's recommendation is to plan plenty of time for a session, even two days, to make sure that nobody is under stress but that everybody feels comfortable with the situation and feels that is feasible to develop creative results. In fact, a Synectics session is like a complex adaptive system whose behavior cannot be predicted (in fact, such behavior is desirable) and as such should be planned with the maximum flexibility. As in the case of brainstorming, challenging problems with a social or human dimension usually improve the group's enthusiasm and help. Also, surprisingly, problems or their initial solutions that are perceived by the group as "funny" have a very positive effect on the group.

The author remembers his experience with Synectics about 15 years ago. He was a member of a group working on a sustainable restaurant (a multistory building with the restaurant on the top, a barn with cattle below, fish tanks on the another level below the barn, and containers with wild mushrooms in the basement). When the group gradually developed this concept of a multistory sustainable restaurant, everybody thought that it was truly funny, although today in the days of sustainability our response may seem odd (it would be much more enthusiastic today). There was a lot of laughter, everybody felt like a small child playing with a crazy idea, and the group entered the state of play. In this way, it is another proof that the fun factor is a key to a successful session.

Another key to a successful session is the selection of the right people to become Synectors. Gordon (1961) believed that the optimal group size is about five people, all volunteers, each carefully chosen on the basis of long interviews (even lasting 6–10 hours) and the thorough analysis of their backgrounds. The reason why volunteers are sought is the same as in the case of brainstorming: finding highly motivated people who are prepared to make sacrifices in order to succeed.

Gordon (1961) recommended a two-stage selection process involving *preliminary* and *final selection* stages, each with its own set of selection criteria, which are particularly meaningful when selecting future Synectors. They have been modified by the author to reflect the present state of the art. These preliminary criteria are briefly discussed in the context of organizing a group of Synectors for a manufacturing company interested in new products. After these criteria are used to conduct the preliminary selection, a different set of nine final selection criteria is used to identify the future Synectors, and these criteria are discussed later. The selection process is extremely important because the method is relatively ill structured, particularly when compared with *morphological analysis* or TRIZ. Therefore, the quality of Synectors is the key to a successful session. Without properly selected and prepared Synectors there is nearly a guarantee that a session will simply end as a waste of time and company's resources.

Preliminary selection criteria are as follows:

1. *"Representation/Professional Diversification" Preliminary Criterion:* About 60% of the group (three out of five) should have knowledge covering all major areas of the company's operations, including research, engineering, production, finances, and so on. The remaining group members, two out of five, should have definitely a nonengineering background, for example in fine art, music, biology, or history. In this way, a relative balance is created between problem-specific knowledge and domains rich in irrelevant knowledge, which is so important for the success of Synectics.

2. *"Energy Level and Attitudes" Preliminary Criterion:* A Synectics session needs to be upbeat, dynamic, and engaging, its participants active and involved. For all these reasons, only people with a high energy are sought. However, sometimes there is a thin boundary between high-energy and dynamic behavior and manic behavior, which could have a devastating impact on a session, and we need to be aware of this issue.

   Equally importantly, all participants should have a positive attitude to life in general and to their work in particular. In other words, they should be practicing *positive psychology* as discussed in Chapter 2, Section 2.2., and they should have significant appreciative intelligence, that is, a rare ability to see mostly positive aspects of any situation and to believe that any situation will ultimately lead to their advancement, no matter what their initial feelings and assessment of the situation at hand. On the other hand, in each organization can be found people who are always complaining about their bad luck in life and about not having a chance. Some such people are highly knowledgeable, brilliant, and, on the surface, ready for action. However, Gordon (1961) discovered that such people are often subconsciously attracted to their bad luck and are trying to prove it. In the case of

Synectics, such people spoil the well with their lack of confidence combined with their secret subconscious desire to fail. Knowing this phenomenon, it is prudent not to invite such people to join a group of Synectors, but eventually to give them a different chance without so many costly consequences of failure.

3. *"Age" Preliminary Criterion:* Practicing Synectics requires a lot of energy and an ability to accept and support all kinds of unusual activities. People who are aged between 25 and 40 years best meet such demands. According to Gordon (1961), the personalities of younger people are still evolving, and that limits their ability to do unusual things. Moreover, such people may not know enough to be effective Synectors. People above 40 may know "too much" and therefore they may easily act as "experts," able and willing to destroy any new ideas. Also, they may have a reduced ability to do unusual things and to accept and expand results of such activities.

There is another reason why the age range within a group should not be too large. The age differentiation brings huge differences in life experience and knowledge, and that subsequently leads to a very undesirable hierarchy emerging within the group. Such a situation transforms group members and partners into hierarchy members who naturally listen to older people and follow their lead instead of being emotionally and intellectually independent and creative.

4. *"Administrative Potential" Preliminary Criterion:* Administrative ability includes, among other things, specific abilities to abstract and generalize that are critical for successful administrators. *Abstraction* is an operation in which quantitative/numerical descriptors of an object are replaced by qualitative/abstract descriptors. For example, the descriptor "weight" = "20 lb" is substituted by "weight" = "light."

Generalization is an operation in which selected descriptors of an object are eliminated, and in this way a given description becomes more general and covers a larger set of objects. For example, the initial description, or the initial set of descriptors, is "structure" = "truss" and "span" = "20 ft." It is substituted by a new more general description in the form of a one-element reduced set "structure" = "truss." This new set obviously describes a much more numerous set of objects.

Both abilities are important for inventors and especially for Synectors. Gordon (1961) therefore argues that Synectors have a great administrative potential. For this reason, one of his selection criteria for Synectors is administrative potential. Unfortunately, the issue is a little more complicated. Administrators can be roughly divided into leaders and managers. Synectors should definitely be good leaders, but managing people and complex engineering systems may not be the best way of utilizing highly creative people at the peak of their inventive potential.

5. *"Entrepreneurship" Preliminary Criterion:* Ideally, in an industrial company, a group of Synectors (and future inventors) should not be focused exclusively on inventing but should also care about the ultimate commercialization of their inventions. It is an interesting situation. Synectors should feel sufficiently independent from the everyday operations of the company to practice inventive engineering without constant pressure about profits. At the same time, the group should be motivated to act as young and hungry entrepreneurs. In this way, the company will benefit from having within its organization a kind of internal entrepreneurship, which is unheard of in well-established and matured companies. For this reason, Gordon (1961) recommended selecting future Synectors using, among others, the entrepreneurship selection criterion.

6. *"Job Background" Preliminary Criterion:* This selection criterion actually reflects two perspectives of the work history of a candidate, including the perspectives of knowledge and personality. In the first case, it is desirable for a candidate to have had many jobs within the company, which should lead to a rich and diversified body of professional knowledge. Usually, administrators do not appreciate *job jumpers* and question their personality. However, this seemingly negative personality trait may simply mean a person who is uncomfortable, bored, or simply impatient with the snail's pace of routine engineering work. When looking for future Synectors, such a person may be the perfect candidate.

7. *"Education" Preliminary Criterion:* The *education* selection criterion surprisingly is not related to any specific area of science or engineering. Instead, it is a measure of intellectual curiosity, reflected in changing fields of study and areas of professional interest. The criterion actually measures the number of fundamental professional shifts in the work history of a given candidate. It is based on the assumption that a Synector should have substantial knowledge in at least two entirely different areas. This knowledge is necessary to create metaphors and analogies, which are so important in Synectics. More fundamental professional shifts are considered better since they all contribute to the growth of the *metaphoric potential*, which consequently has an impact on human creativity.

8. *"The Almost Individual" Preliminary Criterion:* There are highly qualified and talented individuals who for various reasons are not able to produce results corresponding to their potential. For example, they may be shy in an aggressive environment or have a speech impairment yet work in a position requiring making frequent presentations. Gordon (1961) calls such people *almost individuals*. He believes that they could become excellent Synectors, assuming they do not have any serious mental problems, which might limit

their ability to work with a group. Almost individuals may rapidly grow as Synectors and contribute to the group's success in many ways, performing well above expectations based on their prior performance.

Potential Synectors should be selected through a series of interviews conducted by present and past Synectors and by psychologists. Gordon (1961) recommends that several issues be specifically addressed early in the process to prepare the candidate for the interviews but also to ready them for a possible rejection. First, the candidate should learn that his or her company is experimenting with the creation of a problem-solving group, which will be using Synectics. Obviously, the general outline of Synectics should be also revealed. (At this stage, there will be candidates deciding to withdraw from the interview, feeling very uncomfortable with a method that uses their subconscious and metaphorical abilities to produce results.) Next, the candidate should be told the reasons why this activity is considered by the company's leadership as crucial for the company's future. This context for Synectics and the resulting motivation are essential for the effective interview. Also, the candidate should become convinced that the selection process is less based on the evaluation of his/her intellectual abilities and more on the assessment of metaphorical abilities and on a specific combination of knowledge from several domains. In this way, his or her rejection will be easier to rationalize and accept. Finally, it should be made clear that rejection would have no impact on the candidate's position within the company.

The preliminary selection is conducted through lengthy interviews, which gradually reveal the true personality of candidates and help to identify a group of finalists. The final selection is conducted through interviews but also on the basis of observations of individual candidates as they participate in various group activities. These activities force the candidates to show their true face, revealing their various attitudes and behavioral patterns. The activities may include living at a camp site for a day or two, hiking, building a dam on a stream, and so on.

The final selection criteria are intended to identify candidates who will be the "best" Synectors from the group of people who passed the preliminary selection. There are nine such criteria, discussed here following the hierarchy of their importance.

1. *"Metaphoric Capacity" Final Criterion:* During all interviews, the language of candidates is thoroughly monitored for the use of all four analogies (personal, symbolic, direct, and fantasy analogies). In fact, the candidates are requested to speak metaphorically. For some candidates it is easy, but for many it is nearly impossible, and that becomes apparent during the first or second hour of the interview. Such candidates need to be eliminated since the use of analogies is absolutely critical for the success of a group of Synectors.

2. *"Attitude of Assistance" Final Criterion:* Practicing Synectics requires the constant cooperation of all Synectors and their continuous involvement; it requires also that they try not only to respond to the requests coming from the other participants but also to be proactive, that is, to try to actively predict their imminent needs or requests. The best way to determine the *attitude of assistance* is to watch candidates in action when they try as a group to build a gazebo or grill a goat on a fire. Three categories of candidates will be identified: (a) watching, (b) asking if they could help, and (c) suggesting various ways they could help. Obviously, the third category is the best as far as future Synectors are concerned.

3. *"Kinesthetic Coordination" Final Criterion:* There are creative people who are exceptionally disorganized and live surrounded by permanent mess. Their behavior is chaotic, and they seem to be unable to fully control their movements and gestures. When such people need to work as a team, for example, building a dam on a stream, their lack of coordination makes them a serious impediment for the group's effort. Unfortunately, that has also implications for their Synectics potential because the extreme lack of physical/kinesthetic coordination implies a lack of self-confidence, and that may affect their ability to relax and to use their potential. Also, the extreme clumsiness of such people may negatively affect the group's dynamic and may introduce chaotic behaviors, destroying the fluent flow of ideas.

4. *"Risk" Final Criterion:* The interviews should determine if the candidate is risk affinitive or risk aversive. Obviously, the first kind of candidate is preferable. It is important to determine, however, if the candidate takes risks simply for the thrill of doing it and without any benefit analysis, or if he/she is basically a careful person, and only when an extraordinary opportunity arises is he or she able to take calculated high risks associated with great rewards.

5. *"Emotional Maturity" Final Criterion:* Creative people are often like children, asking hundreds of questions all the time, being spontaneous in their reactions, and revealing their feelings in their behavior. That does not mean that they are emotionally immature. It only implies that they are capable of intellectually growing, that is, learning all the time and expanding their understanding of the world. Even more importantly, it shows that they feel comfortable learning from others and revealing their thoughts to others. These are all important qualities sought in Synectors.

6. *"The Capacity to Generalize" Final Criterion:* We have already explained (see Section 4.9) that the process of generalizing, or creating generalizations, involves the elimination of some attributes/descriptors describing a given engineering system, or a few systems, to create a description covering a much larger set of engineering systems. A generalization is therefore a hypothesis of a description of a

class of systems, or a concept of such a class. Creating concepts for engineering systems is definitely difficult and very few engineers are capable of doing it, although this capacity is crucial for Synectors and inventors. As a matter of fact, specialization, or adding descriptors to identify a subclass, is much easier for engineers and the majority of them know how to do it.

7. *"Commitment" Final Criterion:* Using Synectics requires strong motivation and even stronger commitment to succeed. Many creative people are sensitive and often doubtful about their talents and their ability to succeed. That is to be expected, but excessive doubtfulness may "paralyze" Synectors, preventing them from succeeding. To counterbalance this undesired phenomenon, Synectors should be strongly committed to solving the problem in spite of their own doubts or the challenges that emerge during the process.

8. *"Nonstatus Orientation" Final Criterion:* There are traditional power symbols associated with the social or the professional position within a given company, for example the number and size of ferns or carpets in the office. Synectors should not be status oriented because during a session another kind of status will emerge and must be accepted, a status based on contributions and independence, and this is usually entirely different to the traditional status based on the number of Persian carpets in somebody's office.

   During a day-long interview, a true picture of the candidate will emerge. It will reveal if he or she is obsessed and controlled by the traditional status symbols (Persian carpets or Rolex watches) or is open to accepting other creativity- and Synectics-related status symbols.

9. *"Complementary Aspect" Final Criterion:* A group of Synectors should be so much more than a group: It must be a system. Therefore, individual members do not need to score high on all selection criteria, but they definitely need to "fit" the group, and to create the "perfect" combination of personalities, backgrounds, and so on, Gordon (1961) recommends that at first a scholar (and most likely an introvert) and a salesman (and most likely an extrovert) are selected. Next, an *integrator* is selected, a person capable of bringing people together and creating a smoothly working *trojka* ("triple") (an introvert, an extrovert, and an integrator). Only after that, two remaining group members are added to expand the body of knowledge brought by the trojka but also balancing their personalities. Gordon (1961) has listed this criterion as the last one, but from a systems point of view it is most important, as without creating a system nothing will happen.

The selection of a group leader is usually a challenge. Such a person must be particularly knowledgeable and a natural Synector, but without a capable leader, a group of Synectors will never become a system. The leader needs to be an integrator, but in addition to that he or she should have four characteristics that are difficult to find among engineers:

1. *Extreme optimism:* The leader should strongly believe that since his or her group received a given problem to solve, a solution to this problem must exist, and that it needs only to be searched for and found. Also, the leader should be enthusiastic about seeing a given session as an opportunity to make an impact on society and to advance his or her professional career through solving the problem.

    Engineers are usually critical about their work, and this comes with their training and education; this is because self-criticism and being critical about others usually serves them well, preventing mistakes and generally resulting in better and safer products. A Synectics session is fundamentally different in this respect from routine engineering practice, and the leader must be uncritically enthusiastic about the results that emerge.

2. *Total grasp:* Among group members, the leader should have the best understanding of life and of the problem domain. This person should have a lot of experience of dealing with various people in terms of their personalities and backgrounds. Also, he or she should be able to integrate and interpret all the ideas and associations that emerge. Ultimately, the leader is responsible for the integration of knowledge coming from several domains (group members) in order to acquire a transdisciplinary knowledge that will become the foundation for the group's inventions. Therefore, the leader should have some familiarity with these domains, at least on the conceptual level. For all these reasons, the leader should be a *generalist* or a *global thinker* as far as thinking styles are concerned (Sternberg 1997) (see Section 3.2).

3. *Synectics grasp:* The leader should have excellent understanding of Synectics, including its scientific foundation. Individual group members need to know the method's assumptions, procedure, and examples of various analogies, but the leader needs to know much more; he or she should be familiar with the science behind the method and should be able to explain the individual assumptions and interpret them in the context of a given problem. Also, the leader should develop such a deep understanding of the four analogies that he or she should be able to develop on demand problem-related analogies to inspire other group members.

4. *Psychological distance:* The leader must be capable of emotionally distancing himself or herself from the process in order to be able to monitor it and direct it. Such a behavior is very difficult because the leader is probably the best qualified to become a Synector. Therefore, he or she contributes to the session in so many ways that in the process they become emotionally engaged. In fact, this is a contradiction. The leader should simultaneously be emotionally involved so as to

maximize his or her productivity, yet also emotionally distanced from the action so as to remain objective and rational and not be controlled by emotions. Only very few people have the ability to switch from the creative to the rational mode on demand.

Before a session takes place, Synectors must be appropriately prepared. First, they need to become motivated to succeed through learning the social and historical context of Synectics. Based on the author's experience, without this element of preparation, engineering students will never be fully engaged, and this will negatively impact the results.

The best way to motivate the future Synectors is to teach them about human and engineering creativity and its impact on the evolution of our civilization, on individual societies, and on the engineering profession. It must become clear that engineering creativity is the driving force behind progress in modern societies and the key to survival of nations in the international industrial competition. Also, it is the passport to personal fame and fortune for engineers who become inventors.

The historical perspective should include a discussion of the evolution of Western societies and the growing role of inventions, which is absolutely crucial now, in the *age of creativity and innovation* (Pink 2006). An overview of the Renaissance and its creators is particularly effective; it brings human faces into the picture and shows the sophistication of Renaissance creators. They were polymaths and truly transdisciplinary scholars; also they had all psychological and intellectual features still critical for engineering creativity, particularly for Synectics based on using associations. Comparing Renaissance creators with future Synectors often has a magical effect. It usually leads to a boost of their enthusiasm and to equally important changes in their understanding of the role of engineers as creators. They suddenly see themselves as pioneers and the successors of the great minds of the Renaissance—as people ready to change the world.

After the motivation part, training focused on Synectics begins. Gordon (1961) estimated that the entire process may take up to a year, and he believed that this process should not be rushed. It is a process not only of acquiring knowledge but also changing attitudes and at the same time building a team in an industrial company that will operate for many years. Today, there are many short 2–3 day courses on Synectics available, which prepare Synectors in a day or two. Such courses are not necessarily bad, but their objective is more to familiarize people with Synectics than to actually transform them into Synectors.

At first, the trainees should get only a general overview of the method, its assumptions, and the psychological mechanisms involved. A discussion of the four analogies should follow. The training should be provided for all five future team members together, not only to help them to learn about themselves and to build human relationships but also to learn about analogies as a group. In this case, there will be an increased probability that learning will lead not only to an abstract discussion but that the first examples will emerge; initially it will be a trial-and-error process, but the team will be learning fast.

From the beginning, training should be focused on sessions. The first one should be organized pretty early in training as a short one-hour session. The problem should be relatively easy, the trainees frequently but not forcefully guided about the use of mechanisms and analogies, and the entire session recorded. During the next hour, the recording should be played and feedback provided, friendly but substantial. After the feedback is provided, a new one-hour session should begin and the entire process repeated several times. In this way, a learning loop will be created in which the trainees will acquire a lot of practical and methodological knowledge about how to behave during a session, and they will be also gaining confidence.

Finally, after several loops, a concept for a new engineering system is developed. It is properly celebrated in order to recognize the importance of this moment and the emotional and intellectual efforts necessary to create it. At this time, running of sessions is temporarily suspended, and the team builds a model or prototype in order to develop a better understanding of their concept. Next, work on another problem begins and ends with building another prototype. This cycle is repeated many times within a year, with the complexity, difficulty, and importance of the problems growing.

We can distinguish two interrelated learning processes: learning by the entire group and learning by its individual members. We have already briefly discussed learning on the group level. Individual learning is equally important because it helps members to acquire transdisciplinary knowledge, improves their effectiveness during a session, and contributes to the group's cohesion. Team members represent various professional backgrounds, but they need to develop a good mutual understanding or a joint body of transdisciplinary knowledge. It is necessary for them to be able to communicate and to operate as a team. Therefore, they deliberately try to do work that is basically within the area of specialization of the other members.

The training is also about attitudes. In a large traditional engineering organization, the *vector of psychological inertia* (see Chapter 9) is particularly powerful and may significantly slow the pace of all activities, particularly those that are nonroutine or support changes. Employees quietly accept such a situation, but it is simply intolerable for Synectors. For all these reasons, Synectors must develop the attitude "We can" and actually mean and practice it. Creating such an attitude can be done by leading the group through several administrative challenges, which will prove that it is actually possible to do in an hour what traditionally requires days, but only if everybody is motivated and engaged. As a result of such exercises, the group members will develop a perception of their superior ability to deal with all kinds of challenges. That will greatly help them to cope with the actual challenges they will encounter later, no doubt about it.

Synectics training is usually so enjoyable and creates such positive feelings for the exceptionally bright Synectors that they need to cope with an unusual psychological phenomenon: a feeling of guilt that the work on Synectics is neither boring nor tedious. Unfortunately, engineering work is often associated with boredom and is considered hard work. When it is not, engineers subconsciously feel guilty about that. If such a feeling is left unaddressed, it may negatively impact the group's enthusiasm. The best medicine is a small success, which will change the group's dynamics and will subconsciously help them to overcome the feeling of guilt.

8.8 EXAMPLE

Safecity is a small town in Utah, the United States. It is located on a rapidly flowing mountain creek. The creek brings beauty (and tourists) to the town, but it is also dangerous for small children playing on its banks. They occasionally fall into the water and must be rescued by the local fire department, which must respond immediately, otherwise the strong current will take the child away. Unfortunately, that happens two or three times each year, and children are drawn into the pools of deep water or die as a result of serious head injuries and trauma when they are taken by the creek to the rapids located just outside the town.

For a very long time, the city council debated what to do about the problem. Finally, only last year, a popular investigative TV program on the ABC Network, *20/20*, presented a segment on the town showing the problem and the inaction of the city council. Dramatic and emotional interviews with parents who had lost their children as well as conversations with children who had become handicapped as a result of accidents in the creek made a powerful impact on public opinion and brought all kinds of unwanted publicity to the town. Finally, two weeks after the program was broadcast, the city council voted unanimously to hire Successful Solutions LLC, a small consulting company specializing in finding inventive solutions to various socially important problems.

As a result of long deliberations, the city council has defined the problem as follows:

> Find several solutions to the problem of children's accidents caused by the creek. The proposed solutions should not affect the natural beauty of the creek, be easy to implement and relatively inexpensive. They should be found in six months.

The owner and CEO of Successful Solutions LLC, Dr. John Hickory, decided to use the same group of five Synectors who had earlier successfully solved the problem of construction workers falling from elevations and, as a result, had won several patents for the company. The members of the group were

1. Jack, the group leader, a mechanical engineer in his late thirties with a formal background in music (he plays the piano beautifully).
2. Susan, an electrical engineer in her mid-thirties, who is also a certified scuba diver and a passionate tuba player.
3. George, an accountant in his mid-thirties, who lives for numbers and has a great deductive mind, but studied art history and paints abstract landscapes for fun.
4. Milly, an aspiring actress in her late twenties, who studied music and is trying to master yodeling as her hobby.
5. Dolly, a historian in her early thirties, who is interested in the history of engineering but temporarily works as a CAD operator for a local engineering company.

All the Synectors went through the basic and advanced training provided by one of the leading companies in the United States, which specializes in Synectics and offers all kinds of courses on the method. Also, they took an introductory three-day course on TRIZ taught by Ideation International. They completed their limited training in the area of inventive engineering by taking a week-long course taught by Successful Education LLC. As a result of all their training, the Synectors know not only about Synectics but also have some understanding of the other inventive design methods and know about the historical, cultural, and psychological dimension of their work.

The Synectors have been working together for about two years and can be considered reasonably successful, particularly after their recent success with solving the problem of construction accidents. For many people within the company, they are heroes who brought fame and fortune, but there are also employees who consider them parasites for eating up the very limited resources available without an immediate payoff. For this reason, the Synectors are strongly motivated to prove that they are not only the creative core of the company, but that they are also productive and contribute to the company's financial success.

The company has created excellent working conditions for its Synectors. They have a dedicated large room with modern and minimalist furniture and with comfortable leather sofas. The room is painted pale blue and lit by several recessed lights, but there are also two standing lamps by the sofas. A brand new Sonoz music system plays *Love in the Wind*, a soft music for dreaming from Europe, which was purposefully composed to "release you mind to tranquil & dreamlike music created with Angels in mind." All kinds of abstract art is on display, including a sculpture in bronze of a group of Synectors in action. Since it is abstract, it looks like several vaguely defined shapes, reminiscent of eagles taking off. For many, the interpretation of this sculpture is not clear, but the Synectors simply know that it is about them and they love it. The entire space is inviting and soft, and its ambience sends a powerful subconscious message: "Come in and be yourself, be truly creative and change the world"—at least this is the message read by the Synectors.

The Synectors were informed about their new challenge about a week before their session took place. They were provided with a recording of the TV program, with several sets of minutes taken by the city council when firemen and invited experts testified about the accidents, and with all kinds of accident statistics. Also, they were given medical books about head injuries and death by drowning. They were even given several articles about postdeath experience, which occurs particularly often when people are drowning.

In the week before the session, the Synectors read the provided materials, talked about the problem with friends and family members, and became gradually involved in solving the problem on both the emotional and intellectual levels. They were also building confidence that the problem had to be solved and that they would solve it. The question of how and when was not on their minds; it really did not matter, these questions would come much later. All group members found the problem challenging but also potentially highly rewarding, both in professional and in personal or emotional terms. The emotional dimension of the problem was growing and driving their involvement. They all had children, or were planning to have children, and therefore they could easily relate the experience of children dying unnecessarily (and the suffering of their parents) to their own memories from childhood or the early teen years when they put themselves in danger so many times and sometimes only barely survived. The memories of the faces of their parents in pain were still strong, and they would do everything possible to avoid causing so much pain to their parents or to other parents. Now they clearly saw what their mission was, at least in emotional terms.

When the session began, all the Synectors were ready. The author received an audio recording of the session. Its partial record is provided here with short methodological comments to help the reader to follow the process and to see behind the words their intent and the gradual building of various solutions.

JACK: Let me welcome everybody. I am personally very happy that we are working together again and that we were given a chance to solve a truly important problem. Yes, ultimately it is about politics and money, as everything is, but it is also about saving human lives and preventing, or reducing, human suffering. I am sure that a solution, or solutions, exist but we need to find them. Nobody is better qualified than us to solve the problem, and I feel confident that we will solve it. [Jack is building the group's confidence but is also creating the emotional dimension of the problem.]

Let us look at the situation from a distance. [He tries to create the state of detachment first.] We have here three major stakeholders: the city council, the fire department, and parents. The city council wants good publicity and to minimize expenses associated with the creek accidents. The fire department wants more money and power. Parents want their children safe. I am a father; my son nearly died recently when he jumped from the garage roof with small plastic wings while trying to become a Daedalus. Children do stupid things, they also often play too close to the creek; these are facts of life, and we will not change them. But two days ago I had a nightmare and woke up in the middle of the night with a terrible feeling that it was my boy struggling with the powerful current and hitting the rocks. How would you feel if it were your son? [He tries to create the state of involvement.] Yes, I know that we are becoming emotional and we all want to find a solution immediately, just now. I know that each of us has already found many solutions during the last week, but it is too early for solutions. First, we need to develop a good understanding of the entire situation; before we begin finding solutions we need be cold-blooded to conduct a complete analysis of the situation to make sure that we are finding solutions for the right problem. [He tries to create the state of deferment.] When we understand the problem and have our springboard, we will start dreaming about the solutions. To be truthful, I have already had a dream about an angel preventing children from falling into the creek. [He evokes the state of speculation for the first time.] Finally, we must be aware that each child is a mystery; it is an independent entity and we will never fully understand its behavior, not to mention motivation behind it. However, a group of children is more like a swarm of intelligent agents, and their behavior as a group may be somehow predicted, maybe even influenced. [He introduces the state of autonomy of object.] I do not want to talk too much. Let us have fun, let us begin. Let us create our springboard. Susan, you were fantastic last time when you led us through the springboard building process. Please, help us also this time.

DOLLY: Jack, I know that you did not want me to talk, but let me say only that I am here to save the children. I do not care about PR for the city council and I care even less about money for the fire department. I believe that we should learn first why and how accidents happen. When we know that, solutions will come. I have so many ideas; only let me to articulate them.

JACK: Ok, Ok, I also think that our main stakeholders are parents. Remember, though, that firefighters are also important. They are first responders, they are the guys who jump into the creek to pull children out or use their acrobatics on the ladders to help children.

SUSAN: Let us be a little more systematic, at least for now. In fact, I like Dolly's comment. I also like that we have already begun working on the springboard. Let me read you the problem definition provided by the city council. [Reads slowly two times.] Pretty clear, I think. They want to solve the problem of children's accidents, but surprisingly they do not say they want to provide a safe environment for children.

MILLY: Maybe they want to be politically correct and are actually saying "forget children" and find a way to avoid any bad publicity associated with their accidents. A new PR campaign? Safecity is beautiful?

GEORGE: I do not like it. I do not even want to talk about avoiding the real problem and working on a smoke screen. It would be simply wrong.

JACK: George is right. No matter what was the hidden intention of the city council, we must address the problem in the ethical way, no smoke screen. By the way, we have already agreed that our main stakeholders are parents. No parent would support our first idea of a new PR campaign instead of real action and saving their children.

MILLY: Thank you, thank you boss. We are not politicians, we are good people, we are Synectors. No dirty PR tricks please, let us save the children.

SUSAN: We have already come to the conclusion that we actually want to save children, even if it is against the hidden intention of the city council. Let us come back to building our springboard. I have a question for you: What are the most important words or concepts associated with our problem?

MILLY: I know, I know. Obviously the key word "children" comes first, next "creek," followed by "avoiding accidents" and "preserving the natural beauty."

SUSAN: Congratulations, Milly. [Everybody is clapping; Milly stands up and smiles while nodding her head, everybody is laughing, and suddenly they are playing, ready for Synectics.] Any volunteer to write our springboard? George, it must be precise and poetic at the same time. Could you try?

GEORGE: Why not? Let me try. What about "I wish all children all over the world be safe."

JACK: Too general; let us be more specific and focus on Safecity. What about "I wish all children in Safecity be safe from creek accidents while preserving the beauty of the place and spending very little money."

MILLY: Excellent, boss. Does that mean that we will put the creek in a tunnel under the city? In this way the children would be 100 percent safe and we would run water in the creek only when tourists are around.

DOLLY: Or only when they are taking pictures.

GEORGE: We could simply mail them pictures of the creek even before they come here.

MILLY: After getting pictures they would not need to come here. Only to send money to the city council.

JACK: OK, OK. Everybody agrees with our springboard? [He reads it again and everybody says yes.]

MILLY: Yes, but I would formulate it differently: "I wish Safecity to be a place where children can safely enjoy playing on the banks of a beautiful creek."

JACK: In fact I like it more than my own springboard. Apparently yodeling helps you to develop editorial skills. [Everybody laughs while he is writing the springboard on a whiteboard.] We are ready now for finding solutions.

SUSAN: Let us look at a group of children on the bank. They are obviously playing but for an eagle flying above them, they are simply strange objects. [She is making the familiar strange.] He is an old eagle frequently flying above Safecity and he saw several times children walking on the top of a low stone wall separating the park on the creek from the creek itself. He even saw children dancing, racing, and fighting on the wall. He is convinced that the movements of these strange small objects are purely random.

GEORGE: In fact, they are. As a parent and a rational person, I am absolutely convinced that from the system's perspective, [he is also trying to make the familiar strange, moving into the context of systems science] the movements of children, whenever they are playing or not, are random and we must recognize that fact. This observation has clear consequences. Trying to educate children about the dangerous creek will never work. We need to stop them before they enter the danger zone behind the wall.

MILLY: If I were a wall, I would rise high whenever a child approaches me too close. [She is using a personal analogy.]

DOLLY: Fantastic, I love it, a robotic wall. A wall with sensors activating spikes or safety nets emerging from the wall whenever a child approaches it.

GEORGE: Emerging spikes and nets are fine with me, but if I were a wall, I would start screaming at the children; it usually works fine with my boys.

MILLY: If I were one of these children, a screaming wall would not stop me. The wall would need to kick my butt, and do it hard.

GEORGE: You mean an electrified fence? Water hoses built in? High air pressure outlets controlled by sensors? Exploding containers with gas? They all would work, they are fine with me, but they might also stop lovers from sitting on the wall and kissing and that would hurt the image of Safecity, which is supposedly for lovers, like Virginia.

JACK: All children could have transponders, and the safety devices would respond only to people carrying these transponders.

GEORGE: Only boys would need to carry transponders; boys are reckless and stupid and 92 percent of deaths are boys.

MILLY: Sorry, George. It is sexist. Girls have the right to be reckless and stupid too. All children should carry the transponders. No way to exclude girls, it would be discrimination. I would never support such an idea.

GEORGE: I apologize, you are obviously correct; girls have the right to be as reckless and stupid as boys, maybe even more so. Yes, all children should carry the transponders.

JACK: Let us come back to our problem. We have already found many ways to prevent children from climbing the wall. We know for a fact, though, that some children will climb the wall no matter how many sirens will be blasting or how many blinking lights will be working. I think that all these safety measures will only encourage some boys to climb the wall. Children love challenges and fighting restrictions. Let us talk about a hypothetical situation when a child is already in the water and is dramatically screaming for help.

GEORGE: We need to take this child out of the water as soon as possible.

MILLY: I feel his fear; he is fighting for air, trying to grab stones, getting first cuts. [She tries to use a personal analogy again.] Yes, I want to help this blue-eyed boy. I have a vision, I see an angel coming from above, a beautiful angel, she is smiling and smoothly and gently grasping the boy fighting with the current. Now, she is carrying the boy to the bank of the creek. The child is saved.

GEORGE: Angels are foreign to me, they are strange and I do not fully understand their mission on earth.

SUSAN: But I believe in angels, and they have already helped the child in the water. They inspired me to think about the help coming from the air. Let us substitute the strange concept of angels with the familiar concept of flying robotic drones.

MILLY: Do we have angel-sized helicopter drones?

GEORGE: Most likely the army has them. We could also use a swarm of robotic drones pulling children from the water.

JACK: Firemen will love your ideas. Drones are expensive; they require qualified operators and they may bring additional money and positions to the fire department.

MILLY: Do we really need drones? When a firefighter on a ladder grasps a drowning child, the entire arrangement is like a large arm. [She uses a direct analogy.] It has many components: a ladder truck, a ladder, and a big man with his extended arm reaching for the boy.

SUSAN: Could we design a better arm? Without all these components? A single piece?

GEORGE: Why not? It could be a single airbag like in a car. [Again, a direct analogy.] It could be expanded using several small explosives and it would create a huge cantilever or even a bridge above the creek for a firefighter to easily and quickly reach a child.

JACK: We have already been working for more than 4 hours. Let us have lunch. Later, I would like to move in an entirely different direction. I think that we are ready for a small excursion.

JACK: [43 minutes later] I hope that we all enjoyed our lunch. I must share with you a secret. I have talked with James Cook, our administrative director, to make sure that only organic and double organic food is served. Also, I told him that we all like sweets and we needed at least two desserts. I hope that everybody is happy with our order.

MOLLY: I did not know that my oysters were double organic, but they were definitely good. I am ready for action.

GEORGE: Molly, what kind of action are you ready for after eating two servings of oysters?

JACK: George, be politically correct; she meant creative action.

MOLLY: But of course, how could you suspect anything else? [Everybody is laughing.]

JACK: I am glad that we have clarified the matter. Let me invite you for a short excursion. Look, somebody has left a copy of the March 1st, 2015 issue of *Fortune*.

MOLLY: It is not a coincidence; somebody wanted to remind us that we should be concerned not only about our own fame but also about the fortune of our company.

JACK: Yes, it might be a hint. Susan, please go to page 26, row 7 from the top, first word from the right. [Susan opens the magazine and starts reading.]

SUSAN: The word is "pro-golfer."

DOLLY: We need to try again. Pro-golfers do not rescue drowning children in mountain creeks.

MOLLY: Dolly, I love this inspiration. This is a fantastic lead. Pro-golfers are shooting balls, are throwing objects in the air.

GEORGE: Yes, we could create an entire class of objects to be thrown by firefighters to help children. I have several ideas; a large ball with handles, a tube with handles, a small dinghy boat.

SUSAN: Excellent ideas, but your devices with flow with the current. They should be intelligent; they should be robotic devices that would look for children in the water, find them, and keep them in place as long as it is necessary.

SUSAN: I recall that the US Navy is working on a robotic fish; it is intended for all kinds of intelligence work but could be adapted to save children.

JACK: Robotic fish to save children; I like it. Our friends in the fire department will also love it.

GEORGE: I think that the city council will be also impressed; for them, using a robotic fish to save children would create an incredible PR opportunity.

The provided transcript covers only the first several hours of the session, which continued for four days. It has already revealed several interesting features of Synectics. First, to the uninitiated, the session may look like a chaotic conversation between several adults who are pretending to play like children and are trying to solve a problem about which they have no clue. However, for a knowledgeable observer, the session will reveal hidden mechanisms driving the problem-solving process, hidden behavioral patterns of interactions, the various roles of the individual participants, and how various analogies are being used.

The example has also demonstrated the efficacy of Synectics in the development of ideas. In a very short period of time, a large number of ideas were produced, ranging from absolutely crazy to novel and potentially useful. No patentable ideas were found, but the nature of the problem is such that patentable ideas most likely will not emerge; such ideas are much more likely to occur in the case of highly technical and well-defined problems in chemical or mechanical engineering.

8.9 BLACK AND WHITE

8.9.1 White

The method of brainstorming, presented in the previous chapter, has changed the social perception of creativity in the United States, particularly the perception of engineering creativity. However, at the time when Synectics was proposed in 1961, there was a growing wave of criticism of brainstorming. The method was often presented as a silver bullet and used in a way inconsistent with Osborn's guidelines. Also, many charlatans were trying to make easy money out of the method without fully understanding it and, most importantly, not being able to deliver the promised results. Osborn brought hope, but did not entirely deliver. His method is powerful, but it is much more complicated than its proponents want to admit, not to mention as complicated to practice.

When Gordon introduced his method, he was in a good situation because the public already knew that human creativity was not entirely magical and could be practiced by *regular* people. On the other hand, he was also in a difficult situation because of the growing disappointment with brainstorming.

Intellectually, Synectics is a sophisticated method with roots in psychology. All its assumptions, the psychological states it identifies, and particularly its operational psychological mechanisms have been discovered through years of solid repeatable research and are verifiable. For all these reasons, the method is relatively easy to accept for scientists and engineers.

Despite its complexity, nearly 60 years after it was developed, the method is still alive. For example, in the United States the method is taught at several universities, and a number of private companies offer Synectic services and various professional courses. The method is also taught in Europe and in Asia. For example, in at the National University of Kowhsiung in Taiwan, Dr. Po-Jen Shih teaches the method as part of his innovative and popular course on inventive engineering. In Europe, Dr. Sebastian Koziolek also teaches the method in Poland as a part of his unique course on inventive engineering. He has even developed his own methodological contributions to the development of the method (Koziolek et al. 2010), relating it to knowledge engineering and to the various processes of knowledge acquisition.

The author believes that each inventive engineer should learn about Synectics, whether or not he or she is planning to use it in the future. Learning about Synectics is so much more than simply learning about a specific method. First, it is about discovering the power of analogical thinking and seeing the world in all its complexity, which can be only captured if *sfumato*, associated with analogies, is intentionally introduced. Second, it is also about discovering the psychological dimension of engineering creativity. Finally, it is about finding out that to become an inventive engineer, a student needs to undergo a transformation from an analyst to a creative person who has a huge competitive advantage compared to a deductive thinker.

8.9.2 Black

Synectics is strange, complicated, and difficult to use. These three words describe best the reasons why the method is not as popular as it should be, considering its potential efficacy. In other words, the method is not like other engineering methods. Also, it has so many interrelated elements, which must be activated to produce desired results. Finally, the method requires many preparations before it can be effectively used, and each session is a complex operation.

It is possible to learn *about* the method in a lecture or two, but learning *it*, that is, understanding it and knowing how to use it, is much more time consuming and difficult. Synectics can be compared to a complex engineering domain; for example, to structural analysis. A student can learn assumptions in a single lecture but that will not be sufficient to conduct an analysis even of the least complicated structure. Unfortunately, developing an ability to practice structural analysis requires tens of hours of lectures and hundreds more hours of problem solving. Interestingly, students and engineers are willing to invest this kind of effort to learn a traditional domain, but they often believe that they can learn how to practice creativity in one or two lectures.

Synectics is a high-risk–high-payoff method. As with all heuristic methods, results and their timing are unpredictable, but in the case of Synectics even more so. Engineers hate uncertainty and always try to minimize it; in the case of Synectics, this rule often translates into the rejection of the method. Also, the use of the method is simply unacceptable for many engineering administrators who see it as spending company's money in a frivolous way to solve serious problems, which definitely require serious approaches. Even in academia, some administrators (who are traditional engineers) are even concerned about teaching Synectics. The method seems to be entirely incomprehensible for their ultrarational minds and is seen by them as so far away from the traditional engineering methods that teaching it may hurt the rational minds of future engineers.

Probably, the biggest problem with Synectics is the basic contradiction between trying to rationalize it and making it effective. In an effort to familiarize engineers with the method (making the strange familiar), many instructors present it as another engineering method so it will be easy for engineers to understand it and use it. Such a presentation is usually oversimplified, creating a kind of *mechanistic* understanding of the method and, more importantly, minimizing the importance of its emotional dimension, not to mention its psychological roots. As a result of this situation, many potential Synectors are simply not properly prepared for a session and fail miserably. Most likely, this is the key reason why the method is practically unknown and very rarely used in engineering.

Chapter 9

TRIZ

9.1 CREATOR

The Author has a very personal connection to the creator of the Theory of Inventive Problem Solving (TRIZ), Genrich Saulovich Altshuller, and to his method. The Author was in his early twenties and a junior faculty member at the Warsaw University of Technology (it was 1972 or 1973) when he attended, by a pure coincidence, a seminar on TRIZ and unexpectedly discovered a way of thinking entirely different from the traditional analytical thinking of structural engineering. Next came a small book by Altshuller, a translation from the Russian. The translation was bad and the book was written in a Soviet propaganda language that all Soviet writers, including engineers, were required to use, but the message was intriguing and fascinating at the same time: *We can create inventions on demand.* The Author had found his calling. About six months later, the Author became a cofounder of the Heuristics Group of the Polish Cybernetic Society and tried to invite Altshuller to Warsaw. The invitation was sent by a person traveling to Moscow and was mailed from Moscow, otherwise it would never have reached Altshuller (the KGB in action). The reply came in a similarly convoluted way: "I would love to come but I am under a house arrest. A prototype of one of my inventions caught fire during the tests and I was accused of sabotage." (Only much later the Author discovered that this invention was a semiautonomous floating device for cleaning oil spills in ports.) The Author's next personal interaction with Altshuller came more than ten years later (in the mid-1980s), when the Author was working on a talk he had been invited to give on design research in Eastern Europe, and he contacted Altshuller about his contributions to this area. This ultimately led to cooperation with his disciples and with Ideation International Inc., which was cofounded by several close associates of Altshuller and is now the leading TRIZ company in the world. (In fact, the two key TRIZ experts from this company, Alla Zusman and Boris Zlotin, named TRIZ masters by Altshuller, have cooperated with the Author to make sure that the chapter on TRIZ truly represents Altshuller's ideas and the state of the art.) Obviously, this chapter is only an overview of the method, and while it is sufficient for those who want to learn about it on the conceptual level, it is definitely grossly insufficient for becoming a TRIZ expert or even a TRIZ practitioner. Fortunately, there are many books and articles on TRIZ (Altshuller 1984, 1996, 1999; Altshuller et al. 1999; Altshuller and Shulyak 2002; Arciszewski 1988a; Clarke 1997; Orloff 2003), and several companies, including Ideation International Inc., offer all kinds of courses on the topic.

Altshuller (1926–1998) (Figure 9.1) was born in Tashkent in the former Soviet Union. His parents were journalists and intellectuals. He was raised in a family that highly valued learning, knowledge, and books. This was a reflection of his parent's professional activities but also a part of his Jewish heritage. In fact, his Jewish roots were always a factor in his life. On one hand, this made it more challenging for him than for other people also struggling in the Soviet Union. On the other hand, this difficult situation forced him to become more independent and entrepreneurial and definitely helped him to develop his unique personality.

Early in his life, Altshuller became an avid reader, reading mostly science-fiction novels. His passion helped him to develop his extraordinary imagination but also convinced him, at least on the subconscious level, that everything is possible when engineering knowledge and imagination are combined. He read and dreamed about interplanetary travels but also dreamed about becoming a sailor and traveling on this planet. This motive of travel, both imaginary and real, is important to an understanding of his personality and its similarity to the personalities of the great Renaissance creators. For them, travels were also personality forming and strongly contributed to the development of their minds and even more to their transdisciplinary understanding of the world.

Altshuller began working on his dreams of travel when he entered a special naval school after completing Grade 8 in a regular high school. The Second

Figure 9.1 Genrich Altshuller. (With permission. Drawn by Joy E. Tartter.)

World War began, but he was allowed to complete the school's program and after that he was sent to a school for pilots and navigators. He graduated from this school when the war had just ended, so instead of being sent to the front, he was sent at his request to continue his military service with the Soviet fleet located in Baku. Because of his limited but diversified education (naval and flying schools) and his interest in inventions (he had patented his first invention, a scuba-diving system, when he was only 14) he was assigned to the patent department. It was the crucial moment of his life. In his new position, Altshuller had access to patents, and his role was to assist other inventors of all ages to prepare their formal patent applications. He had already developed his investigative skills, but in contrast to the other researchers focused on the psychological aspects of human creativity, he was interested only in engineering creativity, that is, in the development of inventions. He was looking for the *engineering* mechanisms behind inventions and for the "secret" engineering knowledge that could become a foundation of inventions in all areas of engineering. He was simply looking for the holy grail of engineering creativity. Also, as a result of his work with other inventors, he realized that their inventions were accidental and that at this time (the mid-1940s) there was no engineering method in the Soviet Union that could be used in a systematic way to develop inventions when they were necessary.

In 1948, Altshuller was a young man in his early twenties. He was patriotic and confident about his inventive skills, not to mention already a successful inventor. He strongly believed that a new science of inventing could be "invented" and that he was the best person to do it. In fact, at that time, several inventive principles had already been discovered and Altshuller felt comfortable that he could teach his new emerging science. In the tradition of the Soviet Union, he wrote a letter (with a friend, R. Shapiro) to Stalin. In their letter, they criticized the state of innovation in the Soviet Union and offered to develop a Soviet science of inventing. It is not clear if the letter ever reached Stalin's desk. What is clear, however, is that whoever read the letter (most likely KGB operatives) did not like it and that its critical remarks were considered treasonous and anti-Soviet. In 1950, both Altshuller and Shapiro were arrested and tortured, and the courts sentenced Altshuller to 25 years of internal exile; that is, he was sent to a prison camp in the area of the city of Vorkuta in Siberia. He was working in a mine, but because of his technical skills he was given an engineering job and was able to survive several years of the Gulag. The situation had tragic consequences for his family: His father died prematurely and his mother committed a suicide. Only in 1954 was Altshuller rehabilitated and allowed to leave Siberia. Soon afterward, in 1956, he published (together with Shapiro) his first article in the journal *Psychology Issues* on the evolution of engineering and on the general trends behind this evolution. Much later, these ideas evolved into the theory of "Patterns of Evolution" (see Section 9.7).

After his release from the Gulag, Altshuller had three regular jobs. He worked in a cable manufacturing plant, as a journalist, and at the

Construction Ministry of the Azerbaijan Republic. The sequence and nature of his jobs may remind the reader of Alexander Faickney Osborn and his professional evolution (Section 7.1). Most likely, it is not a coincidence that we have just discovered a sequence of jobs providing a body of knowledge necessary and various perspectives required to think big and to support this thinking with a sufficient knowledge to implement great and novel ideas. In terms of thinking styles (see Section 3.2), working in a manufacturing plant expands one's *executive thinking style*, while a journalist uses mostly a *judicial thinking style*, that is, he or she judges various events and shares their opinions with readers. Finally, working in a ministry helps to develop a *legislative thinking style*, since ministries issue all kinds of regulations and codes. In this way Altshuller, like Osborn earlier, had a chance to acquire all three kinds of thinking styles. (This discovery may become another key to the reader's successful career.) During this time period, Altshuller also expanded his education and studied petrochemical engineering. By then he had talents, experience, a large body of diversified knowledge, and had mastered all three thinking styles. He was ready for action, but first he needed at least some independence from the Soviet system, which was brutal and totalitarian and stifled human creativity. He became a science-fiction writer, a very successful one, with books translated into several languages. That gave him some financial freedom and allowed him to go back to his dreams about changing the world.

Altshuller never forgot his calling: the creation of a science of inventing. He continued working on it and tried to get some support and recognition from the Soviet Society of Inventors and Innovators (SSII). This was an organization entirely operated by the Soviet authorities and which actually controlled all innovation-related activities in the country. Without some measure of support or at least acceptance from this organization, it was virtually impossible to do any work in the area of innovation, not to mention to do any fundamental research with potentially great impact. Not surprisingly (see Section 9.4 on the *vector of psychological inertia*), his ideas were ignored for more than 10 years. Only in 1970, did the SSII arrange for Altshuller the first all-union seminar with about 30 participants from various cities. In the same year, the SSII allowed Altshuller to establish the Public Laboratory of Inventive Methodology and, in 1971, the Azerbaijan Institute of Inventive Creativity. Both institutions immediately attracted a number of talented people interested in engineering creativity, and serious research and work on the method began. This work led to the establishment of schools of inventors in many cities in the Soviet Union. A momentum was building that had the potential to significantly change the state of innovation in the country and to improve the Soviet economy. There was only one problem. The entire movement was nearly impossible for the Communist Party to control, and the SSII ordered Altshuller to close his schools. When he disobeyed the order, the SSII closed his laboratory. In this way, the communists proved again that maintaining their power and control was much more important for them than the interests of society, but it is a different story that unfortunately had dramatic consequences for Altshuller. He had to resign along with a group of his close associates from the Azerbaijan Institute of Inventive Creativity and to look for other ways to support their activities.

However, activities in this area continued. In 1974, the second edition of the book *Algorithm of Invention* was published, and a documentary film under the same name about Altshuller and TRIZ was released and became very popular among engineers and inventors. Altshuller's students and people who learned about TRIZ from his books started establishing TRIZ schools in various cities with Altshuller's involvement and support, for example in Moscow, Leningrad, Kiev, Gorky, Perm, and so on. In 1980, the first TRIZ developer and practitioner seminar was organized in Petrozavodsk, and such seminars became regular events that were organized every other year.

In 1989, the Russian TRIZ Association was established and Altshuller became its first president. Ten years later, the International TRIZ Association was established. This organization is still active and has members from all continents. The organization allows the exchange of results among TRIZ scholars and stimulates global research on TRIZ. It also organizes international conferences, which are always big events for TRIZ scholars and practitioners. The organization is still growing, and this proves that Altshuller's ideas are still valid and worth studying.

Altshuller had an outstanding record as an inventor with tens of commercial patents in various areas of engineering; many of them have been used by the industry, thereby validating the method used to produce them. He was also a prolific writer with 14 books and countless articles.

No differently from Osborn or Gordon, Altshuller also became interested in teaching children creativity. In fact, he believed that young minds are much more suitable for learning TRIZ than adults, as well as better at actually incorporating its principles into their thinking. In the mid-1970s, he regularly published monthly inventive contests for children in a popular magazine for children and young teenagers. The response was tremendous and the results of these competitions were used to write the book *And Suddenly an Inventor Appeared* ... . He also developed a creativity program for high-school children and wrote several books on the subject. Unfortunately, the Soviet authorities were highly suspicious of these activities because creativity could be also used to improve the Soviet system, which was unrepairable and could collapse at the first attempt to improve it, as actually happened when Gorbachev introduced *Perestrojka* or "improvements."

Altshuller was a charismatic intellectual leader who created a great following in many countries, which is still growing. All his writings are in Russian. Some of those that have been translated into English suffer from poor translation done without an understanding of Altshuller's cultural and intellectual background, but this may change. It is too early to estimate his global impact on inventive engineering, but there is no question that he has become a cult person and many of his former associates and friends consider him simply a genius. The Author still has one or two letters from Altshuller and they are probably his most precious possessions.

As a man, Altshuller was a surprisingly humble person, almost embarrassed by his talents, knowledge, and fame. His former associates still consider him as a personal friend. The Author also knows that the Altshuller truly cared about the well-being of his associates and always tried to help them. After the fall of the Soviet Union, many of his associates (also Jews) decided to leave the country and make their home in the West. Altshuller contacted the Author several times asking to help them in finding jobs or simply to provide references. There is no question that Altshuller had not only a great and creative mind but also a big heart.

9.2 HISTORY

9.2.1 Introduction

TRIZ is the English acronym of the Russian name "Theory of Inventive Problem Solving." It is the name traditionally associated with a class of methods (a methodology) attributed to Altshuller and to his various associates who contributed to the development of TRIZ and/or developed their own versions. There is no official history of TRIZ or any professional TRIZ historians. In this situation, individual TRIZ scholars and TRIZ-related organizations and institutions offer their own versions of TRIZ history. Therefore, the Author has decided to present here the history of TRIZ as seen by two close associated of Altshuller's: Alla Zusman and Boris Zlotin. Both worked with Altshuller for many years, published books with him, and are considered by the Author as the best and most authentic source of knowledge about TRIZ and its history.

The history of TRIZ is intertwined with Altshuller's personal history (presented in Section 9.1), at least during the first period of classical TRIZ. Its entire history may be divided into three major periods, each with its own stages. This is briefly discussed in Sections 9.2.2, 9.2.3, and 9.2.4 from the perspective of inventive engineering.

9.2.2 Period of classical TRIZ (1945–1985)

9.2.2.1 Initial discoveries (1945–1950)

The roots of TRIZ may be found in Altshuller's original studies of engineering creativity, that is, patents. In the 1940s (1945–1950) he worked in the patent office at the headquarters of the Soviet fleet stationed in Baku. This time period could be called "Experiential Learning" or "Initial

Discoveries." Young Altshuller (then in his early twenties) was fascinated by the variety of patent applications and tried to discover the mysteries behind them. He analyzed hundreds of them, later thousands, and soon discovered that they were mostly accidental. More importantly, he discovered that the key to understanding the process of inventing was not only in psychology, as it was the general belief, but also in the engineering knowledge behind the inventions. Today, when we have the science of knowledge engineering and we routinely build knowledge-based systems, such an observation may seem trivial. However, in the 1940s, about 75 years ago, it was a revolutionary discovery, which immediately led Altshuller to formulate three fundamental questions (presented here in the language of inventive engineering):

- If all inventions have roots in engineering knowledge, do we have any specific knowledge directly related to inventing new engineering systems?
- Is it possible to acquire this knowledge and represent it in a useable form?
- Is it possible to acquire a body of methodological knowledge from patents (an algorithm) that could be used by future inventors?

Altshuller positively answered these questions through the fundamental discovery that inventing is about the elimination of contradictions and that it can be done using universal *inventive principles*. (The concepts of *contradictions* and *inventive principles* are discussed in Section 9.4, "Basic Concepts.") The process of the elimination of contradictions later became the basis for his algorithm of solving inventive problems (ARIZ), and the inventive principles represented universal engineering knowledge, a body of heuristics that have been acquired from thousands of patents. These heuristics were later modified and their set expanded, and many more specialized heuristics have been developed, but the original core of them was actually developed during this period of initial discoveries. Today, we use machine learning, or inductive learning, to acquire knowledge (decision rules or heuristics) from examples. However, in the 1940s, computers did not yet exist, not to mention computer science, knowledge engineering, or artificial intelligence. From this perspective, acquiring inventive principles from patents (examples) was an incredible scientific accomplishment by Altshuller. It could be considered the first known case of a large-scale knowledge acquisition from examples in engineering, a kind of manual inductive learning. Considering the fact that an average human can handle about seven attributes and examples at the same time (due to the limitations of human short-term memory), we can see the significance of Altshuller's discovery. Only an unusual individual with almost superhuman intellectual abilities could produce it—a genius, as many people believe he was.

9.2.2.2 Transdisciplinary learning (1950–1954)

When Altshuller was in the Gulag (1950–1954), he was "lucky" under the circumstances. He was sent to a prison camp that was also the home to many professors, scientists, and artists who were jailed during Stalin's "great purges" in the late 1930s and 1940s. Altshuller had a rare opportunity to interact with all these people and to enthusiastically learn from them. This did not result in any written material (writing in a prison camp was a crime brutally punished by beatings, reduced food rations, and separation). However, Altshuller developed at least a conceptual understanding of various areas of art and science, and that contributed greatly to his improved understanding of his own accomplishments of the previous period.

From the knowledge perspective, he acquired knowledge from various areas and was able to integrate it. In the process, he learned transdisciplinary knowledge, the true foundation for creativity in science and engineering. A psychologist would say that Altshuller's time in the Gulag helped him tremendously to mature his earlier ideas and to prepare him for the future development of TRIZ. A Renaissance historian would say that Stalin unintentionally created an environment in which people could learn from the best minds of their time and acquire a unique combination of knowledge that contributed to their own integrative learning. It was like in the Medici Court, but several centuries later. The living conditions were drastically worse, with people dying every day through starvation or sickness or simply being brutally murdered. However, for all these people, science and art were the only areas where they could preserve their human dignity and maintain their sanity under the extremely harsh living conditions in a prison camp in Siberia. Therefore, they gladly shared their knowledge with those who were willing to listen, and that created incredible intellectual opportunities for young, enthusiastic, and knowledge-hungry people like Altshuller.

9.2.2.3 Developing the algorithm (1954– 1965, approximately)

When Altshuller had his three regular jobs with the government (as all jobs in the Soviet Union practically were) after his release from the Gulag, he continued working on TRIZ. Not surprisingly, during this time he was focused on the development of the algorithm of his method. An algorithm is always a rational part of a method. Therefore, its development appropriately corresponded to his regular jobs, which all required a lot of rational thinking and the following of all kinds of rules.

9.2.2.4 Developing analytical tools (1960–1970, approximately)

When Altshuller became a science-fiction writer, he was focused on abstract thinking and on the generation of all kinds of unusual ideas. That obviously required abductive thinking, which was also used in his work on TRIZ. During this time, he developed several analytical tools that help inventors using TRIZ. Also, he began working on the *patterns of evolutions*, that is, on universal rules describing the evolution of engineering systems over long periods of time (years), and on how these rules could be used to predict possible changes in systems in the future.

9.2.2.5 Creation of scientific foundation (1970–1974)

This was the time when Altshuller established and ran both the Public Laboratory of Inventive Methodology and the Azerbaijan Institute of Inventive Creativity. It was also the time of the rapid development of TRIZ, particularly its methodological foundation. However, it looked like Altshuller felt that his opportunity to work on TRIZ might end soon, or he simply knew the Soviet system. His team continued working practically day and night using a relay system, and those who were unable to follow the pace had to leave. As a result of these several years of "furious" work, TRIZ was finally ready for the engineers to use and for students to learn about in the many schools of inventors emerging in many cities in Russia.

9.2.2.6 Distributed development (1975–1985)

After the Public Laboratory of Inventive Methodology was closed and Altshuller's team left the Azerbaijan Institute of Inventive Creativity, the work on TRIZ continued but in a distributed way. Altshuller was still the leader and coordinator of the research, which was continued by his former associates; by now they had new jobs, but they were still fascinated by TRIZ and continued to do volunteer work on it for several years. Altshuller was able to keep up the momentum until 1985, when his health rapidly deteriorated and he had to practically resign his informal leadership position and pass the baton to his disciples. During this stage of "Distributed Development," many important developments took place. Work was continued on the *algorithm*, on the *separation principles*, and on *natural effects* (all discussed later). Also, the development of a very effective but as well difficult *substance—field analysis* began, although its popularization was later abandoned when it was found to be too difficult to teach and to use by engineers.

9.2.3 Kishinev period (1982–1992)

In 1982, two of Altshuller's leading associates, Alla Zusman and Boris Zlotin, established the Kishinev School of TRIZ in Moldavia in order to work for the industry. In addition, the school also continued TRIZ research. Its major research accomplishments are the continuation of Altshuller's work on patterns of evolution, the creation of lines of evolution, and initial studies of problem identification and formulation.

9.2.4 Ideation period (1992–present)

In 1992, Zion Bar-El, Boris Zlotin, and Alla Zusman, along with two other partners, established Ideation International Inc. in the United States. Zion Bar-El was a very successful entrepreneur and became fascinated with TRIZ. He strongly believed that the next engineering revolution would be the TRIZ revolution, and therefore he decided to invest his life savings in this new enterprise. One of his first decisions was to move the Kishinev School of TRIZ. The Ideation team used their knowledge and deep understanding of TRIZ to create the Americanized version of TRIZ and to use it successfully it for commercial purposes. Also, Ideation International continues the TRIZ research.

Most importantly, Ideation International Inc. has brought information technology to TRIZ and in the process has created I-TRIZ (or Ideation-TRIZ). This is a thoroughly modernized version of TRIZ, a system combining the entire classical TRIZ intellectual foundation with more recent developments and with various sophisticated computer tools. In additional to the software, an entire educational system has been developed with all kinds of courses. The world TRIZ revolution has not yet begun, but its intellectual and software foundation is already in place.

9.3 KNOWLEDGE: THE KEY TO UNDERSTANDING TRIZ

As we have already briefly mentioned in Section 9.2, "History," one of Altshuller's first discoveries was that the key to inventions was engineering knowledge. Stan Kaplan, a TRIZ scholar and author (Kaplan 1996), has best captured this discovery:

> TRIZ is the abstraction of engineering knowledge. It is the abstraction of knowledge from human experience and structuring of that knowledge for efficient and effective use.

Kaplan (1996) has also provided an excellent example of knowledge abstraction. Quadratic equations can always be solved using the method of trial and error. However, each algebra book provides two formulas or abstract solutions that can be used to solve any quadratic equation. We could say that these abstract solutions represent our abstract and universal knowledge about solving quadratic equations; that is, they are applicable to any quadratic equations.

When we consider again acquiring knowledge from examples (e.g., learning from patents), the abstraction of engineering knowledge is a process involving knowledge transformation in which numerical or quantitative attributes are replaced by abstract or qualitative attributes. As a result of this transformation, the transformed knowledge becomes applicable to a much larger class of engineering systems than the original systems that provided the initial knowledge.

In this context, we should assume that TRIZ is a *universal engineering methodology for inventive problem solving*. The key to solving such problems is to use abstract engineering knowledge and follow a simple five-step procedure developed by the Author for his students. This procedure can be called *solving inventive problems by abstraction*, and this name well captures the fundamental nature of TRIZ. It has the following steps:

1. Recognize the nature of your problem
2. Categorize it
3. Find its abstract model
4. Find abstract solutions
5. Specialize abstract solutions for your specific problem

For example, you are asked to solve an inventive problem of a fifth wheel in a light truck (towing heavy trailers). First, you need to realize that your problem means inventing an engineering system capable of carrying large weights and of providing maximum maneuverability at the same time, a system that could work in a light truck. Next, you need to property categorize your problem, that is, to determine that you are dealing with the category of problems called *vehicle weight transfer systems*. In the third step, you need to find a class of abstract models of your system, and in the fourth step you need to find a body of abstract knowledge associated with your abstract models, that is, various abstract solutions for your problem. Finally, in the fifth step, you need to adapt or specialize your abstract solutions to your specific problem of a fifth wheel in a light truck.

The provided procedure significantly differs from the traditional *trial-and-error method*, which is mostly, if not always, used in the case of accidental inventions. Many TRIZ experts would say that their procedure is deterministic (i.e., it always works, producing the same results) while the trial-and-error method is highly probabilistic (i.e., there is only a small possibility that a solution will be found). The Author has a more complex understanding of both processes and believes that the actual situation is more *sfumato*, that is, it is more complicated, and the position of TRIZ experts slightly oversimplifies the issue.

What is most important, however, it is the fact that TRIZ inventive problem solving is best understood as a knowledge-based process of first acquiring abstract engineering knowledge and then adapting and specializing it to solve a specific problem. Everything else is only the logistic and methodological details.

Since knowledge is the key to inventive solutions, an appropriate and sufficient body of knowledge must be always available to an inventor. That requires the realization that a single engineer and a potential inventor is knowledgeable only in his or her area of specialization; that is, a mechanical engineer knows mechanical engineering and a structural engineer knows only structural engineering. The knowledge such a single engineer has may be more than sufficient for solving routine problems in his or her area but usually is grossly insufficient for solving inventive problems. Solving such problems may require knowledge from outside the problem's area, and this limitation may strongly restrict the inventor's ability to invent. TRIZ experts call the described situation *shortage of knowledge* and believe that it is the major impediment to progress in engineering.

In fact, we can explain the entire situation using *knowledge circles*, a system of concentrated circles representing various bodies of knowledge, both engineering and general (Kaplan 1996). The smallest one (No. 1) represents the inventor's personal knowledge, and this is inside a larger circle (No. 2), which represents the knowledge available in a given company. Both circles are inside circle No. 3, representing knowledge available inside a given industry. Outside the three circles already discussed we have circle No. 4, representing all the engineering knowledge in a given country, and this circle resides inside circle No. 5, which represents all available engineering knowledge. This circle is located within an even larger circle, No. 6, which represents all available knowledge.

When our system of knowledge circles is considered from the inventing perspective, we will soon realize that effective inventing requires much more than only circle No. 1, the small body of the personal inventor's knowledge. There is here a fundamental question of how to get access to this all-necessary knowledge outside circle No. 1. TRIZ provides a sophisticated and surprisingly modern answer: Before inventing begins, an inventor must be provided with the all-necessary knowledge that must be acquired earlier and presented in a universal, abstract, and useful form. This answer also provides an excellent explanation of what modern TRIZ is (see Section 9.2, History and Section 9.2.4 about the I-TRIZ).

9.4 BASIC CONCEPTS

In this book, only classical TRIZ is presented, that is, TRIZ as it was developed mainly by Altshuller during the period called classical TRIZ (1945–1985). Classical TRIZ may be called the *method of technical contradictions*, although this name oversimplifies classical TRIZ and reduces it to just a method while it is in fact a knowledge system of which the method is only a part.

The creator of TRIZ was an unusual man who lived in an unusual country (the Soviet Union), one that simply does not exist anymore. It was a totalitarian country driven by the communist ideology and a country that developed its own nearly impenetrable political and scientific culture; it even had its own engineering language, one strongly infested by propaganda. This language is still impossible to simply translate into English without creating confusion and misunderstanding. Even speaking Russian is not sufficient to avoid this confusion.

Creators are products of their times, and they are shaped by their environment in all its complexity, including its historical, cultural, political, and even religious dimensions. Many creators would not agree with this statement, but while the impact of the environment may only be on the subconscious level, it still very important. Our three creators of inventive designing methods are the best example in support of this claim. Both American creators, Osborn and Gordon, lived in a democratic and free society. Therefore, they proposed methods of spontaneous group problem solving through a return to the freedom, play, and fun of childhood. For Altshuller, living in a totalitarian society under a brutal and harsh regime, inventive problem solving was anything but fun. In his letter to Stalin, he wrote about communism and its coming global victory, about the class struggle, and about the national economy. In contrast to human creativity, his method would be systematic and repeatable and would allow the control of engineering creativity, a subconscious reflection of his political environment.

In the early 1970s, the Author was in touch with Professor Johannes Müller of the German Democratic Republic. Müller (1970) created *systematic heuristics* for domestic use in the area of inventive designing. (The Author even wrote a book chapter on the method [Arciszewski 1978].) The name systematic heuristics is a compressed analogy combining two contradictive concepts, and obviously such a science could be developed only in a totalitarian society, providing another proof of the impact of environment on creators.

For all these reasons discussed earlier, to truly understand TRIZ and its roots, we need first to know its original political context, whatever it was. Next, we need to learn several of the basic concepts created by Altshuller and his associates. After that, TRIZ will make a lot of sense to us and we will be able to study it and hopefully also to use it.

9.4.1 Contradiction

If Altshuller were asked what is most important about TRIZ, he would most likely say

> First, using knowledge from previous inventions to develop future inventions; and second, eliminating contradictions as the key to the creation of inventions.

In fact, Altshuller once said

"Every great invention is the result of resolving a contradiction."

Therefore, we need to learn first what contradictions are. The Author used to struggle with various inconsistent definitions of contradictions. Fortunately, after many years of effort to learn from TRIZ experts their true meaning, the Author has finally developed his own understanding of them and will share it with the readers, obviously under the conditions of secrecy.

Basically, Altshuller identified two contradictions:

- Technical contradictions
- Physical contradictions

In casual language, the word *contradiction* describes a situation in which an object may have two entirely different characteristics but only one of them may be associated with the object at a time. For example, black and white, slow and fast, or thick and thin.

In the case of TRIZ, contradictions are defined thus:

A *technical contradiction* is a pair of technical (or qualitative) characteristics of an engineering system, whereby when one of them is improved, the other is worsened.

A *physical contradiction* occurs when a single physical (or quantitative) characteristic of an engineering system should increase and decrease at the same time.

A technical contradiction may be also interpreted as a *technical trade-off*, for example, a trade-off between the safety and the maximum speed of a car. A physical contradiction may be understood as a contradictory requirement, for example, a temperature that is simultaneously > 25 F° and < 14 F°.

Let us consider two examples of pairs of contradictions:

Example No. 1: A family car

A mechanical engineer is designing a small family car. The engineer considers the luggage capacity and the fuel consumption. She knows that these two characteristics are somehow related, although the nature of this relationship is not obvious and not a single function is available that would describe this relationship. She knows that the best way to increase the luggage capacity is by increasing the volume of the luggage space; for example, by increasing both the height and the width of the car. At the same time, she also knows that the best way to improve the fuel consumption is through the reduction of the frontal area, and that requires a decrease in the car's height, width, or both. In the described situation, a technical contradiction clearly exists between the luggage capacity and the fuel consumption—between two different general/qualitative characteristics of our system.

Both car characteristics are dependent on the height and width of the car. When, for example, the car's width is considered, it should be maximized to increase the luggage capacity and should be minimized to improve the fuel consumption. We have here a physical contradiction associated with the car's width, with a single numerical/quantitative characteristic of our system.

Example No. 2: A reinforced concrete beam under bending with a rectangular cross section

A structural engineer is designing a reinforced concrete beam under bending. The cross section of the beam must be rectangular. The engineer considers the flexural rigidity (EI) of the beam and its weight. He is obviously aware that these two characteristics are interrelated. In order to increase the EI, he needs to increase the dimensions of the cross section (its height, width, or both), and that will automatically increase the volume and weight of the beam. In this case, the positive or desired increase in the EI is always accompanied by a negative or undesired increase in the weight of the beam. We have here a technical contradiction between rigidity and weight—a contradiction between two different general/qualitative characteristics of our system.

The same situation may be considered in terms of the structural depth of the beam, that is, in terms of one of the dimensions that is represented by a numerical attribute and is usually measured in inches (or centimeters). In order to increase the rigidity, the designer needs to increase this dimension. However, when he wants to decrease the weight (both desirable results), he needs to decrease the same dimension. We have here a physical contradiction regarding the structural depth (height) of the beam. It is a contradiction associated only with a single and specific/quantitative characteristic of our system.

The concept of contradictions is always difficult for students. Therefore, the Author has prepared a simple comparison of contradictions (Table 9.1) that clarifies many doubts and allows students to better understand these two important concepts.

When Altshuller introduced the original version of TRIZ, he also included the third type of contradictions, *administrative contradictions*. These contradictions identify the contradictive administrative regulations that must be also eliminated, or at least creatively avoided, by inventors. Later, however, these contradictions disappeared from Altshuller's publications. Most likely, they finally caught the attention of censors, who decided that there were no administrative contradictions in the Soviet Union and thus writing about them might give readers the wrong impression as to their existence.

Table 9.1 Comparison of features of technical and physical contradictions

Features	Technical contradiction	Physical contradiction
Level of abstraction	High—abstract	Low—specific
Nature	Qualitative	Quantitative
Attributes used	Symbolic	Numerical
No. of notions	Two different notions	The same notion
Example	Rigidity vs. shape	$V > 30$ and $V < 10$ mph

9.4.2 Inventive problem

Altshuller created a groundbreaking definition of an inventive problem:

An inventive problem is a conceptual designing problem with two characteristics:

1. Solving it requires the elimination of at least one technical contradiction.
2. Its solution represents an invention, that is, an unknown yet feasible and potentially patentable design concept.

This is a revolutionary definition because it defines the inventive problem in methodological terms as the process of the elimination of contradictions. This definition has entirely changed the understanding of inventive designing for so many people and at the same time has created a new conceptual and logistic framework for many others. Whenever a problem cannot be solved using any other inventive designing method, the problem may be reformulated in terms of contradictions, and immediately new inventive opportunities will emerge.

9.4.3 Levels of inventions

Altshuller observed that not all patented inventions are equal. They strongly differ in the levels of their conceptual sophistication and in the ways they were developed. Therefore, he proposed a five-level scale for their classification. These levels are briefly overviewed here.

9.4.3.1 1st level: Apparent solution

A design concept is simply selected from a class of known design concepts in the problem domain. For example, three types of doors are well known, and one of these types is selected and patented.

9.4.3.2 2nd level: Improvement

A design concept is developed through the modification of a known concept or through the combination of two known concepts in the problem domain. For example, two existing car concepts are combined—a concept for a sports car and a concept for a utility vehicle—and a new car concept for a sports utility vehicle (SUV) is developed.

9.4.3.3 3rd level: Invention inside paradigm

A design concept is developed as a combination of a known concept from the problem domain and a known concept from another domain. For example, the known concept for a safety car pillow from the area of crash engineering is combined with the known concept for a controllable balloon from the area of aerospace engineering, and a new concept of an air bag is thus developed.

9.4.3.4 4th level: Invention outside paradigm

A design concept is developed as a combination of a known concept from the problem domain and a new concept coming from a new technology outside the problem domain. For example, the known concept of disc brakes from the problem domain is combined with a new concept for high-strength ceramic materials just developed in the area of ceramic engineering, that is, outside the problem domain. Thus, a concept for ceramic disc brakes is developed.

9.4.3.5 5th level: Discovery

A design concept is developed as a combination of a known concept in the problem domain and a new concept based on a recently discovered new scientific principle. For example, the known concept of a lamp is combined with the just discovered concept of x-rays, and a concept for an x-ray machine is developed.

9.4.4 Basic features of an engineering system

Altshuller's studies of thousands of patents representing various engineering systems led him to the discovery of a *general description of an engineering system.*

> A general description of an engineering system is a collection of 39 descriptors, or Basic Features, which are necessary and sufficient for inventive purposes.

The complete list of descriptors is available in many books on TRIZ (e.g., Altshuller 1984). Here, only the first 11 basic features are provided for illustrative purposes:

1. Weight of moving object
2. Weight of nonmoving object

   3. Length of moving object
   4. Length of nonmoving object
   5. Area of moving object
   6. Area of nonmoving object
   7. Volume of moving object
   8. Volume of stationary object
   9. Volume of nonmoving object
  10. Speed
  11. Force

Unfortunately, the proper use of these basic features requires learning not only their definitions but also their various interpretations, which are necessary for using basic features in a specific inventive situation.

9.4.5 Inventive principles (eliminating technical contradictions)

Studying inventions led Altshuller to two interesting observations. First, the same contradictions could be found in many inventive problems and in various areas of engineering; that is, they were universal, not domain specific. Second, inventors, in no matter what domain, were using the same ways to eliminate the same technical contradictions. These two observations combined led Altshuller to the conclusion that both contradictions and the heuristics for their elimination (called later "Inventive Principles") are entirely domain independent. Next, he had a dream: "If we knew all Inventive Principles, then the life of inventors would be so much easier." He was a man of action and made finding inventive principles one of his top research priorities. Soon, after only several years, he had a collection of 40 inventive principles, which can be described as a group thus:

> "Inventive Principles" is a collection of 40 heuristics that are necessary and sufficient to eliminate all possible technical contradictions in inventive conceptual designing.

"Inventive Principles" can be interpreted as a body of methodological and engineering knowledge about inventing. This body of knowledge has been acquired and generalized for future use in inventing. It comes from many sources, mostly from patents but also from books and other publications. Some of these principles were provided to TRIZ scholars by inventors and even by practicing engineers. Inventive principles are heuristics, and as such they do not guarantee any success in eliminating technical contradictions, but at least they create an opportunity for elimination to happen. Each principle requires learning about it and its interpretations and a lot of practice with using it in various inventive situations. They are discussed with many details in Altshuller (1984) or in Altshuller (1996); Altshuller et al. (1999). The first four principles are

1. Principle of segmentation (how to divide objects)
2. Principle of extraction (how to remove an undesired feature or part of an object)
3. Principle of local quality (how to create the desired quality only locally)
4. Principle of asymmetry (how to use asymmetry)

We will analyze briefly only the first (and one of the most often used) principle, the *principle of segmentation*. In fact, this principle can be understood in three different ways as

1. Dividing an object into independent parts
2. Making an object sectional
3. Increasing the degree of an object's segmentation

In the first case, we conceptually divide our single object into several entirely independent systems. For example, instead of using a transportation plane, we use a large number of drones. In the second case, we divide our system into a number of interrelated components that work together to provide the same function as the original system. For example, a complex electric circuit is divided into a number of interconnected circuits for the simplicity of repairs or replacements. In the third case, when we consider an object that has already been divided into several components, we increase the number of components. For example, when we consider a power station that already has two generators, adding another generator will represent the use of this principle.

9.4.6 Separation principles (eliminating physical contradictions)

Sometimes we simply want to invent through the elimination of a specific physical contradiction. It can be done using 11 "Separation Principles" (Altshuller et al. 1999), which are defined thus:

> Separation Principles is a collection of 11 heuristics that is necessary and sufficient to eliminate all possible Physical Contradictions in inventive conceptual designing.

The complete list of these principles with all methodological guidance is available in Altshuller et al. (1999). According to Clarke (1997), the most useful separation principles are

1. Separation in time
2. Separation in space

3. Separation upon conditions
4. Separation between the parts and the whole

The first principle of separation in time can be used when a process is considered and a specific feature is necessary only during a specific time. For example, a young inventor is working on a concept for a feeding system for a team of cyclists participating in a long multistage race in France. For the best performance, cyclists are supposed to secretly eat sandwiches with fresh oysters every 78 minutes while cycling on the flats and more often, every 49 minutes, when climbing in the mountains. In other words, the cyclists do not need sandwiches all the time but should eat them only at very specific times. Even more, it is impossible to carry sandwiches on the bikes because of their weight and because they would lose their freshness when exposed to the heat of the day in southern France. As an additional constraint, the inventor also knows that his team's secret sandwiches cannot be provided at the regular feeding stations as that might compromise their secret status. The entire situation may be described thus: "Provide sandwiches when they are needed; no sandwiches at any other time."

After a long deliberation, the inventor has decided to use the time separation principle and has chosen to deliver sandwiches exactly when they are needed by means of helicopter drones. As a temporary measure, she will first use the communication drones that are already in use to carry confidential messages among the support team members. Later, special stealth helicopter drones will be ordered (and invented first), which be invisible to other teams and which will fly silently.

The separation in space principle needs to be used when dealing with requirements and constraints related to the spatial aspects of design. For example, an inventor is working on a concept for a snowmobile to be used behind the Arctic Circle to transport frozen seafood. The traditional thinking about designing commercial vehicles for the Arctic Circle is simple: "Insulate the entire vehicle as much as possible." In our case, this heuristic is simply inadequate and we need to use the separation in space principle.

We do not need any insulation around the luggage bay when we transport frozen seafood behind the Arctic Circle, but we need even more insulation around the driver's cabin at such a time. This situation could be described as "Insulation must be in place when needed and somewhere else when not." Therefore, we need to design the vehicle in such a way that three panels with heavy foam insulation surround the luggage bay when it is empty, and they automatically move behind the driver's cabin when the bay is full (maybe even under the weight of the very heavy load of frozen seafood).

The principle of separation under conditions was used by an inventor to develop a choking system or simply a choke. It was simple standard equipment in all cars before automatic injection systems with electronic controls were introduced about 30 or 40 years ago. Before that time, in order to start a cold gasoline engine, the driver needed to use the choke to inject an enriched mixture of gasoline and air into the carburetor. This enriched mixture was required only under the conditions of a cold start, and the driver would deactivate the choke immediately after the engine warmed up a little bit and would start running smoothly.

A military inventor working for one of the NATO armies in Europe was given an urgent order to invent on the same day a new type of a military portable bridge steel truss that would be lighter than the standard and widely used truss XLT 84 (extra light truss, developed in 1984). He was extremely concerned about the feasibility of his order. He was an experienced structural engineer and knew very well that XLT 84 was a product of a long process of shape and section optimization conducted using evolutionary optimization. Therefore, improving it by reducing its total weight seemed to be an impossible task.

The inventor, however, had no choice but to follow his order. As a former student of the Author, he knew about TRIZ. Thus, he began thinking in terms of separation principles. He nearly immediately realized that he could use the principle of separation between the parts and the whole. A steel truss may be understood as a system with three subsystems: members under tension, members under compression, and connections. Members under compression must be designed with consideration for the impact of buckling, and that leads to their relatively larger weight with respect to members under tension carrying similar forces. As a result of this phenomenon, members under compression contribute disproportionally to the total weight of the truss. However, if the principle of separation between the parts and the whole is used, and the members under compression are treated separately from the remaining two subsystems, and if a different and better steel grade is used for their design, an inventive solution will be found. This separation principle–based reasoning led the inventor to the development of the concept of the XXLT15 portable military truss. In such a truss, two separate steel grades are used with a better grade being used for the subsystem of members under compression. This new truss is approximately 4% lighter than its predecessor. That is a big improvement considering the fact that soldiers often carry military portable bridge trusses during the process of bridge assembly, as the Author had to do when he was a student and undergoing military training.

9.4.7 Contradiction table

The *contradiction table* is most likely the first inventive tool developed deliberately for general engineering use. Its function is to help inventors to eliminate all possible technical contradictions as identified by Altshuller.

For any technical contradiction, identified by two conflicting characteristics of an engineering system, the table provides numbers of all inventive principles that could be potentially used to eliminate this contradiction.

For example, when a technical contradiction between the volume of a moving object and its speed is considered, the table offers four numbers of inventive principles, namely 29, 4, 38, and 34. They are listed in order of their diminishing potential usefulness. These numbers identify such inventive principles as

29. Use a pneumatic or hydraulic system.
 4. Use asymmetry:
    4.1. Replace a symmetrical form with an asymmetrical form of the system;
    4.2. If the system is already asymmetrical, increase the degree of asymmetry.
38. Use strong oxidizers:
    38.1. Replace normal air with enriched air;
    38.2. Replace enriched air with oxygen;
    38.3. Treat in air or in oxygen with ionizing radiation;
    38.4. Use ionized oxygen.
34. Rejecting and regenerating parts:
    34.1. After it has completed its function or become useless, reject or modify this part of the system;
    34.2. Restore directly any used up parts of the system.

The first potentially useful principle is clear; the remaining ones may require explanation. These explanations should come from a TRIZ expert who understands all the inventive principles and the science behind them.

The contradiction table can be best explained from the knowledge perspective. In this case, the contradiction table can be interpreted as a certain universal body of inventive knowledge. It is in the form of a collection of decision rules related to all the technical contradictions associated with 39 identified features of an engineering system. Each decision rule has two conditions and at least one recommended decision. Each condition is related to a different feature of an engineering system from a given contradiction. The decision is the recommended action—or more precisely the inventive principle or simply principles—that should be used in this case. When several principles are available, they are listed in the order of diminishing potential usefulness. In the case of the specific contradiction discussed earlier, the decision rule is as follows:

> If condition No. 1 (first feature in the contradiction) is "volume of a moving object" and condition No. 2 (second feature in the contradiction) is "speed" then the decision is to use Inventive Principles No. 29 or 4 or 38 or 34.

The contradiction table is considered the synthesis of inventive knowledge acquired by Altshuller and his associates. Since it is an equivalent of a collection of decision rules, it is a priceless result that may be used for the development of all kinds of computer programs for inventive designing.

9.4.8 Ideal solution

Our master, Leonardo da Vinci, once said, "Think of the end before the beginning." In other words, all inventors should always think not only how to solve a problem but how to develop the ideal solution or a concept for the ideal engineering system in a given design situation.

The concept of the *ideal solution* was initially proposed by Altshuller in his book on *Creativity as an Exact Science* (Altshuller 1984). Surprisingly, in 1973, the Author participated in a workshop organized by Gerald Nadler, an American management and design scholar. It was on his Method of Ideal Solutions (Nadler 1973), which was constructed around the same concept of the ideal solution. Altshuller had no access to foreign literature, and Nadler did not read books in Russian published by the Soviet Radio Publishing House in Moscow. There is no question that two great scholars working in two countries that were so different introduced the same concept independently and that this happened around the same time. Apparently, the time has come for this important concept to emerge. We will understand this concept thus:

> The "Ideal Solution" is a concept of the ideal engineering system in a given design situation, that is, a concept for a system providing all required useful functions, creating no harmful functions, and not using any resources.

We could say that the ideal engineering system performs the required function without actually existing as a separate entity. The function is performed using in an inventive way only the resources that are already available. This is inventive engineering in action: Nothing basically changes; everything remains the same; but the problem is solved. The concept of the ideal solution is so important for building our inventive minds that we will have two examples: a classical example provided by Altshuller (1984) and a second provided by the Author.

In the first example, the inventive challenge is as follows:

> Design a small chamber for testing various metals in acid.

Testing requires providing direct contact between the acid and the metal. Our initial thought is that the metal specimen should be surrounded by acid—submerged in a chamber filled with acid. In fact, our inventive challenge was so formulated that this understanding of the problem is directly implied. However, if we analyze the available resources, we will realize that we have available only the metal to be tested and some acid. If the chamber is designed using materials other than the acid or metal available for testing (e.g., plastic or stone), it will be not the ideal solution because additional resources were used.

In this situation, the chamber should be designed using the metal available for testing. In this way, no other materials will be used and the inventive problem will be solved utilizing only the available resources. We will have the ideal solution.

In the second example, a well-known invention, a self-supporting car body, will be used. This inventive problem emerged in the 1920s as a result of the evolution of cars and the growing problems associated with mass production. The problem was as follows:

> Mass car production requires the minimization of the number of car components. A car is a system with four major subsystems: (1) chassis, (2) frame, (3) engine with transmission and power transfer, and (4) suspension with wheels. From the structural point of view, this system transfers to the wheels the loads applied to the chassis, and all these loads flow from the chassis through the frame to the suspension. The frame also provides both the bending and torsional rigidity to the car. The frame is heavy and difficult to manufacture, and its presence in a car simply reflect the car's roots in the horse carriages; these did not have any chassis, and the frame was absolutely necessary to transfer loads to the suspension and wheels. How to improve the frame?

In fact, many inventors tried to improve the frame design. They considered all kinds of alternatives to the frame's geometry, new cross sections, new materials, and so on. All this work led to gradual improvements but not to a major breakthrough. It came with the realization that the best frame—the ideal frame—is no frame at all. The frame may be eliminated if the available resources are utilized in a new way. In this case, when the transfer of loads from the chassis to suspension is considered, the available resources are in the form of the chassis and the suspension with wheels. These two subsystems may be directly connected and also used to provide the required car rigidity.

In the described situation, the solution is to eliminate the car's frame and use the other existing car subsystems to provide all the frame's required functions. Conceptually, it is the ideal solution, but obviously in this case redesigning the chassis to overtake the frame's functions results in its increased weight and some additional requirements for its design.

Altshuller also introduced the concept of *ideality of an engineering system*, or simply *ideality*, as a quasi-formal measure of the degree to which a given system is ideal. First, we need to introduce the concept of a *useful function* (F_{Ui}), that is, a function that is required of a given system and is the reason that it is being designed. Next, we need the concept of a *harmful function* (F_{Hj}), that is, a functions that the by-product of designing the system and is harmful for its environment. Ideality is defined as the ratio of the sum of useful functions to the sum of harmful functions:

$$\text{Ideality} = \frac{\sum_{i=1}^{m} F_{Ui}}{\sum_{j=1}^{k} F_{Hj}}$$

where:

F_{Ui} is a useful function, i=1, 2, 3, ...

F_{Hj} is a harmful function, j=1, 2, 3, ...

Ideality is basically a qualitative measure and is used to compare the benefits and harm caused by the introduction of a given system. For example, when a gas pipeline in the Shenandoah Valley in Virginia, the United States is considered, its F_{Ui} can be interpreted as the increased energy independence of the country, the increased chemical production, the additional income for the locals, and so on. On the other hand, F_{Hj} may be interpreted as lost valuable agricultural land, harm to the unique views, potential ecological disaster in the case of an explosion and fire, and so on. Such a comparison has purely qualitative nature. However, even in this example, it is entirely possible to assign to all these functions monetary or utility values, and the ideality of a system will be expressed in purely quantitative terms.

Obviously, ideality should be maximized. In the case of the ideal solution, the ideality is equal to infinity. In any other cases, it will be greater than 1 when the value of useful functions prevails over the value of harmful functions (more benefits than harm), and less than 1 when the value of harmful functions prevails over the value of useful functions (more harm than benefits).

When a given inventive problem is considered, the ideal solution is a reference point; that is, all developed inventive concepts are compared with the ideal solution and in this way their relative goodness is determined. Ultimately, the inventor would like to develop a concept representing the ideal solution in her or his case, and this is sometimes possible.

9.4.9 Vector of psychological inertia

Today, and during the last 20 years, so-called design fixation is the subject of systematic studies by cognitive psychologists. According to Youmans and Arciszewski (2014) this term can be defined thus:

Design Fixation identifies the phenomenon when designers limit their creative output because of an overreliance on features of preexisting designs, or more generally, an overreliance on a specific body of knowledge associated with the design problem.

Altshuller was not a cognitive psychologist but an engineer and an inventor. Also, he conducted his studies of design problem solving during the time period from approximately 1945 to 1970. These studies were highly informal from the perspective of the today's experimental cognitive psychology. However, he had incredible intuition and discovered the phenomenon of design fixation about 50 years before American psychologists. He called it even more appropriately *the vector of psychological inertia* and described it in more general terms than design fixation. He presented his term as identifying the natural human tendency to focus on known design concepts and to be extremely reluctant to look for concepts outside one's own domain.

The best example of the vector of psychological inertia comes from the Soviet Union in the 1960s. At that time, the Soviet Union was involved in an extensive program of moon exploration and was planning to send a space probe, Luna 16, to take pictures on the dark side of the moon. Taking pictures obviously would require a lot of light, which had to be provided by a special moon-grade light bulb. It had to survive a trip to the moon on a Soviet rocket, and it should work for at least some time in the moon's nearly perfect vacuum. Unfortunately, Soviet rockets in the 1960s were not known to provide a particularly smooth ride, and the moon bulb would be subjected to all kinds of vibrations and rapid changes in pressure and temperature. This bulb-hostile environment could easily destroy any known seal between the bulb's base and the glass vessel and thus also destroy Soviet ambitions to take a picture of the Soviet flag on the dark side of the moon. This accomplishment might also have had some serious political implications regarding moon colonization and control, but that is a different story.

In the above described situation, the moon bulb became an important scientific but also political matter. A group of the best and most trusted Soviet scientists had got an order to develop a moon bulb that would guarantee success. After a lot of money had been spent and time and efforts invested, finally a truly Soviet moon bulb was developed. It had a secret super seal, which would survive intact all dangers of the trip to the moon and would work at the destination in the vacuum for at least 11 minutes. This development was announced as a major technological success, but a junior scientist then noticed that the entire project, which had been conducted by senior scientists, was simply useless. Unfortunately, whoever gave orders to the scientists was obviously driven by a particularly powerful vector of psychological inertia (most likely he was a high-level administrator in charge of the Soviet space program). He simply followed traditional reasoning regarding the development of light bulbs on Earth and believed that each bulb needed a seal. He forgot that the purpose of the seal between the base and the glass vessel in a bulb is to maintain a vacuum around the lighting element. On the moon, in what is a nearly perfect vacuum, there is no need to have a glass vessel and to maintain a vacuum in the bulb. It will work perfectly well without the vacuum provided by the glass vessel; the lighting element may be simply exposed to the "local" vacuum on the moon.

The existence of the vector of psychological inertia clearly explains why an inventor usually thinks, at least initially, along the existing line of evolution of the system being considered and why his/her reasoning at first brings only very small incremental improvements. It also explains why it is so difficult for inventors to think "outside the box" and to seek knowledge outside their domain. It also provides simple enlightenment as to why the profession, and up to a certain degree society, usually initially reject or delay all new ideas. Even more importantly, knowing about this phenomenon allows inventors trying to commercialize their inventions to consider rejection as a part of objective reality, no matter if they are inventors in Nigeria or in the United States. Rejections should never be expected, but they should definitely not be surprising. There is nothing personal in rejection; it is simply the natural occurrence of an objective phenomenon and nothing can be immediately done about it. Unfortunately, only patience and persistence may help inventors to prevail and implement their concepts.

Finally, an inventor familiar with the phenomenon is much more inclined to address it in a rational way and to try to deliberately overcome its negative impact. In this way, such an inventor will be much more effective in action than the one unaware of the phenomenon. Such an inventor may be simply doomed to make only small quantitative improvements. This will be done exclusively by using knowledge closely related to the problem and without the benefits of introducing radical and qualitative changes, which require overcoming the impact of the vector of psychological inertia.

9.4.10 Patterns of evolution

Altshuller studied all available sources of information about the evolution of engineering systems, including books on the history of technology and patents. His focus was on patents in various engineering domains that were patented in a number of countries over long time periods, in the order of tens of years.

As a result of his studies, he produced three surprising and important results. First, he noticed that engineering systems do not evolve randomly. Second, he observed clear objective patterns behind their evolution. Finally, he discovered that the patterns he found were domain independent. In other words, he discovered that the evolution of all engineering systems was governed by the same set of patterns, which he called

"Patterns of Evolution of Engineering Systems" or simply "Patterns of Evolution."

> Patterns of Evolution is a set of heuristics describing evolution of all engineering systems over the long period of time.

At present, nine patterns of evolution are known. They are presented and briefly described with examples in Section 9.7.

9.5 ASSUMPTIONS

Altshuller and his followers never clearly specified all the assumptions of TRIZ in a single publication. Therefore, the presented assumptions have been compiled from numerous sources, and they are a result of discussions with TRIZ experts. Also, they are presented in the language of inventive engineering.

1. It is a systematic method for the development of inventive design concepts.
2. It is a method with roots in engineering and is intended for engineering applications.
3. The method is intended for a single inventor.
4. Inventive problem solving is about the elimination of technical contradictions.
5. The inventive problem is a conceptual designing problem, the solution for which requires the elimination of at least one technical contradiction and results in a potentially patentable design concept.
6. Each technical contradiction can be transformed into the related physical contradiction.
7. There are five major categories of inventive solutions, classified in terms of the way they have been developed.
8. There is universal inventive knowledge, that is, knowledge applicable to inventive problem solving in all engineering domains.
9. During the history of engineering, a lot of universal inventive knowledge has been accumulated, and this knowledge is sufficient to solve the majority of inventive problems.
10. Patents are the most important source of universal inventive knowledge.
11. Universal inventive knowledge may be presented in the form of heuristics called inventive patterns.
12. The entire available universal inventive knowledge may be presented as a collection of 40 inventive patterns. This collection is necessary and sufficient to solve the majority of inventive problems.

13. There is a set of basic features of engineering systems (39), which are necessary and sufficient to describe any engineering system for inventive purposes.
14. All technical contradictions may be expressed in terms of basic features.
15. Each technical contradiction may be eliminated using at least one inventive principle.
16. The contradiction table is the synthesis of methodological universal inventive knowledge.
17. For each inventive problem, there is an ideal solution.
18. All engineering systems evolve over long periods of time following the same objective patterns of evolution.
19. The ultimate objective of the evolution of any engineering system is the ideal solution.
20. Patterns of evolution are domain independent.
21. There are nine patterns of evolution.
22. Patterns of evolution are intended for the development of engineering systems and for the forecasting of their evolution.

9.6 PROCEDURE AND EXAMPLE

9.6.1 Introduction

The following procedure is mostly based on classical TRIZ, as Altshuller developed it. However, it has been modified by the Author. Also, for the benefit of readers, the use of the *innovation situation questionnaire* (ISQ) has been incorporated into the first stage of the procedure (problem identification), although ISQ was developed much later than Altshuller's original TRIZ. (The ISQ was developed at the Kishinev School of TRIZ in Moldavia in the late 1980s, about 40 years after the initial ideas of TRIZ were formulated.)

The modified TRIZ procedure has six major stages, which are listed and discussed with examples here:

1. Problem identification
2. Problem formulation
3. Development of design concepts
4. Evaluation
5. Methodological knowledge acquisition
6. Final presentation of results

With the exception of the problem identification (discussed with examples in Section 5.3), all stages of the procedure are explained using the same example. Therefore, no separate TRIZ example is provided.

9.6.2 Stage I: Problem identification

Several methods for the identification of problems, including the systems approach, mind mapping, and the ISQ, were discussed in Section 5.3, "Problem Identification." The last one was developed by Altshuller's disciples in the context of TRIZ. Therefore, it is most appropriate to incorporate it into the modified TRIZ procedure, presented here. The ISQ has been discussed with examples before; therefore, it will not be discussed again in this chapter. The reader should remember, however, that when ISQ is performed, it produces

1. Brief (casual) description of the designing problem
2. Detailed description of the designing problem, including
   a. The input and output of the system
   b. The function of the system and functions of its individual elements and subelements
   c. The system's structure, its elements, and their connections
   d. Functioning of the system
   e. The system's environment
   f. Available resources, constraints, and requirements
   g. Allowable changes
   h. Evaluation criteria
   i. Evolution line leading to the present solution

9.6.3 Stage 2: Problem formulation

The objective of the *problem formulation* stage is to formulate the inventive problem in the form appropriate for using TRIZ. Specifically, the results of the problem definition stage must be "translated" into the TRIZ "language" or formulated in TRIZ terms.

There is a three-step process to follow for any given system under consideration:

1. Determination of the basic features
2. Identification of the major technical contradictions
3. Development of the ideal solution

The problem identification stage provides information about the input and output of the system, the function of the system, and the functions of its individual elements and subelements.

The first step requires a careful analysis of these first two pieces of information (input and output) in order to determine the major basic features, that is, those features that are incorporated in the description of the input and output and into the description of the system functions.

For example, let us consider a structural system. If the function of our system is to carry out to supports transverse loading, its input will be vertical forces applied to the system along its length, and its output will be two vertical reactions (forces) applied at its both ends. We are obviously talking here about a structural system; such a system needs to have some required strength, and as a material system it will have some weight. Therefore, reading the problem identification has helped us to determine two basic features: strength (basic feature No. 14) and weight of nonmoving object (basic feature No. 2, called simply "weight" in our chapter).

In the second step we need to identify a technical contradiction involving the basic features that have been distinguished. In our simple example, this is easy: We have a technical contradiction between "strength" and "weight."

In the third step, we need to imagine the ideal solution. This is not easy; usually we are under such a strong influence of the vector of psychological inertia that it requires some effort and dreaming up the ideal solution using, for example, the technique of wishing described as a part of Synectics in Chapter 8. In the case of our example, the ideal solution might be described thus:

> The ideal solution is an engineering system that simply does not exist, but the transverse forces are still transferred to the two support points.

In such a case no weight would be involved, but the required function would still be provided. Obviously, on the surface, our ideal solution does not make any sense, as any structural engineer would tell you. However, if we start thinking the way inventive engineers should (nothing is impossible), we will immediately realize that we know of the example of a self-supporting car chassis in which the car frame is eliminated but its function is still provided. Then, we will discover that we could redesign the super-system of our structural system in such a way that there would be no need for our structural system. How? This is a different question that can be answered only in the context of a specific supersystem. It obviously matters if our structural system is a part of a supersystem in the form of a tall building or an industrial building.

9.6.4 Stage 3: Development of design concepts

This is the most important and also unique and fascinating stage of the entire TRIZ procedure. It is the stage when the entire power of TRIZ is suddenly revealed, and a design concept, or a class of concepts, emerges.

In the previous stage, problem formulation, both the problem's basic features and the technical contradiction have been identified. Now, they will be used to find inventive patterns, which might help to eliminate the identified technical contradiction.

In the case of our example, we have a technical contradiction between the basic features of weight and strength. In general engineering terms, it means that any strength improvement or increase will be associated with the corresponding weight increase. The first one is desirable, the second definitely not. The considered technical contradiction may be easily transformed into a corresponding physical contradiction, for example the height (or depth in the language of structural engineering) of the system under consideration. This dimension should be increased to improve strength and at the same time decreased, or reduced, to improve weight. This is an obvious trade-off, which in traditional engineering would never be eliminated; at best, a compromise would be found.

In the case of our example, the contradiction table (the intersection of the row "weight" and the column "strength") provides three inventive patterns, listed here in the order of the decreasing probability that they would be useful:

1. Pattern No. 28: Replacement of a mechanical system
2. Pattern No. 2: Extraction
3. Pattern No. 10: Prior activation
4. Pattern No. 27: An inexpensive short-life object instead of an expensive durable one

Let us discuss all these inventive patterns and discover how they may be related to finding inventive design concepts in the case of our example.

9.6.4.1 Pattern No. 28: Replacement of a mechanical system

This pattern has four specialized heuristics, each of which may be potentially useful:

1. Replace a mechanical system with an optical, acoustical, or odor system

    Out of these three heuristics, the second one seems to be most feasible. An acoustic system is a system producing sounds or sound waves as a result of a moving membrane, which is driven by a flowing electric current. We can imagine a situation in which an object is placed on a horizontally positioned membrane and a constant current is applied; the object will be raised and will be kept in this new position as long as the constant current is applied.

    Following this reasoning, we can imagine a new system in which under all points of application of our forces, a single speaker-like device is positioned and it is directly supported by the two supports that are supposed to support our system. Whenever forces are applied, they activate the flow of the electric current, which will keep the membrane with the forces sitting on the top in a constant position. We thus have an acoustic system replacing a mechanical system.

2. Use an electrical, magnetic, or electromagnetic field for interaction with the object

The third possibility is definitely the best. Our object is in the form of supports for the forces applied to the system. If these supports are metal plates, for example, a strong electromagnetic field may be used to keep them in the desired positions.

3. Replace fields:

(3.1) stationary fields with moving fields,

(3.2) fixed fields with those changing in time, and

(3.3) random fields with structured fields.

Heuristic No. 3.2 seems to reiterate the previous heuristic, No. 2. It can be interpreted that a field (e.g., an electromagnetic field) may be used, but instead of a fixed field, applied all the time, it should change in time; that is, it should be activated only when it is needed when the external loading is applied.

4. Use a field in conjunction with ferromagnetic particles

This is a useful heuristic applicable when dealing with a timber structure or pieces of timber. Instead of using metal plates supporting applied forces, timber plates carrying these forces could be coated with ferromagnetic particles, and an electromagnetic field could be used to keep them in place.

9.6.4.2 Pattern No. 2: Extraction

This pattern has two specialized heuristics, each briefly discussed here:

1. Extract (remove or separate) a disturbing part or property from an object

This is an outstanding heuristic when dealing with all structural systems carrying transverse loading and, as a result of that, being subjected to bending. In order to increase the strength of such a system, the moment of inertia of its cross section should be increased, preferably without increasing the cross-sectional area leading to the increase in weight. This moment of inertia improvement may be accomplished when the material in a member under bending is moved away from the neutral axis, for example by replacing a member with a rectangular cross section with a member with an I section or by a box member with material removed from the central part of the section.

2. Extract only the necessary part or property

In our case, this heuristic basically only reiterates the first heuristic.

9.6.4.3 Pattern No. 10: Prior activation

Under this pattern, two specialized heuristics are provided and discussed:

1. Carry out all or part of the required action in advance

    This is an excellent heuristic, which may be interpreted as "reduce the loads." How? This is a different question but one with many answers; for example, use lighter materials, use higher-grade steel, reduce dimensions, and so on.

2. Arrange objects so they can go into action in a timely manner and from a convenient position

    This is another excellent heuristic. It provides the suggestion that the configuration of our system should be developed in such a form as to minimize material-hungry bending and replace it, at least partially, by a configuration that mostly uses members under tension, which usually are much more material efficient; for example, a beam could be replaced by a suspended cable structure.

9.6.4.4 Pattern No. 27: An inexpensive short-life object instead of an expensive durable one

This heuristic is the last listed in the contradiction matrix, and that suggests its relatively low usefulness. In our case, this is not exactly so. This heuristic may be interpreted as being related to temporary systems. It may be read that in the case of a temporary system, instead of heavy and expensive metal or concrete structures, a much less expensive pneumatic structure could be used.

9.6.4.5 Design ideas

The stage has produced eight ideas that need to be evaluated in order to decide which of them can be developed into inventive design concepts and eventually used for practical purposes.

In the case of our example, the developed ideas include

1. Gigantic speaker-like device supported at both ends
2. Steel plates kept in place by the electromagnetic field
3. Modification of idea No. 2 but with the electromagnetic field activated only when loads are applied
4. Timber plates coated with ferromagnetic particles kept in place by the electromagnetic field
5. A linear member with an I or hollow cross section
6. Reduction of transverse loads
7. Use of a configuration with mostly members under tension, for example a suspended structure
8. Use of a temporary pneumatic structure

9.6.5 Stage 4: Evaluation

All the ideas developed in the previous stage need to be evaluated using an assumed set of evaluation criteria.

In the case of our example, we will assume three evaluation criteria, which are considered equally important. The value of each criterion will vary within the range 0–1, and any idea getting at least one "0" value will be eliminated. There are several ways to determine the values of the individual criteria for all ideas.

First, the values of the individual evaluation criteria may be arbitrarily assigned by the inventor himself or herself. This is an easy way to obtain the values of the evaluation criteria but is highly subjective, reflecting the understanding and bias of a single person. Second, a team of experts can be created (e.g., from within the inventor's company). In this case, however, the selection of experts may have impact on the results of their evaluation. Selecting traditional engineers, who are usually risk averse, may lead to the elimination of truly novel concepts. On the other hand, selecting only inventors (who are usually risk affinitive) to serve on the evaluation panel may result in a situation in which mostly very novel but marginally feasible ideas are selected. In any case, using an evaluation panel is a good idea, but the selection of experts must reflect the priorities of the company conducting the evaluation.

In the case of our example, the Author arbitrarily assigned the evaluation numbers using his structural engineering experience. All the results of this evaluation are presented in Table 9.2.

The evaluated ideas may be divided into three categories: (1) eliminated ideas, (2) potentially useful ideas, and (3) best ideas.

There are two ideas (Nos. 5 and 6) in the first category of eliminated ideas; three ideas (Nos. 1, 7, and 8) in the second category of potentially useful ideas; and three ideas (Nos. 2, 3, and 4) in the category of best ideas. The best ideas are

Idea No. 2: Steel plates (or plate) kept in place by an electromagnetic field

Idea No. 3: Modification of Idea No. 2, with the electromagnetic field activated only when loads are applied

Table 9.2 Evaluation of TRIZ ideas

| Idea | Evaluation criteria | | | Final evaluation |
	Feasibility 0–1	Novelty 0–1	Patentability 0–1	0–3
1	0.1	1.0	0.9	2.0
2	0.8	1.0	0.9	2.7
3	0.8	1.0	0.9	2.7
4	0.6	1.0	0.95	2.55
5	1.0	0	0	0
6	1.0	0	0	0
7	1.0	0.5	0.8	2.3
8	1.0	0.6	0.8	2.4

>   Idea No. 4: A timber plate coated with ferromagnetic particles kept in
>   place by the electromagnetic field

The obtained results should not be interpreted to mean that the remaining ideas are useless. Quite the contrary; a formal evaluation model has been used that produced a specific set of results mostly driven by the assumption that only patentable design concepts should be considered as meaningful results. In fact, other ideas may be found that are even more attractive and more feasible than the "best ideas." For example, the idea of a temporary pneumatic structure may be developed into a novel and very practical design and may actually be patentable. Inventive engineers should be always aware that initially rejected ideas may still have tremendous potential and that they should be recorded and reconsidered from time to time; they may bring inspiration or may be modified, resulting in valuable design concepts.

The conducted evaluation process has been subjective, and using a different evaluator or group of evaluators might produce different results. However, the evaluation process could be the same.

When the numerical evaluation process is completed, the best ideas should be compared to the ideal solution. In the case of our example, the ideal solution has been described thus:

>   The Ideal Solution is an engineering system that simply does not exist
>   but the transverse forces are still transferred to the two support points.

All best ideas need to be compared to the ideal solution to determine which is the closest one to this reference point. All three of our best ideas (or, precisely speaking, the structural systems based on them) most likely will satisfy the requirement that the transverse forces must be transferred to the two support points. Therefore, the differences among the best ideas will be related to their expected weight and eventually to the expected energy consumption. At this stage of the designing process, only the expected values of weight or energy consumption may be considered because very little is known about the technical details associated with the individual best ideas.

When weight is considered, there will be only a marginal weight difference between No. 2 and No. 3, that is, between the total weight of several small steel plates located under the applied transverse forces and the weight of a single steel plate under all these forces. However, in both these cases, the weight of plates will be significantly smaller than the weight of a traditional structural system supporting transverse forces (i.e., steel or reinforced concrete structural systems), and that means that these two best ideas are relatively close to the ideal solution.

The third *Best Idea* (No. 4) with timber plates under applied forces will be even closer to the Ideal Solution because of the lower weight of timber when compared to steel. Therefore, this best solution may be closer to the ideal solution than No. 2 and No. 3.

Where energy consumption is concerned, No. 3 (the electromagnetic field activated only when loads are applied) is the best, and it is obviously closer to the ideal solution than the remaining best ideas.

The provided overview of the process and the example are sufficient to develop a good understanding of evaluation. However, in the practical applications, the analysis should be much more detailed and extensive. For example, where energy consumption is concerned, specific numbers could be found and the comparison made more quantitative.

9.6.6 Stage 5: Methodological knowledge acquisition

TRIZ has been presented in this chapter as a certain body of methodological knowledge necessary and sufficient to solve a large class of inventive problems. This body of knowledge is not stationary; it is constantly evolving and growing on the level of individual TRIZ practitioners and subsequently on the level of the entire TRIZ community. Each TRIZ user should be aware of this situation and should contribute to this process. For all these reasons, at the end of the TRIZ process, a moment of methodological reflection should be undertaken so as to acquire and record new knowledge, which should be shared with other people.

Specifically, the methodological report should address several issues, which usually cause problems even for experienced TRIZ users:

Interpretation of inventive patterns: The available books do provide examples of how various principles are used, but these examples are grossly insufficient, and there are usually very few in the domain close to the problem domain. Therefore, creating additional examples for private or general use is strongly recommended. The Author also believes that creating mini collections of examples of how a given principle is used in several domains would be most appreciated by the community of TRIZ users.

Assessment of the inventive power of inventive patterns: This is a very subjective and individual matter, but even soft information regarding the inventive power of the individual principles would be most helpful in all situations when it is clear what kind of inventions is desired. For example, sometimes only the most sophisticated inventions are sought because a company is trying to build a patent fence. In other cases, only relatively simple inventions are sought because a company needs patented inventions for immediate use.

Finding inventive patterns that failed: Unfortunately, there are probably are such patterns, which may not work in all domains. Therefore, if a given user operates in a specific domain, any information about patterns that should be avoided would be appreciated.

Interpretation of technical contradictions: Sometimes understanding a technical contradiction in a specific domain is difficult, if not impossible, especially for inexperienced TRIZ users. If such understanding is ultimately acquired, it should be recorded for future use and for sharing with others.

9.6.7 Stage 6: Final presentation of results

Surprisingly, the clarity of the outcome of the *presentation of results* may decide whether the developed design concepts succeed or fail. The results of a TRIZ project are usually prepared by TRIZ experts or by at least TRIZ-knowledgeable engineers. These results, however, most likely will be used and implemented by engineers who know very little, or nothing, about TRIZ. In such a situation, all results contained in the report should be presented in general engineering terms with the minimum use of TRIZ-specific terms. Obviously, a good visualization of the results will significantly improve the clarity of the presentation of results.

The report should be divided into five major parts:

1. Problem identification
2. Problem formulation
3. Developed design concepts
4. Evaluation
5. Acquired methodological knowledge

9.7 PATTERNS OF EVOLUTION

9.7.1 Introduction

In the Section 9.4, the concept of patterns of evolution was introduced. We have at least three important applications for them. First, they are outstanding tools for acquiring learning or knowledge. When studying the historical evolution of an engineering system, patterns of evolution may provide a rare insight why a given system evolved in its specific way. It is also always a surprise to realize that the seemingly random evolution of a system, created by several inventors in various countries over a period of 70 or 80 years, in fact followed the patterns of evolution and up to a certain degree could be predicted. Second, and most likely even more important, the application of patterns of evolution is in the area of the development of future engineering systems. This area is particularly important for inventive engineers, because it allows them to develop inventions in a different way than through the elimination of technical contradictions. Finally, patterns of evolution may be used to forecast lines of evolution of engineering systems and to create a forecast for a specific system for the next 10 or 20 years. All these three applications of patterns of evolution will be briefly discussed.

There are nine patterns of evolution, and they deal with the evolution of engineering systems over time. The time frame depends on the domain. In the area of structural engineering, a period of 30–40 years may reveal the relevance of patterns of evolution, but in the area of electronics, a period of only a few months or years is sufficient to observe these patterns in action. All patterns of evolution are introduced and discussed with examples in the following sections.

9.7.1.1 Evolution based on an S-curve

This is probably the most important and useful pattern of evolution, which has already become a part of universal engineering knowledge and is widely used in many domains. This pattern can be described thus:

> An engineering system evolves following the S-curve pattern. Its evolution begins with birth and ends with death and goes through stages of childhood, growth, maturity, and decline.

Figure 9.2 represents the generic S-curve. It is a function of time and it reveals the nature of this function. It also shows all four stages of evolution.

The concept of the S-curve will be explained in general terms. We will consider here a relationship (a function) between a dependent variable, called simply "variable," and an independent variable, which is in our case "time." The nature of the relationship is the same for several kinds of variable. The variable may represent the performance of the system (for a car: maximum speed, braking distance, fuel consumption, etc.) or its various features (e.g., weight, length, or volume). The variable may also represent the rate of change of the performance or of a feature of the system, particularly when we are interested in the dynamics of change. Finally, "variable" may mean the cumulative value of a specific characteristic of the system, for example, the number of patents associated with the system.

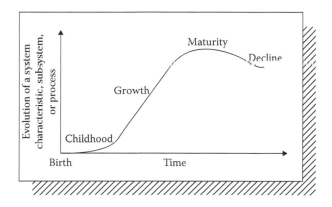

Figure 9.2 S-curve.

Each S-curve begins with *birth*. Birth represents the time when a given system emerges in the market as a new product but may still be under at least some development. However, for a combination of reasons, progress is relatively small during this first period of time, appropriately called *childhood*. It may be so because only limited resources are available during this period or because at the beginning, very little is known about the system and any progress requires a lot of effort.

When the product is accepted and the market begins its growth, the second period, called *growth*, begins. More and more resources become available, some accumulated body of knowledge about the system already exists, and therefore development progress is much faster than during the childhood period. That results in rapid improvements of various characteristics of the system, including its performance.

After a certain period of time, all relatively simple development work is done, there is a large body of available knowledge, and making technological advances becomes increasingly difficult and expensive. Also, the growth of demand is slowing, and the rate of development is going to zero. This could be 25 years for a car or only several months for a smart phone; under such circumstances, performance gradually stagnates and reaches a plateau, while the performance growth rate is very low and finally comes to zero. There is no question that the system is in the period called *maturity*.

When the system is in this state, less and less development work is done, and that obviously results in declining performance growth rates and even in a decline in actual performance. The system begins its stage of *decline* and is approaching the end of its market existence, called *death*, when it is gradually replaced in the market by a product based on a different technology and riding a different S-curve (Figure 9.3); for example, an electric car replaces a car with an internal combustion engine.

In this way, we have found a good explanation as to why we need not only to master existing technologies but to develop new ones. We always need to remember that existing technologies have their natural limits of evolution. Because of that fact, one day we will need to replace existing technologies with new ones simply to facilitate the move of our system from an existing S-curve to a new one, associated with a new technology, if we want to continue the evolution of our system and to improve its performance and its other features.

When working on the development of an invention, it is always important to know where other known contemporary designs could be located on the S-curve. It is usually relatively easy to invent new designs when the system is in the birth or growth stages and the unexplored solution space is still relatively large. On the other hand, when the system is in the maturity stage, inventing becomes much more difficult and usually requires huge resources. Simple, low-level inventions are "gone," and the "remaining" inventions are of a higher level and are most likely quite sophisticated.

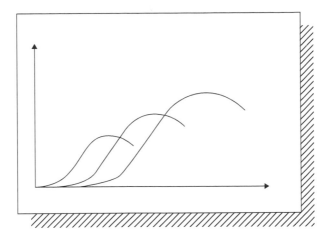

Figure 9.3 Class of S-curves.

9.7.1.2 *Evolution toward increased ideality*

This principle may be interpreted thus:

> An engineering system evolves in such a direction as to optimize the use of available resources and to increase the degree of its ideality.

The principle can be best explained using as an example the internal combustion engine. The first part of the principle is related to the optimization of the use of the available resources. In our case, the available resource is gasoline, and "the optimization of its use" means the reduction or minimization of the fuel consumption over a period of time. There is no question that this phenomenon may be clearly observed over the last 100 years of evolution of the internal combustion engine; such engines have better fuel consumption with each model year.

The second part of the principle regards the evolution in the direction of the improved or increased *degree of ideality*. This increase will be a result of two interrelated processes. First, it will be the growth of the useful functions (the nominator in the *ideality ratio* formula). Second, it will be the parallel reduction of the harmful functions (the denominator in the ideality ratio formula). In the case of internal combustion engines, their useful functions are constantly growing, for example their power and torque, while their harmful functions are decreasing, for example their CO_2 pollution and noise.

9.7.2 Uneven development of system elements

This is an interesting and surprising principle, which can be defined thus:

> An engineering system improves monotonically, but the evolution of its individual subsystems follows their own various S-curves.

The first part of this principle describes how an engineering system improves. It is a monotonic process with periods of nonimprovement, or *stagnation*, when improvements do not temporarily take place; however, the system never *retreats*. During periods of war and recession, the development of many systems is delayed, and there are no improvements. However, we practically never see the worsening of important characteristics of engineering systems, although rare exceptions can be found.

For example, when cars are considered, usually each year we see not only more attractive cars, but also more efficient, safe, and comfortable ones with respect to the previous generation. However, it is not unusual that during times of crisis, less expensive car models are introduced, and these may come with less expensive and sophisticated suspension or tires. That may lead to worsening of the handling characteristics when compared with their predecessors. Obviously, the more expensive and better models will be introduced again as soon as the crisis is over, and the anomaly will disappear.

When a complex engineering device is considered, for example a car, it can be viewed as an entity—as a system providing a single and unique function (the transportation of people and their luggage in our case). Also, when we consider the major components of our device, they may be viewed from a dual perspective. First, they may be viewed as subsystems of our system. Second, they may be seen as complete systems providing various specialized functions. For example, when a car is considered as a system, it will have many components such as suspension, brakes, steering, and so on. All these components can be viewed as the car's subsystems but also as complete systems providing their specific functions; suspension provides comfort, brakes allow speed to be reduced, and so on. This dual nature of the individual system components means that they contribute to the monotonic improvements of the entire system, but at the same time they have their own "private" life, or evolution, which is governed by their own "private" S-curve in accordance to evolution pattern No. 1 (system evolution based on an S-curve).

The described phenomenon can be summarized by saying that the entire system has its own "life" but that all its individual subsystems have their own lives too. There is one S-curve for the entire system and many different S-curves for all subsystems. This pattern has great significance when trying to understand various phenomena from the past but also when trying to prevent their occurrence in future systems.

For example, when the evolution of body armor and helmets is considered, we can distinguish two subsystems: a skull protection system and an eye protection system. Their S-curves could not be more different. In the first case, it can be seen that more than 1,000 years ago, helmets provided excellent skull protection (e.g., Roman helmets), and very little happened later on. That means that the development of helmets has been in the maturity stage for about 1,000 years. (Only in the last century were all kinds of more advanced helmets developed, utilizing new materials and shapes developed as a result of sophisticated computer shape optimization.) However, until recently (in terms of the history of technology), helmets provided none, or at best very limited, eye protection. Obviously, until the last century, the evolution of eye protection systems was in the childhood stage, resulting in all kinds of tragic consequences in this case.

A supertanker is another example of a system whose subsystems have entirely different S-curves. Its propulsion system is a marvel of modern marine technology. Most likely, its evolution is in the advanced stage of growth or even in the maturity stage within the present technology of turbocharged diesel engines, but the evolution of its braking system is still in the childhood stage, since it is nearly impossible to stop a supertanker going at full speed within a distance of less than several miles.

The pattern "uneven development of system elements" provides an unusual insight into the evolution of engineering systems and reveals the importance of the proper selection of their subsystems. The weakest link controls the strength of a chain; similarly, the least developed subsystem is "holding back" the entire system and limits its performance. In extreme situations, this phenomenon may have tragic consequences as it did several years ago with the evolution of the Ford Explorer, a highly successful SUV. In the new model, improved suspension was introduced, but the tires had not been improved, and this combination led to a number of tragic accidents that might have been avoided by knowing about this pattern.

9.7.3 Evolution toward increased dynamism and controllability

This is a complex pattern, which can be described thus:

> An engineering system evolves in such a direction as to maximize its dynamism and controllability.

We can divide engineering systems into *static* and *dynamic*. Static systems do not change their configurations. For example, a steel frame in an industrial building is always static and it never changes its configuration (when this happens, it is the beginning of a plastic collapse and the destruction of the building). Dynamic systems do change their configuration: They may be in one of the several feasible configurations or may continuously adapt their configuration to the changing conditions.

For example, when we consider spoilers in sports cars, we may have small spoilers integrated with the chassis, and these are static systems. There is also a large class of rear spoilers called "wings," which are positioned in a certain distance from the car body and are not integrated with it. We have fixed wings, which are always in the same position, and these are static systems. We have also dynamic wings, which can be adjusted into one of several predefined positions, depending on the expected racing conditions. Finally, we have sophisticated wings controlled by a computer, which are constantly moving and changing their "angle of approach" depending on the speed, the wind, and the car's behavior. When we consider the time line, fixed wings were first, followed by adjustable wings and then by robotic wings. This is our pattern No. 4 in action.

The system's dynamism is directly associated with its controllability, and thus these two notions are covered by a single pattern of evolution. However, many TRIZ experts consider the issues of dynamism and controllability separately and promote two separate patterns.

Controllability and its increase are explained using the same example of wings in sports cars. Once installed, the fixed wings simply cannot be controlled. Adjustable wings may be partially controlled by selecting their desired position before the race. In the third case, robotic wings, we have nearly unlimited ability to control them, restricted only by the sophistication of our computer model and by the available computing power. As our example demonstrates, the increasing dynamism of a given engineering system over time also results in its growing controllability.

Knowing this pattern may help us to develop a concept for a system that will put us ahead of the competition. This could be done by purposely focusing on increasing the system's dynamism and at the same time improving its controllability. Both features are always highly desirable and are good selling points, not only among high-performance drivers, such as the Author.

9.7.4 Increased complexity followed by simplification

This next pattern is not only surprising but also seemingly counterintuitive, since we all basically believe in the ever-growing complexity of engineering systems. However, the actual situation is much more complex, and our belief is based on our observations of evolving systems only over a limited period of time, giving us an incomplete picture. Therefore, knowing and understanding this pattern may become the reader's hidden professional advantage.

> An engineering system evolves through stages of growing complexity followed by stages of growing simplicity.

When a new engineering system is introduced, it begins the *complexity period*. The first of any kind of system in the market usually provides a single function or only a few core functions. They identify the system and were the reason why it was created. Next, when similar systems are introduced by the competition, the first system is expanded (not necessarily improved) to increase its competitive advantage, and various additional subsystems are added that provide additional functions. A *complexity competition* begins, leading to the development of more and more complex systems, overloaded with functions, which in addition to their core functions provide an ever-growing number of additional functions of dubious value. For customers, all these additional functions have only marginal or even negative value, because it requires additional effort to learn how to use new systems, the new functions make the systems confusing, and they make them less reliable because of their unnecessary complexity. At a certain point, the market rejects overcomplicated systems and becomes ready for the *simplification* period. The next winner is the system with a reduced number of functions, which is also much easier to use. A *simplicity competition* begins, which subsequently leads to the development of similarly simplified systems by the competition. This process ultimately leads to the emergence of "barebones" systems, which provide only their core functions. That is the end of the simplification period and the beginning of the second complexity period.

There are many examples showing this pattern in action. It can be observed not only in the evolution of cars but also in the evolution of electronic watches or cellular phones. Knowing this pattern may help the inventive engineer to predict the desired level of complexity of a new system. It can be done when the line of evolution of the system under development is available, and it is possible to determine if the system is in its complexity or simplicity period and how advanced a given period is.

9.7.5 Matching and mismatching of system elements

This is a fascinating pattern, which needs to be understood well before it is used. It can be described thus:

> As an engineering system evolves, its individual subsystems are initially unmatched, then matched, then mismatched, and finally a dynamic process of matching–mismatching emerges.

Let us explain first several terms used in this pattern.

Unmatched systems are two or more entirely incompatible systems put together, for example a diesel fuel system installed in a car with a gasoline internal combustion engine.

Matched systems are two or more compatible systems working together, which were optimized as subsystems of the system they create together, for example, carefully optimized integrated braking and suspension systems in a racing car.

Mismatched systems are two or more systems that were initially matched; however, their evolution followed significantly different S-curves, and as a result of that they are now incompatible. For example, a new sports car is introduced with tires that are precisely optimized for its weight, power, suspension, and so on. Several years later, its new model is introduced. It is heavier and has more power and an entirely different suspension, but the same tires are used. They are incompatible with the car or simply mismatched.

The dynamic process of matching–mismatching occurs when two or more systems are interrelated in such a way that they act like a system in which subsystems communicate through feedbacks, and they appropriately adapt their behavior to the behavior of the remaining parts of the system so as to optimize behavior of the entire system. For example, adaptive suspension in a car communicates with the steering and braking systems and appropriately adjusts its damping characteristics to optimize car handling and road holding.

When Benz built the first automobile in 1885, it had a Benz-designed one-cylinder two-stroke engine, three wire-spoken wheels taken from bicycles, a seat from a horse carriage, and many small parts originally intended not for an automobile but for a horse carriage. All these subsystems of the first automobile were originally unmatched, but soon Benz developed matching subsystems and the evolution of the automobile began. Today, many Benz cars have adaptive suspension, adaptive brakes, and so on. They are clearly using a dynamic process of matching and mismatching. In fact, car evolution is evidently moving in the direction of "smart" automobiles, which will operate like complex adaptive systems and most likely will not even need any human drivers (see Pattern of Evolution No. 8).

9.7.6 Evolution toward microlevel and increased use of fields

This pattern is usually initially seen as confusing, and it requires a good explanation. It can be described thus:

> An engineering system evolves in the direction of using microlevels and the growing number of various fields.

The concept of evolution in the direction of using microlevels may be explained in terms of the evolution of materials used for building walls in civil engineering. In general, it may be described as the evolution from using large-scale natural elements to using smaller and smaller elements, which are highly processed. This line of evolution may include

1. Macro level, large-scale natural elements: whole natural logs
2. Pieces of logs: boards
3. Pieces of boards: particleboards
4. Processed ferrous particles: steel panels
5. Rearranged atoms: nanopanels

This example clearly shows an evolution moving from large natural elements (logs in this case) to the individual atoms, that is, from the log level to the atom level through several intermediate steps. In fact, nanopanels are still not in use, but it is only matter of time before they become commercially available.

The second part of the pattern, that is, the increasing use of fields, is more difficult to explain without referring to field-substance analysis (Altshuller 1984), which is an important part of TRIZ but is not discussed in this book because of its extraordinary complexity. However, the concept of fields can be explained:

In each structural or mechanical system under loading we will find a flow of stresses, and we may say that in both cases will have the material stress fields. When an electric current flows through a conductor, it is surrounded by an electric field; similarly, electromagnets produce an electromagnetic field. In other words, we have all kinds of fields available for various engineering applications. In this context, the second part of the pattern discussed here is simply saying that when a system evolves, the number of fields in use grows.

For example, when the first car transmission was built, it was a purely mechanical system and only a material stress field was used. Next, electric controllers were added with the associated electrical field, which were followed by electromechanical controllers additionally utilizing electromagnetic fields. Later, various sensors monitoring the temperature were incorporated, and they utilize the temperature field. Also, oil-pressure monitoring sensors were added, and they utilize the liquid pressure field.

In the area of electronics, the evolution of computers provides another good example. The first computers used tubes, which were gradually replaced by integrated circuits. Now, within the technology of integrated circuits, we may observe the continuous race toward the microlevel and toward quantum computing, that is, computing on the atomic level.

If we know this pattern, it will stimulate our thinking during inventive designing about considering the move toward the microlevel or about the utilization of various additional fields following the evolution of other systems.

9.7.7 Transition to decreased human involvement

This pattern seems to be nearly trivial for modern man, but it is still important and useful in inventive designing. It can be formulated thus:

> An engineering system evolves in the direction of decreased human involvement.

We can easily provide many examples. One of them is the evolution of heating, which began several thousand years ago. This evolution began with a fire that required a constant attention and human involvement; otherwise it would gradually die or burn the entire forest around us. Therefore, a fireplace represented a huge improvement, although it still required very frequent human attention to control the rate of burning and to reload. Also, heat was produced only when the wood was burning and because the fireplace had very little, if any, thermal inertia. Next, a ceramic room furnace emerged with its heavy ceramic bricks and tiles. It had huge thermal inertia and needed human attention only two or three times a day to reload and to eventually adjust the flow of air. In a large building, there would be hundreds of such ceramic furnaces, and they would still require a small army of people to keep the building heated.

Mainly for this reason, the next step in the evolution was the introduction of the central heating system with its central furnace, usually located in the basement. The heat is distributed through the entire building using air or water as a medium. In fact, the Romans used central air heating systems; so did the Teutonic knights when in the thirteenth century they built their main castle in Malbork in northern Poland, not far from Gdańsk, where the Author was born. Today, such systems are still popular, although in Europe water systems with pipes and radiators located in the rooms are much more popular. In the case of the central heating system, a single person could keep the system running. Such systems used wood at first, later coal, and today natural gas or propane. Today, heating systems practically do not require any human involvement, with the exception of infrequent inspections and occasional adjustments.

Pattern No. 8 is also useful in various inventive design situations because it reminds us that additional automation should always be considered, particularly when we want not only to follow existing trends but also to create them, as should be the mission of all inventive engineers.

9.8 LINES OF EVOLUTION

Patterns of evolution explain the mechanisms of the evolution of engineering systems over time and can be understood as drivers of this evolution. In fact, we need to understand these drivers in order to develop a "big picture" of the evolution of the specific engineering system that is the subject of our interest. Thus, it is time to introduce the concept of *the line of evolution of an engineering system* or simply *line of evolution*. This concept can be defined thus:

> A line of evolution of an engineering system is sequence of design concepts that were used to design the system over the time period.

Usually, we present a list of concepts but also represent a line of evolution in the graphic form, which is much more useful for engineers than a list. This is the same difference as between a list of topics and their mind map. A drawing, with its shapes, words, connections, time lines, and so on, is a powerful tool activating the entire brain, inspiring human creativity and potentially helping inventive engineers to create inventions. Figure 9.4 shows a line of evolution for heating systems as discussed above.

A given line of evolution is developed for a specific feature of the system, for example, the source of energy. Therefore, a system may have several separate lines of evolution associated with its various features. These lines constitute together the *envelope of evolution lines*, which can be described thus:

> An envelope of evolution lines is a graphic representation of the entire class of evolution lines developed for a given engineering system.

An envelope of evolution lines (Figure 9.5) provides the most complete picture of the past and future evolution of a given system. Obviously, we need to understand that it is a *probabilistic* picture as far as its future part is concerned. We are only predicting possible future developments, and sometimes it is even difficult to estimate the probability of their occurrence. Probably, none of these lines will actually and precisely predict future developments. However, the predictive power of an envelope is not in the completeness or correctness of the individual lines; it is in the knowledge, or understanding, they bring together as a system. As much as an individual line is a weak predictor of the future, their entire class, their envelope, is a powerful source of information about the evolution of a given system. There is a high probability that a carefully prepared and complete envelope of evolution lines contains at least parts of the future and actual evolution line, which will take place. For all these reasons, the envelope of evolution

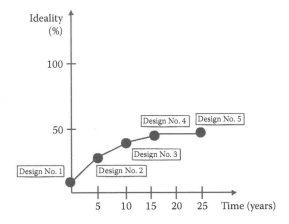

Figure 9.4 Line of evolution for a given product.

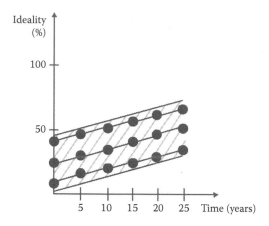

Figure 9.5 Envelope of evolution lines for a given product.

lines should be treated as a secret source of knowledge about the future and as the key to our competitive advantage.

In the late 1980s, the Author conducted with one of his Ph.D. students an analysis of the evolution of joints in steel space structures patented from the 1920s to the 1990s (Arciszewski 1984, Arciszewski and Uduma 1988). This work was actually related to the one of the Author's inventions in the form of a spherical joint in steel space structures (Arciszewski 1989, 1991). The research revealed several independent lines of evolution and a number of specific patterns of evolution controlling this highly specialized domain of joints in space structures. The study was like detective work, looking for the motives and decisions of inventors who worked independently in several countries at different times but whose decisions, taken together, made so much sense; they looked like they were made by a team. The results may be of interest to all structural or mechanical engineering students interested in evolution and are available in Arciszewski (1984).

9.9 BLACK AND WHITE

9.9.1 White

The Author discovered TRIZ in the early 1970s, and it was like a revelation, like a new way of thinking, and it still is. In engineering, we usually strive to find compromises between competing requirements and emerging contradictions, but nobody is happy about the resulting trade-offs. In TRIZ, we eliminate contradictions representing the competing requirements, and in the process we create inventions. This is an entirely different approach and represents the main advantage of an inventive engineer with respect to a traditional engineer focused on using optimization and formal decision making. Moreover, this TRIZ advantage is the key to progress in engineering, which is driven by patented inventions.

Other inventive designing methods, for example brainstorming or Synectics, offer a method-specific body of universal methodological knowledge that can be utilized in various engineering domains. TRIZ is fundamentally different. In addition to a specific body of universal methodological knowledge, it also offers a huge body of universal engineering knowledge, which can be used to support TRIZ inventive problem solving. This represents an enormous advantage for TRIZ with respect to the other methods. It makes TRIZ to a certain degree self-contained and self-sufficient. This is the main reason why a single inventive engineer can use TRIZ and why the use of TRIZ does not require any large professionally differentiated teams, which are necessary in the case of the other methods to provide an appropriately large body of interdisciplinary knowledge.

For all these reasons, an inventive engineer should be at least familiar with the fundamentals of TRIZ and should know that TRIZ can be always used when other methods fail or are simply insufficient for highly sophisticated and knowledge-demanding inventive problems.

As TRIZ has been presented in this chapter, it is a systematic method that can be used in a similarly way to other engineering methods, that is, in a systematic manner following the procedure. This should reduce, if not eliminate, the impact of the "fear factor" still associated with the method. Because of that, TRIZ is appropriate for all inventors working alone, who are usually introverts. For such engineers (probably the majority of all engineers) team problem solving and sessions with several spontaneously behaving people are annoying and very difficult. Therefore, when they are forced to participate, they usually are unable to contribute fully, and their intellectual potential is wasted. Fortunately, TRIZ creates the right opportunity for them.

TRIZ as presented in this chapter looks like a relatively simple method. In fact, TRIZ is so much more. After about 70 years of development and evolution, TRIZ is more like a science of inventive problem solving, as Altshuller envisioned it in the late 1940s (and that put him in prison). There are competing interpretations of classical TRIZ and various versions of contemporary TRIZ, various software packages, numerous courses offered in several countries, and so on. Also, several of Altshuller's former research partners are still active in the global TRIZ community (Fey and Rivin 2005). They have deep understanding of TRIZ and they have been living TRIZ for more than half a century. They, as creative people themselves, are now leaders of the global TRIZ movement; while its time has not yet come, it is only a matter of years.

As a result of this unusual situation, an inventive engineer interested in TRIZ will find a huge body of TRIZ-related knowledge to study, can make a choice as to what is really useful for him or her, and can synthesize this knowledge. This is a dream situation for an ambitious inventive engineer.

The discovery of patterns of evolution has been a fundamental achievement because it has radically changed the significance of the history of technology. Before Altshuller's discovery, the history of technology was considered as a collection of interesting and nice stories and anecdotes, which might help us to develop a better appreciation of the accomplishments of our predecessors and eventually inspire us to reconsider various already forgotten ideas from the past. Today, we should see the history of technology as a source of priceless knowledge about the evolution of engineering systems—a body of knowledge that can be acquired and utilized to develop future systems. In other words, we should see the continuum: a single line of evolution including past designs, today's designs, and future designs. In this context, patterns of evolution may be seen as a reflection of the past and used for the development of future systems, potentially avoiding many random or blind trials.

There is the science of technological forecasting, and its focus is on the methods and results of predicting the future evolution of engineering systems. Patterns of evolution have already been successfully used for forecasting (Clarke 2000), demonstrating their additional importance in engineering.

9.9.2 Black

Unfortunately, TRIZ is still waiting for discovery and reinterpretation by an American or a British design scholar. Yes, there is a huge body of knowledge about TRIZ, but it is available in several countries and in various languages. Even if a given publication is in English, it is most likely a translation from Russian or it has been written by somebody without a full understanding of the Soviet culture and of the Russian language. An American computer scientist, R. Koza, would call this knowledge a *knowledge soup*, since it is not an integrated body of knowledge but only various pieces of knowledge mixed together, which may be understood only if one knows their context. For all these reasons, the available knowledge about TRIZ is difficult to learn and to understand for an American engineer or student. It is possible to learn *about* TRIZ in several hours, but to learn how to *use* it requires a significant time commitment on the level of hundreds of hours. It is not only about learning the assumptions and procedure but also about developing a good understanding of the context of the individual parts of TRIZ.

The chapter has provided only several selected pieces of the TRIZ knowledge soup. These pieces were chosen by the Author to create a simple and interesting method and to reflect the essence of TRIZ. However, TRIZ scholars might reject such presentation as oversimplistic and not reflective of the true beauty and the intellectual richness (if not complexity) of TRIZ. They believe that TRIZ is a scientific discipline and that reducing it to a method is simply wrong, and to a certain extent they are right.

This belief confirms the confusion about TRIZ and about its place within inventive engineering. It also explains the fact that there is no unified and widely accepted version of TRIZ, and that makes learning TRIZ even more difficult.

The fear factor still exists. It is constantly reinforced by various TRIZ experts who, mostly for commercial reasons, claim that TRIZ is so difficult that any practical application requires their involvement, which may be costly and may compromise the intellectual property rights of the potential TRIZ users. The situation somehow reminds us of the situation in the United States in the 1950s, when suddenly many brainstorming pseudo-experts were offering their services and their unsubstantiated claims, and excessive financial demands hurt the image of the method.

TRIZ represents a major development in inventive engineering. But André Gide (a French philosopher and writer, 1869–1951) once said "One doesn't discover new lands without consenting to lose sight of the shore for a very long time." This statement is particularly true in the case of TRIZ. All its eventual users should be therefore warned that they need to first make a strong commitment, and next to invest substantial resources (time, effort, software, patience, and persistence) before any useful results are produced. If they enter the TRIZ world with such a serious attitude and are fully aware of challenges ahead, the Author guarantees that they will not be disappointed and that amazing results will gradually emerge, although this will be a long and challenging process.

Chapter 10

Bioinspiration

10.1 WHY LEARN BIOINSPIRATION?

Our goal in this chapter is to learn about bioinspiration and to realize its importance for inventive engineers. We have at least four major reasons why future inventive engineers need to learn about bioinspiration: *knowledge*, *appreciative intelligence*, the *numerical optimization curse*, and *fun*.

We have already learned that knowledge is one of the keys to inventions. Also, we know that problem domain knowledge is often insufficient, and we need to bring in knowledge from other domains to solve an inventive problem; sometimes, this knowledge must even come from a nonengineering domain. In fact, in Chapter 8 on Synectics, we learned about *excursion*, a method that allows us to acquire knowledge from various unexpected sources in order to utilize it for inventive purposes.

Knowledge may come from sources unusual for engineers, for example from the history of engineering or from nature through the study of biology and botanics. In the first case, about 10 years ago the Author conducted a study of the evolution of body armor in Europe over the period of about 1,000 years (Arciszewski and Cornell 2006). He discovered that behind this evolution were very complex evolution patterns (decision rules), which are also relevant to the development of modern body armor, which unsurprisingly undergoes an evolution of shapes and articulations that is similar to the historic one. In this chapter, however, we will focus on learning from nature.

Nature represents a huge body of *natural knowledge*, that is, knowledge that is "hidden" in trees, in lions, and in our dogs; it is simply knowledge that literally surrounds us in all living organisms and is only waiting for us to discover and use in engineering. Dr. Frank Fish once said to my students that "Nature is our best heuristic tool given by the Universe," and his words precisely capture the importance and essence of the natural knowledge.

Natural knowledge comes in many forms, not only general knowledge in the form of decision rules and heuristics but also various processes (deterministic and probabilistic) and movements, as well as evolution and the creation rules driving these processes. Nature is also an incredible source of facts that could potentially be used in engineering, facts such as specific shapes, colors, sounds, textures, vibrations, and so on that could be directly mapped into various engineering solutions.

All inventive engineers should acquire sophisticated appreciative intelligence (discussed in Section 3.2) and should be masters of reframing, or transforming knowledge from various domains into knowledge that is highly relevant to their inventive problems. Nature is our largest and most important source of inventive knowledge, and therefore acquiring knowledge from nature is a worthwhile goal.

The classical numerical optimization allows the improvement of engineering systems, that is, a reduction in their weight, energy consumption, or pollution. However, it is done within a given design concept, and therefore the expected improvements are usually less than 10%–15%. Such results are still meaningful but obviously grossly insufficient when improvements on the level of 100%–200%, or more, are sought. In such cases, numerical optimization is not the answer because it is "cursed" with its limitations of providing only relatively small numerical improvements and a built-in inability to provide qualitative changes (when a new S-curve is necessary), which could subsequently bring huge quantitative improvements; such improvements took place in the area of computer engineering when moving from a tube to an integrated circuit reduced energy consumption by several orders of magnitude. In all such situations, inventive engineering must be used. Most likely, it will also require looking for inspiration, that is, acquiring and using knowledge from outside the problem domain. Since nature is the best source of inventive knowledge, sometimes even the most traditional engineers may have no choice but go to nature to acquire natural knowledge for solving their problem.

Finally, inventive engineers are obviously creative humans and as such should be motivated not only to work hard but also to have fun. Having fun is good in itself, but more importantly it positively affects human creativity, and that really matters. There is no question that learning about nature and reframing its knowledge (natural knowledge) is always a lot of fun and brings the satisfaction of accomplishing something unusual and meaningful.

In recent years, huge progress in the biological sciences can be observed. Today, biologists have a much better understanding of various phenomena and processes occurring in nature than they had only 20 or 30 years ago. Moreover, they are able to articulate their findings in abstract terms, which are at least partially understandable for engineers. The described situation in turn creates an incredible opportunity for engineers to learn from nature—an opportunity that cannot be lost, at least in the case of inventive engineers.

10.2 WHAT IS BIOINSPIRATION?

As discussed in Chapter 5, the conceptual inventive designing process can be considered as a process of knowledge acquisition, transformation, and utilization in the form of an invention, which is a body of knowledge about a design concept that is unknown, feasible, surprising, and potentially patentable. Considering sources of knowledge to be acquired, we may distinguish between the acquisition of knowledge from within the problem domain and from outside it. We will call this second situation *inspiration in design*, that is, it is a design situation when, in addition to the problem domain knowledge, another body of knowledge from outside the problem domain is acquired and used as inspiration in the development of novel design concepts. For example, a mechanical engineering problem is solved using knowledge from chemical engineering (a problem regarding tubing in a car's cooling system is solved using knowledge associated with the designing of large-scale industrial chemical installations). Therefore, bioinspiration is understood in our book thus:

> Bioinspiration in inventive conceptual designing is a process of acquiring knowledge from biology and appropriately transforming it through integration with the problem domain knowledge in order to utilize it for the development of inventive design concepts.

In this chapter we will use the term "bioinspiration" as described above. However, there are several other practically equivalent terms used to identify the same process. For example, a large community of design scholars associated with the Georgia Institute of Technology (the leading research center in this area in the United States) promotes the terms *biologically inspired design* or *bioinspired design* (BID) (Goel 2013; Stone et al. 2014). Also, in a recent publication (Hoeller et al. 2013), a group of leading bioinspiration scholars has used the term "B³D," that is, biomimicry, biomimetics, and biologically inspired design. The first term, *biomimicry*, is understood as the mechanistic imitation of forms and colors used in nature, and it is explained in Section 10.4 on "Visual Inspiration." *Biomimetics* is the area of study of the natural chemistry and processes occurring in nature in order to develop artificial imitations of these. It is more general in focus than biomimicry, but these two terms are sometimes used as equivalent, depending on their interpretation.

When bioinspiration is considered, three major areas of inquiry can be distinguished. These areas include

1. Visual inspiration or biomimicry
2. Conceptual inspiration or biomimetics
3. Computational inspiration

All three areas of bioinspiration are discussed in Sections 10.4 through 10.6.

All inventive engineers practice appreciative intelligence, and the majority of them are naturally enthusiastic about the future and coming inventions. However, in the case of bioinspiration, this unconstrained enthusiasm should be combined with an understanding of the limitations of biology, their new heuristic tool. All tools have their limitations, and we obviously need to know them to avoid frustration and costly mistakes.

Both natural and engineered systems operate in the same environments and are subject to the same laws of physics and to the same forces. In many cases, they also provide the same functions. For all these reasons, natural knowledge associated with natural systems seems to be very attractive for designing engineered systems. Unfortunately, the situation is a little more complicated, and there are significant differences between natural and engineered systems. Most importantly, natural systems use organic materials, while engineered systems use mostly nonorganic materials; however, there are more significant differences. Last year, Dr. Fish prepared for my students a more detailed comparison of natural and engineered systems using seven key attributes (see Table 10.1). This comparison clearly shows fundamental differences, which obviously must be remembered, particularly when using biomimicry.

Size matters. Engineered systems are relatively large, while natural systems (animals) are relatively small. When we compare a modern nuclear submarine (Fish 2009) with a whale, both systems possess a fusiform shape to reduce drag and energy expenditure. Both also operate in the same environment, that is, underwater, and they are subjected to the same forces of nature. The blue whale is a huge animal, growing up to 33 meters in length, but it is still tiny when compared with a modern submarine, which may be

Table 10.1 Comparison of natural systems and engineered systems

Attribute number	Attribute	Natural systems	Engineered systems
1	Size	Generally small	Relatively large
2	Materials deformability	Compliant	Rigid
3	Source of motion	Translational movements produced by muscle (biochemical motors)	Rotational motors
4	Control	Complex neural networks with multiple sensory input	Computer systems with limited sensory feedback
5	Functions	Limited number	Functionally multifaceted
6	Energy storage	Continues energy supply	Limited capacity
7	Rebuilding when damaged	Self-healing	None

300 meters long; we have here two systems differing in length by an order of magnitude. Most whales, dolphins, and porpoises are less than 5 meters in length, and fish may be two orders smaller still. Scaling is always difficult, and it is usually not a linear process, particularly when dealing with such complex phenomena like fluid dynamics. Therefore, fluid dynamics finite-elements models that sufficiently describe the flow of water around a whale may be grossly insufficient for a submarine. Even more, when scaling up the legs of an ant and building an artificial ant the size of a horse, we may be unpleasantly surprised to discover that the legs of our robotic ant will collapse under the weight of the robot in the same way a stone beam that is too long will collapse under its own weight, as was once discovered by the Egyptian engineers.

When materials used in building natural and engineered systems are compared, we have other significant differences. Engineering systems are built from nonorganic materials, which are relatively rigid or have limited deformability, like metals, concrete, hard plastics, or ceramics. We could call them *dry and rigid*. Animals are built using compliant, or highly deformable, materials of fibrous organic compounds with a high water content. Such materials might be called *wet and flexible*. In general, biological materials are weaker and much more deformable that materials created by humans. These differences translate into substantial dissimilarities in how, for example, the sufficient strength and flexural and torsional rigidity of the entire body must be created in natural and engineered systems. As a result, natural solutions may be unnecessarily complicated and costly for engineers, who can use artificial materials that are superior, at least as far as the mechanical properties of those materials are concerned.

The source of movement is entirely different in animals than in engineered systems. In the case of animals, movements are generated through forceful contractions of the muscles, which are transmitted to an articulated skeleton by tendonous connections (Fish 2009). Most importantly, animals do not use rotational movements, and therefore the energy efficiency of their propulsion systems is generally lower than in the case of human-built systems.

There is a fundamental difference between how natural and artificial systems are controlled. In the first case, it is done by complex neural networks with many various sensors. Such systems exhibit a very high degree of redundancy but at the cost of high complexity. Artificial systems usually have a very small number of absolutely necessary sensors, which provide feedbacks to a computer that runs a program controlling the system. Such an approach is optimal from the point of energy consumption and costs but may be much less reliable when one or more sensors are damaged or provide the wrong readings, not to mention in the case of a computer hardware failure.

The complexity of control is directly related to the next difference: Natural systems are functionally multifaceted (they move, feed, and reproduce), while machines are usually designed to perform a single clearly specified function (and this function may be obviously divided into several interrelated separate subfunctions). Therefore, a mechanical system may be optimized in our quest to create an ideal solution in accordance with the patterns of evolution (Chapter 9). Natural systems do not look for ideality; they do not even look for optimality. Their goal is survival, and that means that all functions must be performed in a way that is at least satisfactory or adequate, and none of them must be neglected; that would have catastrophic consequences. A nonfeeding or immobile animal would soon die; a nonreproducing animal will live, but a nonreproducing species would simply disappear, as has happened so many times as a result of climate cooling or warming or for various other reasons.

Probably one of the most difficult challenges in using bioinspiration is the energy challenge. Animals can control energy input and can feed when they need more energy. For this reason, they can operate practically indefinitely, and this fact is reflected in how their various systems are built. On the other hand, engineered systems usually have only a limited operating time. They have a tank of gas or a load of coal, and after this energy is used and if it is not immediately replaced, their ability to operate ceases.

Finally, injured animals have at least some ability to self-heal, while engineered systems do not yet self-repair.

We should also remember that the ultimate objective of an animal is to survive, and that may lead to solutions that are not strictly optimal. For example, when the propulsive systems of fishes are considered, they are definitely not optimal from the energy-efficiency point of view, particularly for routine swimming when a fish is cruising or migrating.

Muscle mass is composed of two kinds of fibers: red and white. Red fibers are highly energy efficient and can be used over long periods of time, but their contractions are relatively slow. White fibers contract much faster than red fibers but can be used only over short time periods, and they are much less energy efficient than red fibers. Surprisingly, the majority of muscle mass is made up of white fibers. This is definitely not the optimal solution from the engineering point of view when weight and energy efficiency are considered. This situation can be compared to the Author's practice of driving two miles to George Mason University in very heavy and slow traffic (typical for northern Virginia in the United States). This commute was recently done in a Ford Mustang GT with a V8 425 HP engine and a manual transition. Even the Author, an inventive engineer and inventor, could not find any justification for this choice. This muscle car was obviously not the optimal commuter car. Unfortunately, its choice could be only explained by the fact that all boys like driving powerful and noisy sports cars, no matter what their age.

Fortunately, fishes have a better justification for having so much muscle. There are so many life-and-death situations when a fish must rapidly move and accelerate to catch their prey or to avoid becoming prey. Such situations justify the creation of propulsion systems that are not strictly optimal but which improve the survivability of the fish. On the other hand, many engineered systems also have a dual nature, and their design could be improved by learning from nature. A muscle car, for example, can be used as a commuter vehicle but may be also occasionally driven on a road racetrack.

10.3   WHAT IS THE HISTORY OF BIOINSPIRATION?

Engineers were probably the first professionals on this planet, and they have sought ideas from nature for many millennia. A fallen log may have inspired the first single-span beam bridge, and a rock avalanche may have led to the building of various grain-crushing devises and ultimately to the emergence of the modern agricultural industry. The great inventors of the past millennium, like Leonardo Da Vinci or the Wright Brothers, were also inspired by their observations of natural phenomena.

Surprisingly, only recently—during the last 20 or so years—research on learning from nature has gained momentum, and today we have a large global community of scholars active in this area, with members not only in the United States but also in Europe and Asia. Many scholars believe that the publication of the book *Biomimicry, Innovation Inspired by Nature* (Benyus 1997) has created this momentum and contributed to the gradual transformation of various "learning from nature" studies into research conducted with the discipline of a scientific inquiry. There is a movement (Stone et al. 2014) led by Dr. Ashok Goel, an engineer and one of the leading bioinspiration scholars, toward the creation of a separate research program called Bio-inspired Design at the USA National Science Foundation, and this will be the last step in the emergence of a new science of bioinspired design.

10.4   VISUAL INSPIRATION

Visual inspiration, or biomimicry, can be described thus:

> Visual inspiration in the inventive conceptual designing is a process of direct mapping concepts from nature into engineering design concepts.

In more pragmatic terms, visual inspiration can be understood as a process of using various observations from nature, particularly those observations related to the form of living organisms, their shapes or colors (Vogel 1998). Pictures (visuals) of plants or animals, or their organs, are used to develop similar-looking engineering systems or their components.

In biology, the concept of *mimicry* exists. This is an evolutionary strategy, and plants or animals using this strategy evolve to look or behave like another species in order to improve their chances of survival. Animals often advertise their unpalatability with a warning color or a pattern. Gradually, predators learn to associate that message with an unpleasant experience and seldom attack these prey. The Batesian mimicry system is the best-known example of this phenomenon; a palatable prey species protects itself from predation by masquerading as a toxic species, and this protection is obtained through visual mimicry.

For centuries, humans used concepts found in nature for two purposes. First, these concepts would become incorporated in the myths created by humans, and second, they would be used for design purposes. In the first case, for example, medieval knights in Europe, or warriors of all kinds in Asia or Africa, always wanted to frighten their opponents on the battle-field. Therefore, they would cover their bodies with skins of dangerous animals, like lions or leopards, or even use animal masks to instill fear. They did not want to mimic animals but only the desirable concepts associated with these animals: concepts of danger, death, or cruelty.

Another good example of myth-building biomimicry is from Poland. In this country for several centuries the core of the cavalry were the *Huzars*, a light cavalry believed to have the mysterious powers of Polish cougars: stealth, lightning speed, and beauty in action. They were elite troops with excellent training and a big heart for a fight. Many historians believe, however, that the reason they won so many battles, sometimes in situations that made victory practically impossible, was also because of the myths surrounding them. Figure 10.1 shows the beautiful armor and leopard skin of a Huzar. The photograph does not show, however, the large feathered "wings" that were connected to his armor to produce terrifying sounds during an attack in a similar way that attacking animals also do; this is another example of biomimicry in action, this time in the area of sound.

Once the Author had a workshop on Heron Island off the Australian coast, where he took a picture of a magnificent "mama" sea turtle after she had laid her eggs in a nest and was slowly and painfully creeping back to the ocean (Figure 10.2a). Several years later, he attended a conference in Singapore where he took a picture of a new concert hall (Figure 10.2b). There is no doubt that the designers of the building were inspired by the beautiful smooth look of the sea turtle's shell. We might even claim that they were inspired by the concept of using a shell as a structure for a building that would be exposed to strong winds and frequent hurricanes coming from the nearby ocean. In fact, this example shows the complexity of issues associated with using visual inspiration in designing; a shape is mimicked, but we simply do not know all the reasons why it has been done. It could have been done purely for esthetic reasons or because the designers wanted to use this shape because of the myths associated with sea turtles—myths about the legendary strength of their shells and their aerodynamic perfection, or myths about their survivability. All these myths perfectly fit into the present campaign of building myths (or perceptions) about Singaporean

Figure 10.1 Polish light cavalry officer (Huzar).

Figure 10.2 (a) Sea turtle, (b) concert hall in Singapore.

society, which is still very "young" in historical terms and is undergoing a process of integration and of finding its national identity. (It is still composed of separate Chinese, Malaysian, and Indian communities with their different cultures, religions, value systems, etc.) For a structural engineer, the shell as a type of a structural system is not only beautiful but also perfect for a large-span roof structure subject to huge wind and earthquake forces. It provides a smooth flow of internal forces and small deformations under loading. Therefore, we could even claim that this is an example of a *conceptual inspiration*, discussed in Section 10.5.

The use of visual inspiration is relatively simple, but unfortunately this kind of inspiration is only skin deep because concepts are *mechanistically* moved from nature to engineering systems. The situation is particularly confusing in the case of complex natural systems that are copied to inspire the development of intricate engineering systems. To avoid potentially dangerous mistakes, a designer needs to develop a full understanding of the functions of the natural system so as to mimic this system in a meaningful way, that is, in such a way that the shape and essential functions of the natural system are preserved while secondary, redundant, or unnecessary functions or features may be eliminated, if desired. There is always a danger that the final product will preserve all the secondary functions of the natural system, which are unnecessary in the engineering system, and that this will dramatically reduce the effectiveness of using visual inspiration. Even worse, when a shape from nature is used and it is done mechanistically, it may bring unknown or not fully understood dangers into the human design (Arciszewski and Cornell 2006). For example, about 5,000 years ago, Egyptian engineers were inspired by logs and began using stone beams for various purposes, mostly as beams in their monumental stone buildings. When they initially copied the proportion of a log (the ratio of the span to the cross-section height) and used it for relatively short beams, their stone beams worked properly. Later, however, they began expanding the span of their stone beams and they discovered that the beams were breaking, apparently under their own weight. Obviously, in this case there was a dangerous limit for using biomimicry. This limit could be easily explained by the modern theory of elastic bending, but obviously this theory was unknown 5,000 years ago. (According to this theory, when a beam under bending is considered, the relationship between the maximum bending stress and the span is not linear. When the span is increased by the factor of two, the maximum bending stress increases four times for the uniformly distributed loading.) Today, we know how to deal with beams, but we are also trying to copy from nature's much more complicated forms and mechanisms. In this situation, all inventive engineers should still be aware that the described danger of oversimplification driven by a lack of understanding or knowledge is still with us.

10.5 CONCEPTUAL INSPIRATION

Conceptual inspiration, as understood by the Author, can be described thus:

> Conceptual inspiration in inventive conceptual designing is a process of acquiring knowledge from nature and transforming it into abstract knowledge useful for a large class of inventive problem.

The previous section confirmed once again that knowledge (or understanding) is the key to inventive designing, and that biomimicry, although useful in many situations, may be dangerous when used in a mechanistic way and without fully understanding the design situation. Fortunately, conceptual inspiration represents a significant improvement in addressing this major deficiency of visual inspiration. Conceptual inspiration is about acquiring abstract knowledge, not individual pictures or pieces of information, as in the case of visual inspiration. Such abstract knowledge is in the form of decision rules or heuristics and represents our understanding of nature. Therefore, it is much more universal and reliable than visuals, and thus it is associated with a much lower probability that its use may cause dangerous consequences, a concern that is always important in engineering.

Conceptual inspiration is particularly attractive for inventive engineers because its use has the greatest potential for acquiring knowledge that will lead to inventions and subsequently to fame and fortune, the hidden dream of all inventors. Also, it has a great academic value because it produces not only immediate benefits in the form of knowledge for solving a specific problem but also contributes to the advancement of science since the acquired abstract knowledge may be added to the state of the art.

Dr. Frank Fish is a biologist at West Chester University in West Chester, Pennsylvania, and is probably the greatest scholar in the area of experiential bioinspiration (bioinspiration directly driven by experimental studies and supported by the theoretical analysis of experimental results). He has developed a fascinating research program and for the last 20 or so years has been studying the behavior and anatomy of fish in their natural environment in order to learn their secrets. He has helped the Author to share with you two of these secrets: the magic of the tubercles on humpback whales' flippers and the mystery of Batoid fishes (Miklosovic et al. 2004; Fish 2009, 2013; Fish et al. 2011a,b; Moored et al. 2011).

10.5.1 Magic of the tubercles on humpback whales' flippers

The humpback whale feeds on a variety of foods, including Antarctic krill and schooling fish. Whales hunt by encircling a shoal of prey in gradually tightening circles while striking the water surface with their flukes and emitting air bubbles when they are underneath the prey. In this way, they build a *bubble net* surrounding the prey and forcing it to form a giant living *prey ball,* which is a delicious meal for a whale. Such preying on fish or krill requires an incredible ability to perform nearly acrobatic turning maneuvers, which seem literally impossible for a gigantic whale. This outstanding maneuverability has caught the eye of researchers who hoped to discover the magic behind it and eventually use it for engineering purposes.

The ability to make such rapid and tight turns is unique among whales and is associated with two facts. First, the humpback whale has the longest flipper of any cetacean (a family of fish composed of whales, dolphins, and

Figure 10.3 Humpback whale's flipper with tubercles. (With permission. Drawn by Joy E. Tartter.)

porpoises). Second, there are large rounded tubercles along the leading edge of the flippers (Figure 10.3), which give the flippers a scalloped appearance. Both experimental and analytical research (Fish et al. 2011a,b; Miklosovic et al. 2004) has concluded that it is mostly the tubercles that are responsible for modifying the hydrodynamic characteristics of the flippers. They contribute to the exceptional effectiveness of the flippers in turning, not to mention increasing their lift while reducing the drag (hydrodynamic resistance). It has been discovered that all these benefits are caused by the vortexes (Figure 10.3) that are produced by the tubercles. These vortexes provide additional support in the water for the flippers and thus improve their performance.

For many years, all the phenomena associated with the tubercles have been the subject of both the experimental and analytical research conducted mostly at the West Chester University and at the US Naval Academy. The research included recording the behavior of whales in their natural environment when they perform their acrobatic maneuvers and demonstrate their unusual turning skills. It has also included building various analytical finite-elements models of growing complexity, which describe the hydrodynamic behavior of flippers with tubercles when a whale performs tight turns. As a result of all this research conducted by biologists and hydrodynamics experts, a large body of natural knowledge has been acquired and transformed into advanced formal scientific knowledge, which also includes a formal analytical hydrodynamic model of a flipper with tubercles. This knowledge represents the present excellent understanding of all issues associated with the whale's flippers with tubercles and their behavior during the whale's maneuvers.

This is an outstanding example of conceptual inspiration in action. The entire effort was focused on acquiring natural knowledge and transforming it into a deep understanding of the natural system. In our example, only after this advanced stage of understanding was reached did Dr. Fish begin working on various potential engineering applications. He is a biologist and the brain and heart behind the reported research. He has also obtained an international patent (with Stephen Dewar and Philip Watts) for a "Turbine & Compressor Employing Tubercle Leading Edge Rotor Design," the best proof of the novelty and value of the acquired knowledge. The described process is fundamentally different from biomimicry, which is a simple process of mapping pieces of information from nature directly into design concepts. Such a transfer of natural knowledge is easy and immediate but is devoid of any deep understanding of the phenomena behind this knowledge, and that is very dangerous, as the history of technology has proved so many times.

Today, tens of patent applications related to the phenomena associated with tubercles are in various stages of development in the United States and in Europe (Fish et al. 2011a,b). They may be roughly divided into aquatic and aerial applications. In the first area of aquatic applications, the patent has already been commercially used for designing a surfing board with a keel incorporating tubercles. This product is available for all inventive engineers and surfers who want to win any surfing competition in which maneuverability is a factor. Similar applications of the invention are being considered for keels in small sailing boats and for all control surfaces, including rudders and keels, in very large ocean-going superyachts. New designs for masts with tubercles are also being considered for sailing boats, yachts, and even for ice gliders (Figure 10.4). Recently, a group of students from Sussex County Technical High School in Sparta, New Jersey, United States, designed and built a small experimental submarine with various control surfaces with tubercles. The submarine proved to be highly maneuverable, mostly as a result of the use of tubercles. Finally, new types of

Figure 10.4 Ice glider using tubercles.

marine propellers are being investigated that have tubercles on the attack edges of their blades.

The power of conceptual inspiration in the case of this research is best demonstrated by the emergence of a new company in Canada, appropriately called WhalePower Corporation. Its sole mission is to develop an entire engineering technology based on the tubercles phenomenon and to use it for various large-scale industrial applications. For example, the company has been working on a new generation of wind-turbine blades with tubercles. Initial studies indicate that the new blades will allow a wind turbine to capture more of the wind's energy at a much lower wind speed while producing less noise that the blades in use today. The new generation of blades could revolutionize the entire wind-power industry, not only improving the economics of wind power but also allowing the construction of wind farms in locations with weak winds where the present generation of wind turbines with blades without tubercles cannot be located.

Blades with tubercles are already in use in large industrial and agricultural ceiling fans (Figure 10.5) manufactured by Envira-North Systems Ltd. Their use represents a significant improvement with respect to fans with traditional blades, that is, without tubercles. First, four WhalePower blades produce the same ventilation effect as 10 traditional flat blades, and that obviously leads to huge material cost savings. Second, replacing 10 unmodified blades with four blades with tubercles leads to a noise reduction from 64 db to 56 db, and that is a huge difference in comfort for factory workers or animals living on large industrialized farms.

In Germany, the DLR Institute of Aerodynamics and Flow Technology has been involved in research on the impact of tubercles on the performance of helicopter blades. In fact, the research has already resulted in a patent on "Leading-Edge Vortex Generators," and the invention has been used to design and build various experimental helicopter blades, which are undergoing tests. If all the results of these tests are positive, as it is expected, a

Figure 10.5 Industrial fan with WhalePower blades.

new generation of helicopter blades with tubercles will be created, resulting in increased speed and reduced fuel consumption.

10.5.2 Mystery of batoid fishes

Batoid fishes are a group of about 500 species that have evolved dorso-ventrally flattened bodies with a whip-like tail and expanded pectoral fins that are fused to the head forming a wide flat disc (Rosenberg 2001). The unusual but also beautiful form of their bodies (Figure 10.6) is a result of an adaptation to living on the ocean floor. The manta ray is probably the best-known batoid fish. Seeing it from the perspective of a scuba diver as it is majestically moving through the water is probably one of the most unforgettable and spectacular views nature may offer humans, as your Author may confirm after his experience of diving with manta rays off Heron Island in Australia.

The batoids swim solely by the movement of their greatly expanded pectoral fins, which are flapped vertically in a way similar to the flight of birds. This kind of movement is called *oscillatory locomotion* and is considered extremely energy efficient. The manta ray is also considered the most evolutionarily advanced batoid fish. Also, manta rays may be over 6 meters wide and weigh over 1580 kg. These parameters make them comparable in size with autonomous underwater vehicles (AUVs) and thus eliminate the problems associated with scalability. For these three reasons, manta rays have caught the attention of researchers, just as humpback whales became the focus of research because of their extreme maneuverability.

We are witnessing a true robotic revolution. Applications of industrial robots are rapidly increasing, but robots are also used for other purposes, including various military applications. Particularly well known are drones or flying robots, which are remotely controlled or can be entirely autonomous. Drones are widely and successfully flown not only on reconnaissance missions but are also used for targeting and for the actual combat

Figure 10.6 Batoid fish.

purposes. Much less known is the area of AUVs, where a lot of research and development work is focused on building AUVs that could be deployed quickly and used for a variety of missions, including surveillance, search and rescue, sentry duty, logistical support, or detection of chemical and biological agents (Fish 2013). There are at least eight major design requirements for future AUVs:

1. Energy efficiency
2. Maneuverability
3. Stability in high-energy environments
4. Ability to operate in the littoral zone and open ocean
5. Station holding
6. Ability to follow bottom terrain
7. Rigid hull
8. Minimum of control and propulsive surfaces

With the exception of No. 7, these requirements are very similar to the requirements used by nature to develop the family of Batoid fishes, which live mostly in the littoral zone (close to the coast), as future AUVs will also operate in this zone. In this context, it is not surprising that the manta ray, the most evolutionarily advanced and sophisticated batoid, has been chosen by researchers as a source of bioinspiration to develop the next generation of bioinspired autonomous underwater vehicles or BAUVs.

The research on manta rays has included recording them swimming not only in the ocean but also in a large tank with a glass wall in the marine laboratory of the US Navy Research Center. The recordings were analyzed in order to understand the kinetics of the manta ray's swimming and to build a kinetic model capturing the dominant features of their motion. Interestingly, the model may be adapted for many species in the batoid family. Comparison of the recorded movements and those produced by the model has shown the remarkable accuracy of the model. Therefore, it has been used for studying the swimming of batoid fishes when using oscillatory propulsion. Also, a sophisticated finite-elements model of a manta ray has been developed and used for the numerical analysis of fluid–fish surface interactions.

As a result of all these experimental and analytical studies, a good understanding of batoid fishes has been developed, as it is a requirement for the successful use of conceptual inspiration. Recently, this inspiration in the form of acquired natural knowledge has been used to develop two BAUVs, appropriately called "MantaBots" (Fish 2014). The first MantaBot was developed by a group of students at Princeton University and the second one at the University of Virginia.

The main difference between these two MantaBots is in the way their pectoral fins are actuated. In the Princeton design, four metal roads are

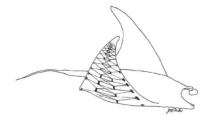

Figure 10.7 Tensegrity structure with actuators in the pectoral fin of a MantaBot. (With permission. Drawn by Joy E. Tartter.)

shifted by electric servomotors, and these provide the movements of the fins. In the University of Virginia design, the pectoral fins are activated by mobile tensegrity structures located inside the fins (Figure 10.7). A mobile tensegrity structure is a three-dimensional truss in which the members under tension are in the form of wires, which can be contracted or expanded by individual actuators, thereby creating movements of the entire structure.

Both designs were used to build experimental BAUVs. They were tested and compared in a competition organized in 2011. Both BAUVs could swim and perform complex maneuvers, but the propulsive movements of the pectoral fins were merely similar to those of manta ray, not identical. This was the main reason why their performance was worse than the real fish. Nevertheless, this was a successful demonstration of the power of conceptual inspiration and a confirmation that natural knowledge can be acquired and transformed into useful engineering knowledge, ready for design purposes.

10.5.3 Learning from natural evolution

About 10 years ago, the Author conducted with his daughter Joanna (PhD in biology) a study about acquiring design knowledge from history and nature. The subject was plate body armor, its evolution, and the design principles driving this evolution, while the goal was to look for conceptual inspiration, which might eventually be used for inventing new human body armor. There are many obvious similarities between human and animal body armor, and looking for inspiration in nature seemed to be a good idea, as has been proven (Arciszewski and Cornell 2006). We will review here some of the results regarding knowledge acquired from nature.

When body armor is considered, no matter if it is for humans or animals, its ultimate objective is clear: maximize survivability. This objective may be also reformulated as a technical contradiction in accordance with TRIZ (see Chapter 9): maximize body protection and maximize mobility. Since the same force (maximization of survivability) drives the evolution of both kinds of body armor, knowledge about animal body armor evolution is relevant to designing human body armor. It may be an oversimplification of a complex issue, but poorly designed human armor results in higher mortality rates, and poorly evolved natural armor does the same and can even lead to the extinction of a species.

In nature, various different species use plate armor for protection, with differentiated adaptations for movement. Surprisingly, in nature, separate lines of evolution have emerged, each focused only on a specific requirement: maximization of protection, maximization of mobility, or finding a balance between protection and mobility.

The gopher turtle is an excellent example of the first line of evolution. Its body is nearly completely protected by a plated shell. Even its legs are armored, and all major and vital organs are protected underneath and on top with very strong, heavy plates. As a matter of fact, for centuries, humans have imitated this kind of full protection. Moreover, about 2,000 years ago, the Romans developed a military formation called "the turtle," obviously inspired by the turtle's body armor (Figure 10.8). In this case, marching soldiers would form a rectangular formation with the leading soldiers holding their shields in front, those on the side holding their shields to the side, and the soldiers in the middle holding their shields over their heads. In this way, a huge marching turtle armor would be formed with all the soldiers safely protected inside.

The second line of evolution represents a balance between protection and mobility, and its best example is the armadillo (Figure 10.9). Armadillos can move quickly and are relatively well protected by an armor-like shell from head to toe, except for their underbelly, which has a thick skin covered with coarse hair. Armadillos evolved over the last 10,000 years, and today they are much smaller than their predecessors. Probably, a smaller size has helped them to survive by lightening the weight of their bodies and increasing their mobility.

The third line of evolution is focused on the maximization of mobility, with canines and felines being good examples of this line. They use speed, strength, and intelligence to survive without any armoring. Biologists believe that armor is exponentially more costly if the species is high on the

Figure 10.8 Roman "turtle" formation.

Figure 10.9 Armadillo.

food chain, and this may explain the fact that lions or horses do not have any armor.

The study of evolution of turtles has revealed several design heuristics that are directly applicable to the development of new types of human body armor:

1. Maximize size and volume of body armor
2. Create smooth surfaces
3. Create multilayered body armor
4. Introduce shock-absorbing layers
5. Minimize weight
6. Maximize articulation

The first heuristic, "Maximize size and volume of body armor," is consistent with our understanding that increasing mass and volume of a solid body increases its energy absorption capability and provides increased safety; it simply delays the moment when the body breaks. The second heuristic, "Create smooth surfaces," is most likely a reflection of the fact that a smooth surface may save our life when a beast tries to grab us with its fangs or nails. Also, a smooth surface may create light reflections, confusing an attacker, and may help us to slide down a hill if a beast is chasing us. Finally, a smooth surface may reflect flying arrows or other projectiles when we come under attack.

The third heuristic, "Create multilayered body armor," can be explained by the fact that combining materials may create unique features of armor: impact damping, thermal insulation, penetration resistance, lightness, and so on. The fourth heuristic, "Introduce shock-absorbing layers," only adds specificity and sophistication to the previous one and reminds inventors that impact protection is extremely important and that this can be improved using a layer of fat, for example, to protect a brain against impact.

Heuristic No. 5, "Minimize weight," can be understood as concern about mobility, even among turtles and particularly among humans, but it is also about minimizing the energy and food required to build and maintain heavy armor. Finally, the last heuristic, No. 6, "Maximize articulation," is naturally about mobility and the need to maximize it while also knowing that the maximization of articulation is the best way to improve mobility without compromising protection.

The use of conceptual inspiration is never easy; it is always a challenge. First, an engineer must be trained in abstract thinking in terms of decision rules and heuristics. (As the Author's experience with learning and teaching Synectics indicates, such training, although possible, is time consuming and difficult. However, the student may benefit from using various computer tools like "Gymnasium," which is a part of MindLink software.) Second, using conceptual inspiration is time consuming because it requires studying the problem area and the related area of biology before working on understanding both areas and finally on the integration of this knowledge. Third, little is known about the methodology of conceptual inspiration use, and that makes the process even more difficult and sometimes frustrating. Finally, the challenge of size and performance mismatching is always present. However, in this case we know what to do (Fish et al. 2011a,b). To avoid this challenge, bioinspiration should be used preferably only in cases of overlapping scale and performance; that is, natural knowledge associated with whales could be used for inventing new types of medium-size boats but not for inventing steering systems for aircraft carriers. Similarly, in terms of performance, natural knowledge hidden in migrant birds could be used for inventing new systems for long-distance transportation planes but not for jet fighters.

The three challenges presented here constitute together a high price for success, but this price is entirely justified by the expected results. The process activates the entire power of the human mind, particularly its creative part, since the generation of decision rules is an abductive process. Thus, the final results may be groundbreaking and lead to a variety of inventions.

10.6 COMPUTATIONAL INSPIRATION

10.6.1 Justification: Why do we need it?

We have already learned about visual inspiration (biomimicry) and about conceptual inspiration (biomimetics). The first is basically a mapping process from nature to engineering, while the second is learning from nature, a complex *manual* process of natural knowledge acquisition and the transformation of this knowledge into scientific and engineering knowledge. Both forms of bioinspiration have been successfully used many times and have proved their value for inventors. Visual inspiration does not involve any use of computers, and in the case of conceptual inspiration, computers are used but only in a supportive role and only for analytical purposes, like conducting a finite-elements fluid–fish interaction analysis in the research on the manta ray propulsive system.

In both cases we use a *static*, or *stationary*, approach to nature; that is, we learn from it as it is *now*. It is like taking a single picture of a natural system at a given time and learning only from this single picture. Obviously, there is nothing wrong with such an approach, which has proved to be successful. However, if we want to learn much more about a given natural system, instead of taking a single picture, we take many pictures over a long period of time—over thousands or even millions of years. Such an approach can be called *dynamic* and will allow us to understand the process of evolution of our natural system.

The dynamic approach may help us to acquire natural knowledge about the evolution and growth of natural systems. Evolution and growth are two different concepts. The first describes the process of gradual changes in natural systems, their evolution from generation to generation, which occurs over long time periods. The second, growth, describes a process used by nature to grow a single natural system from a seed into a complete plant or animal. Learning about these two processes will allow us to develop a deep insight into them and, most importantly, to discover the mechanisms driving them. In other words, the dynamic approach may reveal nature's top secrets about how various plants and animals evolve with time and how they grow. Such an understanding cannot be developed using visual or conceptual inspiration.

Knowing nature's mechanisms behind evolution and growth is potentially priceless for inventors. These mechanisms may be used to simulate on computers the various processes of evolution or growth of engineered systems over long periods of time but also during a conceptual designing process. Mechanisms may be relatively simple, but repeating them millions of times may produce results that could never be predicted using natural knowledge that is acquired using only the static approach. In fact, the computer simulation of the evolution of engineered systems occurring over long periods of time and the growth, or emergence, of the individual systems is the most fascinating part of modern computing. It is probably also the most challenging part of computing, but its inventive potential cannot be overestimated, as various examples will later demonstrate. For all these reasons, *computational inspiration* supplements the two other forms of inspiration and is a subject of growing interest for both scholars and practitioners.

10.6.2 What is computational inspiration?

The area of computing that deals with various activities usually associated with humans is sometimes called *artificial intelligence* (AI). This unfortunate term was introduced in the 1950s, when there were high hopes that a computer program could be built that would exhibit intelligence comparable to the human intelligence or one that would be even more advanced. These expectations were infeasible 50 years ago, and they are still today. The term AI has created a lot of damage to the reputation of the field of advanced computing with its premature and unjustified implication that AI is possible. Therefore, today we prefer to use other terms that do not have the negative connotations associated with the term "AI," as you will learn in the following sections on *evolutionary computation* and on *cellular automata*.

We will describe computation inspiration thus:

> Computational inspiration is a process of using natural evolution and growth processes to develop similar computational processes that can be used in computer programs for the automated development of design concepts, including inventive design concepts.

We will focus on two specific forms of computational inspiration that are particularly promising for inventive engineers. First, we will discuss evolutionary computation (EC), which is inspired by the mechanisms of the evolution of natural systems. Next, we will overview cellular automata, which are inspired by the mechanisms of growth in natural systems.

10.6.3 Evolutionary computation

The term EC describes a class of modern computational methods inspired by the process of natural evolution. These methods incorporate computational models of mechanisms of reproduction, natural selection, and genetic inheritance in order to solve problems in many fields of engineering and science. They are also applicable to conceptual designing and can be used for searching large design representation spaces to find the best design concepts, including inventive concepts. The best way to understand EC is to learn about it in the context of the natural evolution. Therefore, a brief overview of natural evolution and the emergence of EC will be provided here, followed by a discussion of EC in designing and examples produced by the Author's research team working for NASA several years ago.

In the nineteenth century, Darwin proposed his famous theory of evolution (Darwin 1859), which explained how various features of a living organism are passed from generation to generation. The core of this theory is a mechanism called *natural selection*. According to Fish (2013), Darwin was not familiar with Gregor Mendel's work on the genetic basis of inheritance; only later did contributions by R.A. Fisher, J.B. Haldane, Sewell Wright, and T.H. Huxley integrate Darwinian evolution with the principles of genetic inheritance, and neo-Darwinism, or the *synthetic theory of evolution*, was created.

The synthetic theory of evolution is based on the two fundamental notions of *natural selection* and *inheritance*. It is also based on the assumption that the natural evolution occurs on the level of the population. Natural selection determines the composition of the next population. Inheritance means that members of a population reproduce, producing children or offspring that inherit some features of their parents. Interestingly, from the perspective of inventive engineering, each organism participating in the natural evolution process may be considered as an equivalent to a design concept in the process of conceptual designing and could be even called a *natural design concept*.

In inventive engineering, we use such phrases as "design concept for a steel structural system in an industrial hall" and "steel structural system in an industrial hall." The first one means an abstract description (in the form of attributes and their values) of this specific structural system. In the EC area, an abstract description of a system (natural or engineered), the equivalent of a design concept, is called a *genotype* or *genome*. The genome is an organized collection of *genes,* which are the equivalent of symbolic attributes. Values of genes are called *allies* and they are equivalent to values of symbolic attributes. Examples of a genome and allies are provided in Figure 10.10.

The phrase "steel structural system in an industrial hall" means the actual system as it can be shown in a technical drawing with all the necessary engineering details. Its equivalent in biology and in the area of EC is a *phenotype*, which represents the actual system, natural or engineered, associated with its genome (Figure 10.10).

Natural evolution occurs on the level of population, but natural selection is done on the level of individual organisms. It is a process in which organisms with favorable features leave more offspring than organisms with less favorable features. In this way, the frequency of favorable features is increased in the next population. The selection is done on the basis of the ability of the individual offspring to survive in their environment, that is, its *survivability*, and it is the ultimate measure of the fitness of a given organism for its environment. Therefore, the selection criterion is appropriately called *fitness* and the natural selection can be described as *survival of the fittest* (Darwin 1859). Therefore, fitness is a relative measure and is valid only within a given population of competing organisms. Fitness can be also understood as a measure of a specific organism's ability to pass its features on to the next generation (Fish 2013). Fitness is always assessed on the level of phenotypes; that is, actual organisms and their performance are evaluated. Therefore, we should consider two levels of natural evolution: genotypes and phenotypes. Evolution occurs on the level of genomes, but the competition for survival, decided by fitness, occurs on the level of phenotypes. This duality is similar to the inventive process in the context of society: Inventions are ideas, that is, inventive design concepts, but their

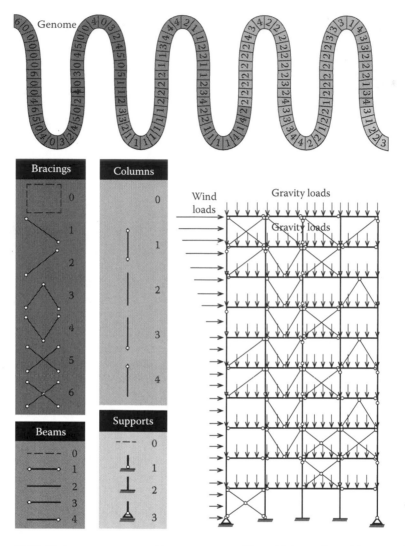

Figure 10.10 Skeleton structure and its components. (From Kicinger, R. and Arciszewski, T., *American Scientist*, 2007, 95(6), 504.)

actual value (fitness) is determined in the market when they become real products competing against similar products.

From an engineering perspective, natural selection can be explained as a multicriteria selection process involving criteria like energy consumption, weight, size, developmental costs, speed, body protection, maneuverability, and so on. All these criteria contribute in a complex way to the survivability or to the fitness of a given offspring. The other engineering interpretation of natural selection is in the context of utility theory (Keeney and Raiffa 1993). In this case, utility functions are established for all-important features of a living organism. A utility function allows the conversion of a value of a given feature into a value of *utility*, that is, into a value of a measure representing a "generalized goodness" of this value. Next, the *total utility function* can be created as the sum of these individual utility functions. Thus, the total utility of a given natural design concept can be determined and used within a given population as a measure of the goodness of a given natural design concept and as the selection criterion. In fact, the total utility can also be interpreted as a relative measure of the survivability of a given natural design concept versus other competing natural design concepts—the ultimate measure of its fitness.

The selected natural design concept may not be optimal in the context of a single selection criterion but it is best from the perspective of survivability. (The natural design concept must also satisfy the strict requirements associated with survivability, that is, the ability to reproduce, to move, and to feed.) In the case of engineered systems, the single selection criterion is usually imposed or it is relatively easy to identify. Usually, such selection criteria as weight or cost are used, but obviously other criteria may be used. In fact, there is an emerging research field dealing with multicriteria evolutionary optimization (Kicinger and Arciszewski 2004, Obayashi 2015) and researchers are investigating various models of how to use of many selection criteria in the evolutionary computational process.

Natural selection will work effectively only when a sufficient variety of organisms exists within a population. This variety of organisms is called *genetic variation*, and nature has several reproduction mechanisms that contribute to this variation. They include *mutation, genetic recombination*, and bringing in genetic material from a different population. All these mechanisms operate on the level of genotypes and allies.

In the case of mutation, a genotype is simply modified through changing one or more values of genes, that is, changing allies. Genetic recombination, or sexual reproduction, means creating a new genotype of an offspring as a result of using genetic material from two organisms and mixing together their genotypes. Finally, bringing in genetic material from a different population corresponds to the situation, for example, when two populations of animals live on two adjacent islands and occasionally animals from one island swim to the other, bringing with them their genetic material.

In Chapter 4, the concept of a design representation space was introduced. When a given design situation is considered, this space can be understood as the body of knowledge related to our design problem. Such a space may contain a huge number of potential design concepts, even on the level of billions, since each point in this multidimensional space represents a potential design concept. The best design concept must be found on the basis of its *optimality*, which is measured by an assumed optimality criterion. Therefore, we have to search the design representation space for the best design concept, and for this purpose various search methods can be used. EC provides a class of search methods that in this case are inspired by the natural evolutionary process adapted from nature. In all these methods, the optimality of design concepts is measured by the fitness function, a concept that has already been introduced and discussed. From the engineering perspective, the introduced process can be understood as a process of search and optimization in which a population of design concepts gradually evolves through many generations and undergoes incremental changes. The objective of such evolution is to optimize the fitness function, that is, to minimize it or maximize it depending on the nature of the problem.

During more than 50 years of EC research, a large variety of evolutionary algorithms (EAs) has been developed. They all are based on the fundamental concepts of evolution and inheritance but differ in the reproduction strategies and/or representation on which they operate (De Jong 2005). Three major EAs include evolution strategies (ES), evolutionary programming (EP) (Schwefel 1965, Fogel et al. 1966), and genetic algorithms (GAs) (Holland 1975). There are also many hybrid algorithms incorporating various features of these three major categories of evolutionary algorithms. All these algorithms have been developed mostly for evolving solutions to parameterized problems, that is, problems for which a representation space with attributes and their values can be constructed. Recently, a fourth major EA has been developed and is called genetic programming (GP) (Koza 1992). In this case, actual computer programs are evolved to solve various computational tasks (Luke 2000). An excellent introduction to the entire field of EC can be found in De Jong (2005). Other books also provide good overviews of the fundamentals, for example Goldberg (1989), Bentley (1999), and Kumar and Bentley (2003).

Research on the engineering applications of EC has a relatively long history, which was initiated in Europe in the early 1970s by Rechenberg (1965). In the United States, Goldberg (1987) was first to use GAs in the optimization area to optimize complex gas pipeline systems. In the area of structural engineering, Hajela and Lamb (1986), Hajela (1992), and Hajela and Lee (1995, 1996), were the first to work on the numerical optimization of structural systems (Hajela 1997). Your Author has also done some research on the applications of EC in the area of inventive conceptual designing (Arciszewski and DeJong 2001; Arciszewski and Kicinger 2005).

GAs are particularly popular among engineers and have been the subject of the Author's research for several years. Therefore, a conceptual design process driven by a GA will be briefly overviewed here to help readers to develop at least a conceptual understanding of EAs. First, however, the basic terminology needs to be introduced. In the case of conceptual designing, we will be using the following terms:

> A design concept will be called an *individual* when it is viewed as a member of a population, but it will be called an *offspring* or *child* when it is considered as a product of the reproduction process.

A *population* means a set of individuals for a given generation in the evolutionary process. When the current population is replaced by offspring, the new population is called a *new generation*. Traditionally, the initial population associated with the first stage of the evolution process, or with the first generation of this process, is numbered as the "0" population.

Each stage in the evolution process is called a *generation*.

A *reproduction process* is a process of creating offspring or children. There are two kinds of reproduction process: mutation and crossover.

A *genome* is the EA abstract representation of a design concept. Each genome consists of a sequence of *genes,* which correspond to attributes that describe a design concept. A value of a gene is called *ally*.

A *phenotype* is the actual engineering system represented on the abstract/conceptual level by a genotype.

Mutation is the process of creating offspring through changes in the genome of a single parent. (Figure 10.11)

Crossover is the process of creating offspring through combining part of the genomes of two parents. (Figure 10.11)

Fitness is a quantitative optimality measure of an individual.

A *stopping criterion* is a criterion used to determine if the evolutionary process needs to be continued or if its goal has been reached and the process needs to be terminated.

We will discuss an evolutionary conceptual design process conducted for a specific engineering system for which we know dimensions, loads, location, and so on. It will be a process with n generations, that is, we will have a sequence of $n-1$ populations, excluding the initial or 0 population. We will also make an assumption that the size of all populations is the same and each has p individuals.

10.6.3.1 Preparation

Before the actual evolutionary conceptual design process begins, some initial preparations must be done. Specifically, all known assumptions regarding the engineering system under consideration must be specified, including its dimensions, location, loading, design codes, and so on. Particularly important is the selection of the fitness, although in structural applications, weight is usually a good approximate measure of both the size and complexity of a structural system and of its expected cost.

Also, certain assumptions regarding the evolution process have to be made, like the population size, number of generations to be used (or another stopping criterion), specific assumptions regarding the selection mechanisms and the used mutation and crossover, number of offspring per generation, and so on. Finally, an automated design and optimization computer program must be selected and integrated with the evolutionary

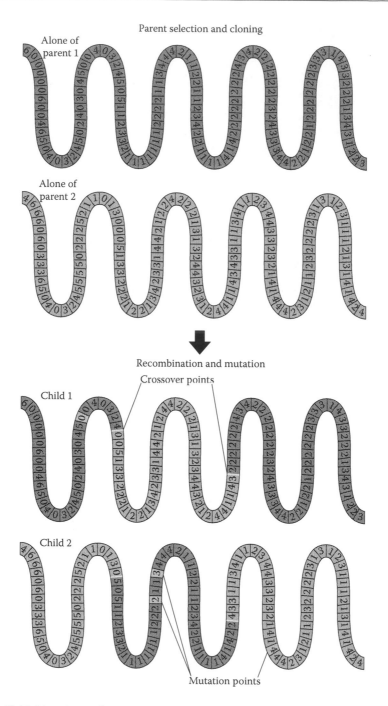

Figure 10.11 Mutation and crossover. (From Kicinger, R. and Arciszewski, T., *American Scientist*, 2007, 95(6), 505.)

program so as to be able to determine the fitness of the individuals in the evolving population.

10.6.3.2 Generation No. 1, initial population No. 0

The process begins with the initial population, or generation No. 1. The initial population will have p individuals. They may be randomly generated, or known design concepts (or patented inventions) may be used as initial individuals. Both approaches have advantages and disadvantages. The Author's experience indicates that using known design concepts leads to good results faster than using randomly generated individuals. There is probably a cost associated with such an approach in the form of the somewhat reduced novelty of the final results. The Author consulted about this issue with his former PhD student, Dr. Rafal Kicinger, now a well-established EC scholar and practitioner. He confirmed that there are no available quantitative experimental results regarding the impact of the selection of initial individuals on the novelty of the final results. Interestingly, he also warned that using known design concepts must be done carefully and these concepts should be sufficiently differentiated to avoid the lurking danger of the local optimization, that is, searching for the best design concept only in one or several regions of the design representation space. However, using randomly generated initial individuals means that they will most likely come from various parts of the representation space, and therefore there is a smaller probability that some regions of this space will not be searched and potentially attractive design concepts will be missed. This is a practical way to avoid local optimization.

When the initial population is available and it is composed of randomly generated individuals, obviously not all of them represent feasible design concepts. Some individuals may represent geometrically variable, or unstable, structures. Others may represent infeasible design concepts because of the proposed material, for example wood or paper for columns in a tall building, and so on. When dealing with the potential design concepts only in the form of genomes, an inventive engineer may conduct such feasibility analysis manually, but it is difficult and extremely time consuming, as the Author has experienced once or twice. A much better approach is to send all genomes to an automated design and optimization system (like SODA [structural optimization in design and analysis] in the area of steel structures [Grierson and Cameron 1989]) that will use genomes and the other input data (dimensions, gravity, wind loads, etc.) and transform this input into a set of complete detailed designs. As the first step in this process of detailed designing, the automated design system will analyze the feasibility of the provided genomes and will simply reject those that are infeasible.

When the feasibility of all individuals in the initial population is confirmed or not, and all infeasible genomes are replaced by feasible ones, the evaluation of all individuals begins. Again, an engineer may manually prepare detailed designs based on all genomes and determine their fitness using their actual characteristics like weight, cost, or constructability. It is possible but definitely not recommended, particularly in the case of complex engineering systems like tall buildings or tunnels. As in the case of feasibility analysis, practically the only possible approach is to use an automated design and optimization system that will use as input genomes and the other required data for detailed designing and transform this input into complete detailed designs.

When a set of detailed designs based on all initial individuals is available, the fitness for all individuals must be determined using the fitness criterion assumed earlier, like weight, speed, cost, time, and so on. Obviously the nature of this criterion depends on the kind of design problem involved. In the case of a car, it could be "maximum speed," but in the case of a submarine, it could be "stealth."

The fitness of all individuals will be used to determine which ones will be allowed to have offspring and how many. When the most fitted individuals are identified, their copies are produced, and they will participate in reproduction and produce the desired number of offspring. The reproduction, or breeding, will be done using mutation and/or crossover. The number of the produced offspring may be much larger than the size p of the next population because many offspring may be infeasible or their fitness is very low. Now we have a collection of offspring, or children, which will be used to create the next population for generation No. 2. Therefore, the feasibility of all offspring must be assessed (in the same way as happened to their parents) and the feasible ones must replace obviously infeasible offspring. When this is done, the fitness of all the offspring must be calculated (the same way as for their parents).

Now a collection of evaluated offspring is available and ready for the selection of those offspring that will become individuals in the next population. Various selection methods can be used; for example, p offspring with the highest values of fitness are selected. Also, it is possible that parents will compete with their offspring, and the next population No. 1 will be a mix of p surviving parents and their offspring.

10.6.3.3 Generation No. 2, population No. 1

As a result of evolution taking place during the first stage of the entire multistage process, population No. 1 has emerged. We know all p individuals in this population and we also know fitness values for these individuals. The evolution that is taking place during the second stage, or in generation No. 2, is similar to the previous stage.

First, the desired number of the individuals with the best fitness values is selected from the population and they are copied. These copies are used for the purposes of reproduction, which is conducted as in the previous stage. Next, the feasibility of all offspring is determined, and for the feasible ones their fitness is determined. Infeasible offspring are replaced by feasible ones, which also get their fitness values calculated. Finally, offspring compete—eventually with their parents, who are members of population No. 1—and population No. 2 emerges.

When population No. 2 is available, we need for the first time to determine if the entire evolution process needs to be continued or terminated. In other words, we need to use the so-called stopping criterion. This is assumed before the evolutionary designing begins. We may terminate the process after, for example, 10,000 generations or after we have improved the fitness by, for example, 10%, 20%, or 30% with respect to the best individual in our initial population of generation No. 1. Let us assume that in our hypothetical process the stopping criterion has not been met and we are moving to the next stage, stage No. 3, with our population No. 2.

10.6.3.4 Generation No. 3, population No. 2

This stage begins when population No. 2 is available with all its individuals and their fitness values. All the activities at this stage are identical to the activities of the previous stage with one obvious difference: they are done on population No. 2 instead of population No. 1 as in the previous stage. If the stopping criterion is satisfied, the evolution process is over; population No. 3 becomes our final population, and its individuals constitute our final result. If the stopping criterion is not satisfied, population No. 3 becomes the population of parents for the fourth stage.

We have described three stages of a hypothetical evolutionary designing process. This description should provide a good understanding of the nature of the entire process. Eventually, readers are encouraged to learn more on the subject from such sources as De Jong (2005) and Goldberg (1989).

Very rarely, if ever, only three stages are sufficient to produce meaningful results. Usually, several hundred stages are sufficient, but sometimes tens of thousands of stages are necessary for complex design problems associated with a large design representation space that needs to be searched.

10.6.3.5 Example

Around the turn of millennium, a group led by the Author conducted a NASA-sponsored research project on evolutionary designing. As a result of this project, a computer system was developed, appropriately called Inventor (the actual software design and programming was done by Dr. Krzysztof Murawski, at that time a visiting scholar at George Mason University from the Warsaw Military University of Technology) (Murawski et al. 2000).

Inventor produced complete designs of steel skeleton structural systems in tall buildings. The system had two main integrated components. The first was an evolutionary component for the development of design concepts, while the second was a structural design component for the production of complete designs utilizing these concepts. The evolutionary component was a unique computer program developed at George Mason University and was based on genetic algorithms. The structural design component was the commercial computer program called SODA, which has been adapted and integrated with the evolutionary program. SODA conducts the fully automated design and optimization of steel structural systems. Therefore, it was a perfect tool for converting design concepts (genomes) into complete designs (phenotypes) in order to determine the weight of the designed structural systems (but also the weight of the individual structural members, their number, number of types of structural members, etc.), which was used as a measure of their fitness (Grierson and Cameron 1989).

Inventor works with genomes, and all feasible genomes represent various design concepts of structural systems. When designing a steel structural system in a tall building, decisions have to be taken to specify components of four subsystems: bracings, beams, columns, and support footing. Therefore, each genome has four parts corresponding to the descriptions of the individual subsystems.

In a genome, each part of it—that is, each gene—is related to a different structural member or connection in the structural system described by this genome, while the value (ally) of this gene specifies the nature of this member or connection. In the case of Inventor, genes may have several numerical values (allies) representing the nature of used structural members or connections.

Let us consider a single *cell* in a steel skeleton structural system. By the term "cell" we mean a part of a structural system that is contained between two adjacent columns and two adjacent beams (see Figure 10.10). When the subsystem "bracings" is considered for our cell, the nature of the bracing in this specific cell is identified by a single gene. When the value of this gene is 0, there is no bracing in our cell, like in a rigid steel frame, which is not braced. The value of "1" means a diagonal bracing in our cell, and a value of "6" represents the situation when an X-bracing is used, and so on. Thus, we have here a one-to-one mapping between genotypes and phenotypes—between the values of attributes identifying our design concept and the actual structural members and connections used in the structural system.

Before the automated evolutionary designing process begins, the user must provide information about the building for which various structural systems will be designed—about the location of the building, wind and gravity loads, and so on. Also, the user may decide if he or she wants a symmetrical structure or if symmetry is not required. Usually, allowing asymmetry leads to more "interesting" design concepts, although they may not be ultimately acceptable because of their low constructability. However, such design concepts may inspire the human designer, who will use them to develop concepts that are a little more traditional but feasible and still novel. In this context, Inventor should be understood not only as a structural design tool but also as a source of inspiration. Ultimately, it is the human who becomes an inventor and all kinds of sources of his or her inspiration are desired.

The user may provide a population of known design concepts (represented by a collection of corresponding genomes), but it is not required. Inventor may start the process by randomly generating a set of genomes. In this case, they are simply sequences of numbers, and their feasibility must be checked (or more precisely the feasibility of structural systems represented by them), infeasible genomes replaced, and the initial population created. Next, Inventor conducts the entire evolutionary designing process (described earlier) until the stopping criterion is satisfied and the final results are presented to the user. These come in the form of simple drawings clearly showing the entire design configuration, including members and connections; and detailed data regarding individual members, their cross section, dimensions, weight, and so on.

Several examples of structural systems generated for a skeleton structure in a tall building are shown in Figure 10.12. There are two numbers below each drawing. The top number represents the total weight of the steel structure (in pounds) of the building under consideration, and the bottom number is the maximum horizontal displacement of the top floor (or sway), which is in inches. Both weight and sway are important characteristics of structural systems in tall buildings, and they should be minimized.

Probably no structural engineer would produce such chaotic configurations, which are feasible but have very low constructability. Also, non-engineers could be confused by the chaotic nature of these designs and might suspect that there is something wrong with the designs and that they may be even dangerous. On the other hand, architects truly love these

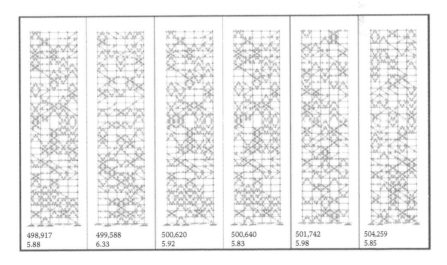

| 498,917 | 499,588 | 500,620 | 500,640 | 501,742 | 504,259 |
| 5.88 | 6.33 | 5.92 | 5.83 | 5.98 | 5.85 |

Figure 10.12 Structural designs produced by Inventor.

configurations, which are obviously not boring and are considered by architects as inspiring and bringing beauty to the structural designing of tall buildings, which usually results in monotonous, simple, and repetitive arrangements of structural members.

10.6.4 Cellular automata

The nature-inspired concept of cellular automata (CAs) was proposed in the 1950s by von Neumann (1951). CAs may be understood as simple mathematical representations of complex systems and their behavior. Since their introduction, CAs have been successfully used in science to model complex systems and processes driven by a large number of interacting, simple, and identical components. However, in the twentieth century, no conceptual designing applications took place.

In 2002, the true renaissance of CAs began with the publication by Steven Wolfram of his groundbreaking book *A New Kind of Science* (Wolfram 2002), which was intended to entirely change the dominant understanding of nature and science. Wolfram believes that all processes in nature, science, and engineering can be explained using the CAs approach. Even more, he claims that, using this approach, a new, if not complete, understanding of the world may be developed. His bold claims and enthusiasm, not to mention the solid mathematical foundation of his work and his outstanding reputation as a pioneer of computing, have attracted the attention of one of the Author's PhD students, Rafal Kicinger (already mentioned in the previous section). Kicinger became the first engineer selected by Wolfram to attend his first summer school in 2003 on A New Kind of Science as a part of a very small group of PhD students from all areas of science. After completing the program, Kicinger became probably the first engineer with an excellent modern understanding of CAs and their enormous potential. Next, he did what all inventive engineers need to do: He took high professional risks and focused part of his doctoral studies on the applications of CAs in engineering design. The results were the best award for risk taking and obviously also for his hard work, not to mention talents and excellent background. The results were presented in a number of publications (Kicinger 2004; Kicinger et al. 2005a–c), including a popular article in the *American Scientist* (Kicinger and Arciszewski 2007), which is particularly recommended for all inventive engineers looking for a short, clear introduction to CAs in inventive designing. The following part of this section is at least partially based on Kicinger's findings and resulting publications.

CAs can best be explained best in the context of evolutionary designing because both fields are closely related, and EC and CAs could be used in a complementary way. Natural evolution occurs on the level of a population, which gradually evolves to improve the fitness of its members, or individuals, and in this way to improve the chances of survival of the entire population. As the result of reproduction, new individuals are born and they emerge as already entirely formed organisms, like small baby elephants or baby whales. Similarly, in the case of computational evolution in engineering designing, as a result of mutation or crossover new complete design concepts emerge—concepts for a new type of bridge or a new type of tunnel.

There is no question that natural evolution is a powerful source of computational inspiration for all inventive engineers. Unfortunately, computation programs utilizing the principles of natural evolution have various limitations in terms of conceptual designing (Kicinger and Arciszewski 2009), and their design products are not always good enough for engineering purposes (as briefly discussed in the previous section). Fortunately, biology provides another amazing process with huge engineering potential, which is called *development*. In contrast to natural evolution, which occurs on the level of the population, development occurs on the level of a single organism.

Development is a process driving the emergence of an organism from the initial very small number of simple cells, called the *embryo*. This is a multistage process in various parts of an organism gradually emerge. It is like watching growth of an entire tree from its root. First, the trunk emerges, next the main branches, followed by secondary branches and leaves at the end. This growth is driven by "secret" rules of growth associated with a given embryo.

Development can be also used to develop various design concepts. In this case, a part (or a subsystem) of an engineering system is provided, and it is used to develop the entire system using development rules, which are specified. For example, the development of a concept for a steel skeleton structural system in a tall building may be understood as a process of *growing* the structure from the foundation level (Figure 10.13). The structural system for the ground floor is provided to initiate the growth, but the development rules are used to create the second-floor structure, and they use this structure to create the third-floor structure, and so on, until the top structure is created and the design concept for the entire structural system is known.

There is another fundamental difference between evolutionary designing and development. In the first case, we have a direct mapping between individual genes (and their allies) and the corresponding elements of the considered engineering system, that is, a direct mapping between a genome and the related phenotype. When development is used, we do not have this direct mapping for the entire engineering system. The picture is a little more complicated. We also have a genome, but it does not represent the entire engineering system as in the case of evolutionary designing.

The genome used in development is like a recipe for the future engineering system and contains all the necessary information to develop this system from an embryo. Therefore, the entire genome has two distinctive parts: the embryo and the decision rule. The embryo is a sequence of genes and their values that describes a part of the future engineering system

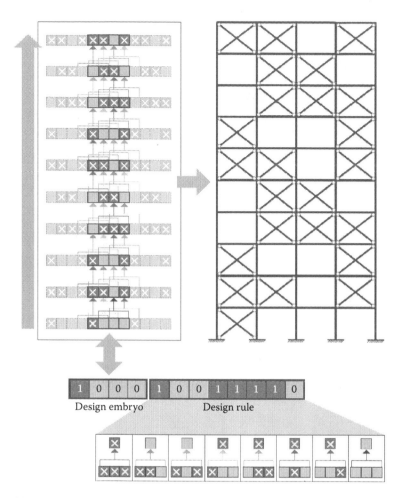

Figure 10.13 Cellular automata and a skeleton structure. (From Kicinger, R. and Arciszewski, T., *American Scientist*, 2007, 95(6), 506.)

(or its subsystem), which will be used to develop the entire system. The design rule is in fact a collection of rules for the development of the embryo into the entire system. The design rule is also called cellular automaton and may be understood as a formal computational model of development.

When a specific genome is considered, its *embryo part* is represented by the first several genes and their values that together identify the embryo. Its *cellular automaton part* is represented by the remaining several genes in the genome, and it contains all the developmental rules. These rules are known as a sequence of numbered rules. The outcome of the first rule is represented by the first gene of the cellular automaton part. Similarly, the outcome of the second rule is represented by the second gene, and so on. How these developmental rules are understood and how they are implemented is a little bit complex, and this will be explained next by considering a specific design case.

When a genome is given, developing the entire engineering system is a simple computational process, which can be conducted in a mechanistic way. The novelty and quality of the developed engineering system obviously depend on the genome used in a given case, that is, on the embryo and the cellular automaton, which are a recipe for the design. Therefore, the genome may be improved, if not optimized, using GAs and by running the evolutionary design process for many generations, as described in the previous section.

We will learn about cellular automata in action by considering the process of designing a planar (two-dimensional) structural system for a tall building (Figure 10.13). We will make a number of assumptions to simplify the problem for the sake of clarity. However, the problem will still be realistic to a certain degree while allowing us to understand the developmental process.

Our structural system will be a four-bay braced rigid frame. This frame will be designed on a rectangular grid, which in our case will be defined by columns and beams whose locations are already known. Also, in this case, each two adjacent beams and two adjacent columns identify a *cell* in our structural system. Our challenge is to develop a system of wind bracings; that is, we need to determine which cells should be braced. Moreover, we will assume that only one type of bracings can be used, the so-called X-bracings. Thus, for each cell we have a simple question: to brace it or not. If we brace it, the corresponding gene will have the value "1." If we do not brace it, the value of the same gene will be "0."

In the case of our four-bay structure, the embryo is in the form of the bracings of the first-floor structure. Since we have four bays, only four genes will be necessary to identify the embryo. The first cell on the left (see Figure 10.13) is braced, and therefore the value of the first gene on the left (corresponding to this cell) is "1." All three remaining cells on this level are unbraced, (empty) and all genes corresponding to these cells have the same value, "0." The entire embryo is specified by a sequence of values of the individual genes; that is, in our case it will be the sequence 1, 0, 0, 0 (Figure 10.13).

When the embryo is known, that is, the first-floor structure and the corresponding content of the all cells in the first row of cells, we are able to initiate the development of the second-floor structure.

However, before we begin the actual development process, we need to specify the design rule. In our case, this design rule will be a collection of eight specific rules, as shown in Figure 10.13.

When a specific cell in the second row of cells is considered (second-floor structure), the decision to brace it or not will be governed by the content of the three cells located below this cell. A cell directly below the considered cell will be examined as well as two cells to the left and right of this cell. All these three cells below constitute a *local neighborhood* of the considered cell in the row above. When three cells are analyzed and each may be braced or not, we have a total of eight combinations or eight decision situations. Naturally, for each of them a specific rule needs to be known, simply arbitrarily assumed, or a product of evolutionary optimization.

Before we discuss examples, we need to learn another secret of cellular automata. This is the *secret rule of cellular automata*, and it says that

> If the cell below a considered cell is on the end of the row, the value of the cell on the other end is used to fill the missing neighborhood slot. In this way, the rule always applies to three cells in a given row.

As an example, let us discuss the development process for two cells in the second row of the grid: one in the middle (second from the left end) and one located on the left end. In the first case, the cell directly below the considered cell is unbraced or empty, but the cell on the left is braced, while the cell on the right is unbraced. We have a sequence of values for these three cells: "1," "0," "0." Figure 10.13 provides the specific rule for this situation, which is the fourth rule from the left, and it says that the considered cell should be braced, as is shown in the drawing of the structure on the right side of Figure 10.15. When the first cell on the left in the second row is considered (second case), we need to use the secret rule of cellular automata, which has just been introduced. The cell below our considered cell is braced, the cell on the right is empty, and there is no cell on the left (the end of the row). Therefore, according to the secret rule of cellular automata, the conjugate cell needs to be examined, that is, the cell on the right end of the row below the considered cell. This conjugate cell is empty and its value is "0." In this way, we have identified a combination of three cells in the local neighborhood and a unique combination of their values: "0," "1," "0." This situation is identified by the specific rule that is sixth from the left, and it says that the considered cell should be braced (1), as shown again in the structural drawing in Figure 10.13. Obviously, similar analysis has to be conducted for all the cells in the second row. When this is done, we will know the values of all cells for the second row. These values (and the configuration of bracings of the second floor) will be used as the embryo for the third floor, and so on, until the top floor bracing is known and the entire structural system has been developed from the bracings of the first floor.

The described process is fascinating both conceptually and in its nature-like behavior. It is driven by simple rules, but they may lead to self-organizing behavior and the emergence of very complex patterns. These patterns are often entirely unexpected and may provide a unique understanding of engineering systems. It is like discovering the hidden secrets of engineering systems and finding complex processes governing their design. For all these reasons, the idea of using cellular automata in conceptual designing has attracted the attention of Rafal Kicinger (already mentioned). He has expanded his doctoral studies and has also focused on the developmental conceptual designing in addition to his studies on evolutionary designing. Moreover, he has decided to integrate these two domains and to study a designing process in which both cellular automata (CAs) and evolutionary computation (EC) are used. As a result of his studies, an experimental computer program has been developed called Emergent Designer (Kicinger 2004). The name reflects the fact that the program *produces* design concepts in which new and fascinating patterns may emerge. The program has been developed for conducting various experiments with EC and CAs and their complete statistical analysis. The program also facilitates an experimental integrated designing process for designing steel structural systems in tall buildings. In this case, CAs are used, but the genome is not arbitrarily assumed or randomly generated. EC is used to produce the genome through a multigenerational evolutionary process. Emergent Designer has been successfully used for designing numerous structural systems in tall buildings, and the produced results have provided a unique understanding of CAs versus EC in design. Examples of designs produced by Emergent are shown in Figure 10.14.

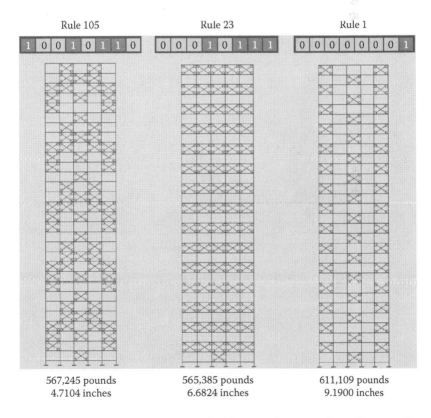

Rule 105	Rule 23	Rule 1
1 0 0 1 0 1 1 0	0 0 0 1 0 1 1 1	0 0 0 0 0 0 0 1

| 567,245 pounds | 565,385 pounds | 611,109 pounds |
| 4.7104 inches | 6.6824 inches | 9.1900 inches |

Figure 10.14 Structural designs produced by Emergent Designer. (From Kicinger, R. and Arciszewski, T., *American Scientist*, 2007, 95(6), 507.)

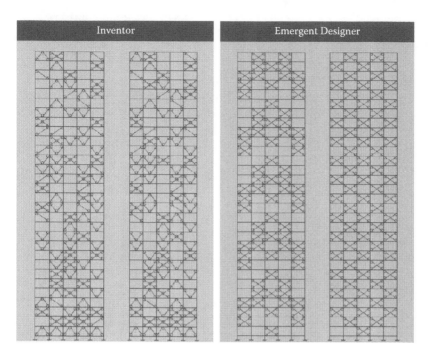

Figure 10.15 Comparison of structural designs produced by Inventor and Emergent Designer. (From Kicinger, R. and Arciszewski, T., *American Scientist*, 2007, 95(6), 508.)

Figure 10.15 shows a comparison of structural designs produced by Inventor and by Emergent Designer. All the designs are feasible, but there are huge differences in the produced configurations. As discussed in the previous section, the designs produced by Inventor, and based on EC, seem to be chaotic in terms of the distribution of bracings throughout the structural system, and no patterns can be found. Surprisingly, the evolution process does not converge into symmetrical designs, as was expected. However, the designs produced by Emergent Designer are simple and "elegant." Moreover, the majority of them are symmetrical, and this feature is highly desirable as it reduces the number of different structural members and usually contributes to a smooth flow of internal forces. Some of the designs have another "human" feature. They look like they are designed on two scales: on the global scale of the entire building and on a local scale. That means that a regular geometry for the entire structure may be observed, and this is augmented by local patterns used in various critical parts of the structure to reinforce it, as a human designer would do (see the left drawing of a structure produced by Emergent Designer shown in Figure 10.15.)

The structural designing of steel skeleton structures in tall buildings is one of the most challenging areas of structural engineering. We want such structures to be light (to minimize their cost), and at the same time we want them rigid; that is, their deformations under wind or earthquake should be very limited so as to maintain the comfort of people living or working in tall buildings. When a specific weight-optimized structure is considered within its assumed design concept, increasing rigidity naturally leads to increased weight. So we have here a technical contradiction: rigidity versus weight. For this reasons, structural designers of tall buildings are always looking for novel design concepts as a way to resolve this contradiction.

During the "Golden Age" of tall buildings in the United States (the 1960s and 1970s), Fazlur Khan was probably the most creative designer and had a number of groundbreaking ideas. One of them was the use of so-called macrobracings, which are gigantic diagonal cross bracings in the outer walls of a tall building and span large areas of the building. He discovered a new specialized "inventive pattern": "Use macro-bracings if you want to resolve the ever-present contradiction 'weight' versus 'rigidity' in tall buildings." Macrobracings force the redistribution of wind forces toward outside columns and other outside structural members, and in this way they reduce internal forces in the entire structural system. Most importantly, their use leads to the reduction of weight (compared to traditional rigid frames) while maintaining the sufficient rigidity of a skeleton structure. Macrobracings were used first by Khan in the John Hancock Tower in Chicago in the late 1960s and more recently in the Bank of China Tower in Hong Kong (for photographs of both buildings, see Figure 10.16).

Kicinger was extremely happy and proud when he discovered that his computer program (based on his evolutionary developmental algorithm) has developed several designs in which macrobracings were used. In this way, the creative power of computational bioinspiration was demonstrated,

Figure 10.16 Macrobracings in existing tall buildings and in designs produced by Emergent Designer. (From Kicinger, R. and Arciszewski, T., *American Scientist*, 2007, 95(6), 507.)

if not proved. However, the area of computational bioinspiration is still grossly unexplored. The very few results that are currently available have already revealed the potential of bioinspiration, particularly when we are looking for entirely new design concepts and engineering systems. But it is like the tip of the iceberg, and so much more has to be discovered and made available for practicing engineers. Therefore, if future inventive engineers are looking for challenges and opportunities to build their research careers, this is the area waiting for you.

References

Alligood, K.T., Sauer, T., and Yorke, J.A. (2000). *Chaos*, New York: Springer.

Altshuller, G. (1984). *Creativity as an Exact Science*, Philadelphia, PA: Gordon and Breach, Science.

Altshuller, G. (1996). *And Suddenly the Inventor Appeared: TRIZ, the Theory of Inventive Problem Solving*, Worcester, MA: Technical Innovation Center.

Altshuller, G. (1999). *The Innovation Algorithm*, Worcester, MA: Technical Innovation Center.

Altshuller, G. and Shulyak, L. (2002). *40 Principles, TRIZ Keys to Technical Innovation*, Worcester, MA: Technical Innovation Center.

Altshuller, G., Zlotin, B., Zusman, A., and Philatov, V. (1999). *Tools of Classical TRIZ*, Dearborn, MI: Ideation International.

Arafat, G., Goodman, B., and Arciszewski, T. (1992). Ramzes: A knowledge-based system for structural concepts evaluation, Special issue on artificial intelligence in civil and structural engineering, *International Journal on Computing Systems in Engineering*, 4(2–3), 211–221.

Arciszewski, T. (1976). A systems approach to structural shaping of linear elements under bending, *Proceedings of 2nd Conference on Heuristic Methods*, pp. 92–98, Warsaw.

Arciszewski, T. (1977a). Simplified multi-variant preliminary design process (in Polish), *Proceedings of 5th Scientific Conference on Computer Systems in Design*, Bialystok.

Arciszewski, T. (1977b). Systems approach to the analysis of steel space structures (in Polish), *Proceedings of the 3rd Conference on Heuristic Methods*, pp. 101–110, Warsaw.

Arciszewski, T. (1978). Systematic heuristic, In A. Goralski (Ed.), *Problem, Method, Solution. Techniques of Creative Thinking*, pp. 150–165. Warsaw: Polish Scientific-Technical.

Arciszewski, T. (1984). Design of joints in steel space structures, *Space Structures*, pp. 695–700, London, England: Applied Science.

Arciszewski, T. (1985). Decision making parameters and their computer-aided analysis for wind bracings in steel skeleton structures, *Advances in Tall Buildings*, pp. 281–299, New York: Van Nostrand.

Arciszewski, T. (1988a). ARIZ 77—An innovative design method, *Journal Design Methods and Theories*, 22(2), 796–820.

381

Arciszewski, T. (1988b). Stochastic form optimization, *Journal of Engineering Optimization*, 13(1), 17–33.

Arciszewski, T. (1989). Joint for space frame, USA, Patent No. 4,866,902.

Arciszewski, T. (1991). Joint for space frame, Canada, Patent No. 1,291,626.

Arciszewski, T. (2009). *Successful Education. How to Educate Creative Engineers*, p. 200, Fairfax, VA: Successful Education LLC.

Arciszewski, T. and Cornell, J. (2006). Bio-inspiration: Learning creative design principia, In I. Smith (Ed.), *Intelligent Computing in Engineering and Architecture*, pp. 32–53, Berlin, Germany: Springer.

Arciszewski, T. and DeJong, K. (2001). Evolutionary computation in civil engineering: Research frontiers, In B.H.V. Topping (Ed.), *Civil and Structural Engineering Computing 2001*, pp. 161–185, Fairfax, VA: George Mason University.

Arciszewski, T., Dybala, T., and Wnek, T. (1992). A method for evaluation of learning systems, Special issue on machine learning in knowledge acquisition, *Journal of Knowledge Engineering Heuristics*, 2(4), 22–31.

Arciszewski, T. and Kicinger, R. (2005). Structural design inspired by nature, In B.H.V. Topping (Ed.), *Innovation in Civil and Structural Engineering Computing*, pp. 25–48, Stirling, Scotland: Saxe-Coburg.

Arciszewski, T. and Kisielnicka, J. (1977). Morphological analysis, *Part I, Problem, Method, Solution. Techniques of Creative Thinking*, pp. 76–96. Warsaw: Polish Scientific Publishers PWN.

Arciszewski, T., Michalski, R.S., and Wnek, J. (1995). Constructive induction: The key to design creativity, *Proceedings of the Third International Round-Table Conference on Computational Models of Creative Design*, pp. 397–426, Heron Island, Australia.

Arciszewski, T. and Pancewicz, Z. (1976). New type of connection in skeleton structures, Poland, Patent No. 171367.

Arciszewski, T. and Rebolj, D. (2008). Civil engineering education: Coming challenges, *International Journal of Design Science and Technology*, 14(1), 53–61.

Arciszewski, T. and Rossman, L. (Eds) (1992). *Knowledge Acquisition in Civil Engineering*, p. 217, Reston, VA: American Society of Civil Engineering.

Arciszewski, T. and Russell, J. (2013). Change demands renaissance in civil engineering education, *Journal of Structure and Environment*, 3(4), 5–13.

Arciszewski, T., Sauer, T., and Schum, D. (2003). Conceptual designing: Chaos-based approach, *Journal of Intelligent & Fuzzy Systems*, 13(1), 45–50.

Arciszewski, T. and Uduma, K. (1988). Shaping of spherical joints in space structures, *Journal of Space Structures*, 3(3), 171–182.

Argule, M. (1987). *The Psychology of Happiness*, London, England: Methuen.

Arnheim, R. (1969). *Visual Thinking*, Berkeley, CA: University of California Press.

Bayazit, N. (2004). Investigating design: A review of forty years of design research, *Design Issues*, 20(1), Boston, MA: MIT Press.

Beazley, M. (2003). *From the Dark Ages to the Renaissance: 700–1599 AD (History of Europe)*, London, England: Mitchell Beazley.

Bedau, A.M. and Humphreys, P. (2008). *Emergence, Contemporary Readings in Philosophy and Science*, Cambridge, MA: The MIT Press.

Bentley, P.J. (Ed.) (1999). *Evolutionary Design by Computers*, San Francisco, CA: Morgan Kaufmann.

Benyus, J.M. (1997). *Biomimicry: Innovation Inspired by Nature*, New York: Perennial.

Blitz, D. (1992). *Emergent Evolution: Qualitative Novelty and the Levels of Reality*, Dordrecht, the Netherlands: Kluwer Academic.

Bloom, B.S., Englehart, M.D., Furst, E.J., Hill, W.H., and Krathwohl, D. (1956). *Taxonomy of Educational Objectives, the Classification of Educational Goals, Handbook I: Cognitive Domain*, New York: David McKay.

Bowles, S. and Gintis, H. (2002). The inheritance of inequality, *Journal of Economic Perspectives*, 16(3), 3–30.

Buzan, T. (2010). *Mind Maps for Business*, p. 254, London, England: Pearson.

Buzan, T. and Buzan, B. (1994). *The Mind Map Book*, New York: E.P. Dutton.

Cameron, K. (2008). *Positive Leadership. Strategies for Extraordinary Performance*, p. 130, San Francisco, CA: Berrett-Koehler.

Clarke, D. (1997). *TRIZ: Through the Eyes of an American TRIZ Specialist*, Detroit, MI: Ideation International.

Clarke, D. (2000). Strategically evolving the future: Directed evolution and technological systems development, *Technological Forecasting and Social Change*, 64, 133–154.

Corning, P.A. (2002). The re-emergence of "emergence": A venerable concept in search of a theory, *Complexity*, 7(6), 18–30.

Darwin, C. (1859). *On the Origin of Species by Means of Natural Selection, or the Preservation of Favoured Races in the Struggle for Life*, London, England: John Murray.

De Jong, K.A. (2005). *Evolutionary Computation: A Unified Approach*, Cambridge, MA: MIT Press.

Descartes, R. and Lafleur, L.J. (Trans.) (1960). *Discourse on Method and Meditations*. New York: The Liberal Arts Press.

Diener, E. (1984). Subjective well-being. *Psychological Bulletin*, 95, 542–575.

Diener, E. (2009). *The Science of Well-Being*, Springer Science+Business Media.

Dietrich, A. (2004). The cognitive neuroscience of creativity, *Psychonomic Bulletin & Review*, 11(6), 1011–1026.

Duckworth, A.L., Peterson, C., Matthews, M.D., and Kelly, D.R. (2007). Grit: Perseverance and passion for long-term goals, *Journal of Personality and Social Psychology*, 92(6), 1087–1101.

Fey, V. and Rivin, E. (2005). *Innovation on Demand: New Product Development Using TRIZ*, New York: Cambridge University Press.

Fish, F.F. (2009). Biomimetics: Determining engineering opportunities from nature, biomimetics and bioinspiration, In R.J. Martin-Palma and A. Lakhatia (Eds.), *Proceedings of SPIE*, Vol. 7401, Bellingham, Washington.

Fish, F.F. (2013). Advantages of natural propulsive systems, *Marine Technology Society Journal*, 47(5), 37–44.

Fish, F.F. (2014). Evolution and bio-inspired design: Natural limitation, In A.K. Goel, D.A. McAdams, and R.B. Stone (Eds.), *Biologically Inspired Design*, London, England: Springer-Verlag.

Fish, F.F., Weber, P.W., Murray, M.M., and Howle, L.E. (2011a). The tubercles on humpback whales' flippers: Applications of bio-inspired technology, *Integrative and Comparative Biology*, 51(1), 203–213.

Fish, F.F., Weber, P.W., Murray, M.M., and Howle, L.E. (2011b). Marine applications of the biomimetic humpback whale flipper, *Marine Technology Society Journal*, 45(4), 198–207.

Florida, R. (2002). *The Rise of the Creative Class*, New York: Basic Books.

Fogel, L.J., Owens, A.J., and Walsh, M.J. (1966). *Artificial Intelligence through Simulated Evolution*, Chichester, England: John Wiley.

Fredrickson, B.L. (2009). *Positivity*, p. 277, New York: Three Rivers Press.

Freud, S. (1930). *A General Introduction to Psychoanalysis*, New York: Perma Books.

Fulbright, R. (2011). I-TRIZ: Anyone can innovate on demand, *International Journal of Innovation Science*, 3(2), 41–45.

Gelb, M.J. (1998). *How to Think Like Leonardo da Vinci*, New York: Random House.

Gelb, M.J. (1999). *How to Think Like Leonardo da Vinci, Workbook*, New York: Random House.

Gelb, M.J. (2004). *Da Vinci Decoded: Discovering the Spiritual Secrets of Leonardo's Seven Principles*, New York: Bantam Dell.

Gelb, M.J. and Miller, C.S. (2007). *Innovate Like Edison—The Success System of America's Greatest Inventor*, p. 299, New York: Dutton Penguin.

Gero, J.S. (2007). Situated design computing: Principles, In B.H.V. Topping (Ed.), *Civil Engineering Computations: Tools and Techniques*, pp. 25–35, Stirlingshire, Scotland: Saxe-Coburg.

Gero, J. and Schnier, T. (1995). Evolving representations of design cases and their use in creative design, In J. Gero, M.L. Maher, and F. Sudweek, (Eds.), *Preprints of Computational Models of Creative Design*, pp. 343–368, Boston, MA: Key Center of Design Computing.

Goel, A.K. (2013). Biologically inspired design: A new program for computational sustainability, *Intelligent Systems*, 28(3), 80–84.

Goethe, J. (1995). *Scientific Studies, Goethe: The Collected Works*, p. 57, Princeton, NJ: Princeton University Press.

Goldberg, D.E. (1987). Computer-aided gas pipeline operation using genetic algorithms and rule learning, part I: Genetic algorithm in pipeline optimization, *Engineering with Computers*, 3, 47–58.

Goldberg, D.E. (1989). *Genetic Algorithms in Search, Optimization & Machine Learning*, Boston, MA: Addison-Wesley.

Goldstein, J. (1999). Emergence as a construct: History and issues, *Emergence: Complexity and Organization*, 1(1), 49–72.

Gordon, W.J.J. (1961). *Synectics: The Development of Creative Capacity*, New York: Harper & Row.

Grierson, D.E. and Cameron, G.E. (1989). Microcomputer-based optimization of steel structures in professional practice, *Journal of Microcomputers in Civil Engineering*, 4(4), 289–296.

Gruber, T.R. (1993). A translation approach to portable ontologies, *Journal of Knowledge Acquisition*, 5(2), 199–220.

Hajdo, P. and Arciszewski, T. (1991). Computer generation of structural concepts: A knowledge-based approach, *Proceedings of the ASCE Seventh Conference on Computing in Civil Engineering*, pp. 278–287, Washington, DC.

Hajela, P. (1992). Genetic algorithms in automated structural synthesis, In B.H.V. Topping (Ed.), Vol. 1. *Optimization and Artificial Intelligence in Civil and Structural Engineering*, Dordrecht, the Netherlands: Kluwer Academic Press.

Hajela, P. (1997). Stochastic search in discrete structural optimization—Simulated annealing, genetic algorithms and neural networks, In W. Gutkowski (Ed.), *Discrete Structural Optimization*. CISM International Centre for Mechanical Sciences. Vol. 373.

Hajela, P. and Lamb, A. (1986). Automated structural synthesis for nondeterministic loads, *Computer Methods in Applied Mechanics and Engineering*, 57(1), 25–36.

Hajela, P. and Lee, E. (1995). Genetic algorithms in truss topological optimization, *Journal of Solids and Structures*, 32(22), 3341–3357.

Hajela, P. and Lee, E. (1996). Constrained genetic search via schema adaptation: An immune network solution, *Structural Optimization*, 12(1), 11–15.

Harrison, J. (2001). *Synaesthesia: The Strangest Thing*, Oxford: Oxford University Press.

Hazelrigg, G. (2012). *Fundamentals of Decision Making for Engineering Design and Systems Engineering*, Washington, DC: George Hazelrigg.

Hoeller, N., Goel, A., Freixas, C., Anway, R., Upward, A., Salustri, F., McDougall, J., and Miteva, K. (2013). Proposal developing common ground for learning from nature, *Zygote Quarterly*, 7, 134–145.

Hoffman, D.D. (1998). *Visual Intelligence*, New York: W.W. Norton.

Holland, J.H. (1975). *Adaptation in Natural and Artificial Systems*, Ann Arbor, MI: University of Michigan Press.

Holland, J.H. (1998). *Emergence from Chaos to Order*, London, England: Oxford University Press.

Johansson, F. (2004). *The Medici Effect*, Cambridge, MA: Harvard Business School Press.

Kaplan, S. (1996). *An Introduction to TRIZ, the Russian Theory of Inventive Problem Solving*, Dearborn, MI: Ideation International.

Kashdan, T.B. (2009). *Curious?* p. 336, New York: Harper-Collins.

Kashdan, T.B. and Ciarrochi, J. (Eds) (2013). *Mindfulness, Acceptance and Positive Psychology*, p. 335, Oakland, CA: Context Press.

Keeney, R.L. and Raiffa, H. (1993). *Decisions with Multiple Objectives: Preferences and Value Tradeoffs*, New York: Cambridge University Press.

Kicinger, R. (2004). *Emergent Engineering Design: Design Creativity and Optimality Inspired by Nature*, PhD Dissertation, George Mason University, Fairfax, VA.

Kicinger, R. and Arciszewski, T. (2004). Multiobjective evolutionary design of steel structures in tall buildings, *Proceedings of the AIAA 1st Intelligent Systems Technical Conference*, Chicago, IL, September 20–23, American Institute of Aeronautics and Astronautics Press, Reston, VA, AIAA, 2004–64.

Kicinger, R. and Arciszewski, T. (2007). Breeding better buildings: Engineers design improved structures by borrowing from genetics, *American Scientist*, 95(6), 502–508.

Kicinger, R. and Arciszewski, T. (2009). Bio-inspired computational framework for enhancing creativity, optimality and robustness in design, *Journal of Computing in Civil Engineering*, 23(1), 22–23.

Kicinger, R., Arciszewski, T., and De Jong, K.A. (2005a). Emergent designer: An integrated research and design support tool based on models of complex systems, *Journal of Information Technology in Construction*, ITcon, 10, 329–347.

Kicinger, R., Arciszewski, T., and De Jong, K.A. (2005b). Evolutionary computation and structural design: A survey of the state of the art, *Journal of Computers & Structures*, 83, 1943–1978.

Kicinger, R., Arciszewski, T., and De Jong, K.A. (2005c). Evolutionary designing of steel structures in tall buildings, *Journal of Computing in Civil Engineering*, 19(3), 223–238.

Koza, J. (1992). *Genetic Programming: On Programming of Computer by Means of Natural Selection*, Cambridge, MA: MIT Press.

Koza, J. (2003). *Knowledge Soup, Lecture Notes*, Washington, DC: Smithsonian Institute.

Koziolek, S., Rusinski, E., and Jamroziak, K. (2010). Critical to quality factors of engineering design process of armoured vehicles, *Solid State Phenomena*, 165, 280–284.

Kumar, S. and Bentley, P.J. (Eds) (2003). *On Growth, Form and Computers*, London, England: Academic Press.

Lopez, S.J. (2013), *Making Hope Happen: Create the Future You Want for Yourself and Others*, p. 260, New York: Atria Books.

Luke, S. (2000). *Issues in Scaling Genetic Programming: Breeding Strategies, Tree Generation, and Code Bloat*, PhD Dissertation, Department of Computer Science, University of Maryland, College Park, Maryland.

Lyobomirsky, S. (2013). *The Myth of Happiness*, New York: Penguin Books.

MacKenzie, C. (1928). *Alexander Graham Bell*, pp. 72–73, New York: Houghton Mifflin.

Maier, M.W. and Rechtin, E. (2000). *The Art of Systems Architecting*, Washington, DC: CRC Press.

McManus, J. (2008). Da Vinci's parachute: Field tested 523 years later! dailygalaxy.com. April 29. Accessed October 31, 2015.

Menzies, G. (2009). *The Year a Magnificent Chinese Fleet Sailed to Italy and Ignited the Renaissance*, New York: Harper Perennial.

Miklosovic, D.S., Murray, M.M., Howle, L.E., and Fish, F.F. (2004). Leading-edge tubercles delay stall on humpback whale, *Physics of Fluids*, 16(5), 39–42.

Miller, K. (2009). *St. Peter's*, Wonders of the World Series, London: Profile Books.

Moored, K.W., Fish, F.F., Kemp, T.H., and Bart-Smith, H. (2011). Batoid fishes: Inspiration for the next generation of underwater robots, *Marine Technology Society Journal*, 45(4), 99–109.

Morowitz, H.J. (2002). *The Emergence of Everything: How the World Became Complex*, London, England: Oxford University Press.

Müller, J. (1970). *Grundlagen der Systematischen Heuristic*, Berlin: Diez Verlag.

Murawski, K., Arciszewski, T., and De Jong, K. (2000). Evolutionary computation in structural design, *Journal of Engineering with Computers*, 16, 275–286.

Nadler, G. (1973). *Production Systems Design with the Ideals Concept*, London, England: Industrial and Commercial Techniques.

Neisser, U., Boodoo, G., Bouchard, T.J., Jr., Boykin, A.W., Brody, N., Ceci, S.J., Halpern, D.F. et al. (1996). Intelligence: Knowns and unknowns, *American Psychologist*, 51(2), 77–101.

Obayashi, S. (2015). A study on many-objective optimization using the Kriging surrogate-based evolutionary algorithm maximizing expected hypervolume improvement, *Mathematical Problems in Engineering*, 4, 1–15.

Oguejiofor, E., Kicinger, R., Popovici, E., Arciszewski, T., and DeJong, K. (2004). Intelligent tutoring systems: An ontology based approach, *International Journal of Computing in Architecture, Engineering, and Construction*, 2(2), 115–128.

Orloff, M.A. (2003). *Inventive Thinking Through TRIZ*, New York: Springer.

Osborn, A.F. (1921). *A Short Course in Advertising*, New York: Scribners.

Osborn, A.F. (1952a). *Wake Up Your Mind: 101 Ways to Develop Creativeness*, New York: Scribners.

Osborn, A.F. (1952b). *Your Creative Power. How to Use Imagination*, New York: Scribners.

Osborn, A.F (1953). *Applied Imagination: Principles and Procedures of Creative Thinking*, p. 317, Oxford: Scribners.

Osborn, A.F. (1962). Developments in creative education, In S.J. Parnes and H.F. Harding (Eds.), *A Source Book for Creative Thinking*, pp. 19–29, New York: Scribners.

Panek, R. (2009). The father of dark matter still gets no respect, *Discover*, January issue.

Pawlak, Z. (1991). *Rough Sets: Theoretical Aspects of Reasoning About Data*, Dordrecht, the Netherlands: Kluwer Academic.

Pink, D.H. (2006). *A Whole New Mind. Why Right-Brainers will Rule the Future*, New York: Riverhead Books.

Quillian, M. (1968). Semantic memory, In M. Minsky (Ed.), pp. 227–270, *Semantic Information Processing*, Boston: MIT Press.

Rechenberg, I. (1965). *Cybernetic Solution Path of an Experimental Problem* (Vol. Library Translation 1122). Farnborough: Royal Aircraft Establishment.

Rechtin, E. (1991). *Systems Architecting. Creating & Building Complex Systems*, Upper Saddle River, NJ: Prentice Hall.

Ritchey, T. (2010). *Wicked Problems—Social Messes: Decision Supporting Modelling with Morphological Analysis*, Stockholm: Swedish Morphological Society.

Rittel, H. (1972). *On the Planning Crisis: Systems Analysis of the 'First and Second Generation*, pp. 390–396, Berkeley, CA: University of California.

Rittel, H. and Webber, M. (1973). *Dilemmas in a General Theory of Planning*, pp. 155–159, Amsterdam, the Netherlands: Elsevier Scientific.

Roam, D. (2008). *The Back of the Napkin: Solving Problems and Selling Ideas with Pictures*, New York: Penguin Group.

Robertson, L. and Sagiv, N. (Eds) (2005). *Synesthesia: Perspectives from Cognitive Neuroscience*, Oxford: Oxford University Press.

Rosenberg, L.J. (2001). Pectoral fin locomotion in batoid fishes: Undulation versus oscillation, *Journal of Experimental Biology*, 2042, 374–394.

Ross, T. (1995). *Fuzzy Logic for Engineering Applications*, New York: McGraw-Hill.

Sage, A.P. (2000). Transdisciplinarity perspectives in systems engineering and management, In M.A. Somerville and D. Rapport (Eds.), *Transdisciplinarity: Recreating Integrated Knowledge*, pp. 158–169, Oxford, England: EOLSS.

Schwefel, H.P. (1965). Kybernetische Evolution als Strategie der experimentelen Forschung in der Stromungstechnik. Master's thesis, Hermann Föttinger Institute for Hydrodynamics, Technical University of Berlin, Berlin.

Shelton, K. (2007). *Design for Robustness of Unique, Multi-Component Engineering Systems*, PhD Dissertation, George Mason University, Fairfax, VA.

Shelton, K. and Arciszewski, T. (2008). Formal innovation criteria, *International Journal of Computer Applications in Technology*, 30(1,2), 21–32.

Shimoff, M. (2008). *Happy for No Reason: 7 Steps to Being Happy from the Inside Out*, p. 320. New York: Free Press.

Simon, H. (1969). *The Sciences of Artificial*, Cambridge, MA: MIT Press.

Sternberg, R.J. (1985). *Beyond IQ: A Triarchic Theory of Intelligence*, Cambridge, MA: Cambridge University Press.

Sternberg, R.J. (1996), *Successful Intelligence*, New York: Simon & Shuster.

Sternberg, R.J. (1997), *Thinking Styles*, Cambridge, MA: Cambridge University Press.

Stone, R.B., Goel, A.K., and McAdams, D.A. (2014). Charting a course for computer-aided bio-inspired design, In A.K. Goel et al. (Eds.), *Biologically Inspired Design*, pp. 1–16, London, England: Springer-Verlag.

Terninko, J., Zusman, A., and Zlotin, B. (1998). *Systematic Innovation: An Introduction to TRIZ*, Boca Raton, FL: St. Lucie Press.

Thatchenkery, T. and Metzker, C. (2006). *Appreciative Intelligence: Seeing the Mighty Oak in the Acorn*, p. 211, San Francisco, CA: Berrett-Koehler.

Urgessa, G.S. and Arciszewski, T. (2011). Blast response comparison of multiple steel frame connections, *Journal of Finite Elements in Analysis and Design*, 47(6), 668–675.

Valler-Radot, R. (1902). *The Life of Pasteur*, Westminster, London: Archibald Constable.

Vogel, S. (1998). *Cat's Paws and Catapults*, New York: W.W. Norton.

Von Neumann, J. (1951). The general and logical theory of automata, In L.A. Jeffress (Ed.), *Cerebral Mechanisms in Behavior: The Hixon Symposium*, pp. 1–32, New York: Wiley.

Ward, J. (2008). *The Frog Who Croaked Blue: Synesthesia and the Mixing of the Senses*, London, England: Routledge.

Wheeler, R.A. (2014). Alex F. Osborn: The father of brainstorming. Seen at russellawheeler.com on October 27, 2015.

Wiener, N. (1950). Cybernetics, *Bulletin of the American Academy of Arts and Sciences*, 3(7), 2–4.

Wolfram, S. (2002). *A New Kind of Science*, Champaign, IL: Wolfram Media.

Wrey, W. (2005). *Leonardo Da Vinci in His Own Words*, New York: Gramercy Books.

Youmans, R. and Arciszewski, T. (2014). Design fixation: A cloak of many colors, In J. Gero (Ed.), *Design Computing and Cognition's 12*, pp. 123–140, New York: Springer Science and Business Media B.V.

Zadeh, L.A. (1965). Fuzzy sets, *Journal of Information and Control*, 8(3), 338–353.

Zlotin, B. and Zusman, A. (2006). *Directed Evolution: Philosophy, Theory, and Practice*, Dearborn, MI: Ideation International.

Zwicky, F. (1969). *Discovery, Invention, Research through the Morphological Approach*, London, England: MacMillan.

Index